Oil and Urbanization on the Pacific Coast

Energy and Society
Brian Black, Series Editor

Other Titles in the Series
Oil and Nation: A History of Bolivia's Petroleum Sector
Stephen C. Cote

Oil and Urbanization on the Pacific Coast

Ralph Bramel Lloyd and the Shaping of the Urban West

Michael R. Adamson

West Virginia University Press
Morgantown 2018

First edition published 2018 by West Virginia University Press
Printed in the United States of America

ISBN:
Cloth 978-1-946684-43-1
Paper 978-1-946684-36-3
Ebook 978-1-946684-44-8

Library of Congress Cataloging-in-Publication Data
Names: Adamson, Michael R., author.
Title: Oil and urbanization on the Pacific Coast : Ralph Bramel Lloyd and the shaping of the urban West / Michael R. Adamson.
Description: First edition. | Morgantown : West Virginia University Press, 2018. | Series: Energy and society | Includes bibliographical references and index.
Identifiers: LCCN 2018010576| ISBN 9781946684431 (cloth) | ISBN 9781946684363 (pbk.)
Subjects: LCSH: Lloyd, Ralph Bramel. | Businessmen--Pacific States--Biography. | Petroleum industry and trade--Pacific States--History. | Real estate development--Pacific States--History.
Classification: LCC HC102.5.L56 A33 2018 | DDC 333.33092 [B] --dc23
LC record available at https://lccn.loc.gov/2018010576

Cover design by Than Saffel / WVU Press
Cover image: Lloyd Corporation, "Lloyd Center Properties," undated, Homer D. Crotty Papers and Addenda, box 33 (Addenda), folder 19, The Huntington Library, San Marino, California.

Contents

Figures

Tables

Acknowledgments

My research into the role of independent operators in petroleum exploration and production began more than two decades ago on a project funded by the U.S. Minerals Management Service (MMS Cooperative Agreement No. 14–35–0001–30796). I thank Randy Bergstrom and Harvey Molotch for inviting me to join an intellectually rigorous, multidisciplinary team that investigated the industrial history of petroleum extraction across California's Coastal Region. Both on the project and in Harvey Molotch's graduate seminar, I benefited from critical and constructive discussions on the local social and economic impacts of the oil industry with Thomas Beamish, Leonard Nevarez, Krista Paulsen, and Jaqueline Romo.

My research on the project included interviews with active independent operators and petroleum consultants, including Cecil O. Basenberg, Floyd Clawson, H. Lynn Hall, Jr., W. J. Lovingfoss, Ted Off, the late Robert L. Richardson, and Clif Simonson. Their experiences and insights captivated my attention and motivated me to continue my study of the role of the independent upon the completion of the project.

In the course of this investigation, I learned of the Lloyd Corporation Archive at the Huntington Library. I first researched this vast trove of material as part of an investigation funded by MMS (MMS Cooperative Agreement No. 14–35–01–00-CA-31063). I thank James T. Lima for identifying the need for a one-volume study that made the findings of prior research into the Coastal Region impacts of the oil industry more accessible to the public and policymakers, which I filled by completing the contract. I also thank Russell Schmitt and Jenifer Dugan of the Marine Science Institute at the University of California, Santa Barbara, who served as the project's principal investigators.

I completed work under this contract as a Visiting Scholar at the Office for History of Science at the University of California, Berkeley. I thank Cathryn Carson and Diana Wear for providing encouragement, resources, and a vibrant intellectual environment for my research.

The project took shape when I researched Ralph Lloyd's real estate development activities with support from the Fletcher Jones and Andrew W. Mellon Foundations. I am fortunate that the Lloyd Corporation and others associated with the Lloyd family donated papers and other material to the Huntington Library. I can imagine no more intellectually stimulating and supportive setting for scholarly investigation. I am grateful to William Frank for graciously helping me to negotiate the Lloyd Corporation Archive and bringing to my attention additional materials that informed my understanding of Ralph Lloyd's early career, in particular. I also thank myriad staff who, over the years, facilitated my research by patiently pulling hundreds of boxes on my behalf, and the photographers who provided many of the images contained in this book.

Other archivists, librarians, and staff facilitated my research into other collections. I thank Special Collections Librarian Bruce Tabb for facilitating my research into the John C. Ainsworth Papers from afar. Charles Johnson of the Ventura County Museum pointed me to invaluable oral histories, photographs, and other materials related to Ventura, California's oil boom and the development of the Ventura Avenue field. Municipal staff made available the records of the cities of Los Angeles, Portland, and Ventura.

During 2008, the aforementioned "wildcatter," Clif Simonson, was drilling for oil on the Sexton Ranch in the hills behind the city of Ventura. I thank him for showing me the active drill site and giving me a tour by Jeep of the shut-in Lloyd Corporation Lease in the Ventura Avenue field.

In researching and writing this book, I benefited from thoughtful responses of many who read or reviewed conference papers, articles, and chapters. For their contributions, I thank Carl Abbott, Sean Patrick Adams, Randy Bergstrom, Chris Castaneda, Peter Coclanis, Stephanie Dyer, the late William Freudenburg, Margaret Garb, Diana Hinton, Robert Lifset, Harvey Molotch, Krista Paulsen, Joseph Pratt, and Ty Priest. I am grateful to Louis Galambos for inviting me to present a draft chapter to his postdoctoral seminar at Johns Hopkins University.

I owe a particular debt of gratitude to W. Elliot Brownlee, my dissertation adviser, who instilled in me an appreciation of the value in pursuing topics that can be supported in large part by archival sources in one's backyard, so to speak. Our discussions of California business and economic history have been points of departure for several productive lines of inquiry that have indirectly or directly informed my understanding of Ralph Lloyd's career.

Portions of the Introduction and Chapters 1 and 5 draw on material published in my article "The Role of the Independent: Ralph B. Lloyd and the Development of California's Coastal Oil Region, 1900–1940," *Business History Review* 84 (Summer 2010): 301–28. Chapter 2 is adapted from my article "Oil Booms and Boosterism: Local Elites, Outside Companies, and the Growth of Ventura, California," *Journal of Urban History* 35 (November 2008): 150–77. As cited in the notes, various discussions of project delivery methods, such as design-bid-build and design-build, first appeared in my book *A Better Way to Build: A History of the Pankow Companies*, which Purdue University Press published in 2013. The material appears courtesy of Purdue University Press.

I am privileged to have had Derek Krissoff and Sara Georgi as editors. I am grateful for their enthusiastic support of the project. I also thank Allison Scott for copyediting the manuscript with care and the entire production staff at the Press for seeing this book to the finish line. Finally, I thank Martin Melosi for his ongoing interest in my work and suggesting that I bring the project to the Press.

Abbreviations

AF	*Architectural Forum*
API	American Petroleum Institute
AR	*Architectural Record*
Associated	Associated Oil Company
BD&C	*Building Design & Construction*
BI	Proposed Buildings and Installations, Lloyd Corporation Archive, Henry E. Huntington Library, San Marino, California
BOE	Barrel of Oil Equivalent
BPA	Bonneville Power Administration
CCCOP	Conservation Committee of California Oil Producers
CMP	Controlled Materials Plan
CPA	City of Portland, Oregon, Archives and Records Center
DMV	California Division of Motor Vehicles
Fed	Federal Reserve Board
FHA	Federal Housing Administration
FTC	Federal Trade Commission
GDP	U.S. Gross Domestic Product
GNP	U.S. Gross National Product
GP	General Petroleum Company
GPO	U.S. Government Printing Office
GSA	U.S. Government Services Administration
JCA	John C. Ainsworth Papers, University of Oregon Libraries, Special Collections and University Archives
LAE	*Los Angeles Examiner*
LAH	*Los Angeles Evening Herald*
LAT	*Los Angeles Times*
LC	Lloyd Center records, Lloyd Corporation Archive, Henry E. Huntington Library, San Marino, California
LCA	Lloyd Corporation Archive, Henry E. Huntington Library, San Marino, California
LCL	Lloyd Corporation Letters, Lloyd Corporation Archive, Henry E. Huntington Library, San Marino, California

LCR	Lloyd Corporation Real Estate documents, Lloyd Corporation Archive, Henry E. Huntington Library, San Marino, California
NPA	National Production Authority
NRA	National Recovery Administration
Photos	Photographs, Lloyd Corporation Archive, Henry E. Huntington Library, San Marino, California
Port	Portland file, Lloyd Corporation Archive, Henry E. Huntington Library, San Marino, California
PVA	Park View Apartments file, Lloyd Corporation Archive, Henry E. Huntington Library, San Marino, California
PWA	Public Works Administration
RBL	Ralph B. Lloyd Letters, Lloyd Corporation Archive, Henry E. Huntington Library, San Marino, California
RE	Real Estate files, Lloyd Corporation Archive, Henry E. Huntington Library, San Marino, California
RFC	Reconstruction Finance Corporation
SBMP	*Santa Barbara Morning Press*
Shell	Shell Company of California; Shell Oil Company
Standard	Standard Oil Company of California
Union	Union Oil Company of California
VCM	Museum of Ventura County
VCS	*Ventura County Star*
VFP	*Ventura Free Press*
VL&W	Ventura Land and Water Company
VLW	Ventura Land and Water Company file, Lloyd Corporation Archive, Henry E. Huntington Library, San Marino, California
VS-FP	*Ventura Star-Free Press*
VWP	*Ventura Weekly Post*
VOAHA	Virtual Oral/Aural History Archive, California State University, Long Beach, Library

Introduction

The Role of the Independent Oil Operator

Every man that has a piece of land with oil on it can drill a well, as they have been doing throughout the state.

—Charles A. Canfield, 1905

Ralph Bramel Lloyd was a small businessman who made a large and lasting regional impact through two of the industries in which he operated. As an "independent" oilman, he was instrumental in the development of one of America's largest onshore fields, which continues to produce millions of barrels of crude oil per year. As a real estate developer, Lloyd catalyzed the materialization of several commercial districts, in particular on the East Side of Portland, Oregon, where an entire district bears his name. This book describes and explains the achievements and travails that Lloyd experienced in both industries with particular attention paid to capital flows that linked the two. His multifaceted career illustrates the integral role that small businesses have played in sectors that have received comparatively little attention from scholars. Significantly, Lloyd thrived not by establishing a niche market through specialty product or service provision but by competing against and cooperating with big businesses.[1]

As a "citizen outside the government," to use William Issel's phrase, Ralph Lloyd lobbied municipal government to approve policies and undertake projects that he deemed essential to the viability of his projects.[2] Because he operated in three cities of dissimilar sizes, namely Los Angeles and Ventura, California, and Portland, Oregon, Lloyd's career also offers an opportunity to examine and understand in comparative terms the role of small business owners in shaping urban places through political action.

To be sure, Ralph Lloyd was driven by self-interest, as he frankly conceded to his correspondents.[3] At the same time, he strongly identified with the communities in which he operated. Moreover, he approached commercial real estate as a long-term investment. Rarely did he acquire a property only to "flip" it for short-term profit. Lloyd's willingness to hold property more or less indefinitely and his abiding interest in the cities in which he invested were mutually reinforcing. Further, his Lloyd Corporation continued to develop his real estate holdings, if not his oil properties, for some three decades after his death in 1953. Together, these factors helped to sustain the material impact of Ralph Lloyd's career as a developer.

The California Oil Industry: The Role of the Independent

Ralph Lloyd enjoyed his greatest success in the oil business. More than three decades ago, Roger Olien and Diana Davids Olien noted how little had been written about how nonintegrated oil companies competed in an oligopolistic industry dominated by vertically integrated corporations, or "majors."[4] During the twentieth century, hundreds of these "independents" operated in California in the so-called upstream segment of the industry, which includes exploration and production. Yet we know little more about their influence on the Pacific Coast industry than we did in 1970 when historian Gerald D. Nash called on his colleagues to devote more attention to their study.[5] In an often-uneasy symbiotic relationship with two of California's major oil companies, Ralph Lloyd played a crucial and pivotal role in developing one of America's greatest onshore oil fields. As a result, his Lloyd Corporation became a leading producer in the state.

During the early twentieth century, independents became an increasingly important factor in the search for, and development of, crude oil reserves in California. Regional market structures of the upstream segment of the industry remained open to new entrants, even as rationalization occurred "downstream," in the transportation, refining, marketing, and distribution of petroleum products.

California's independents successfully cooperated with, and competed against, majors within evolving upstream market structures. Ralph Lloyd's career neatly encapsulates the ways in which they did so and illustrates the reasons why distinctive market structures characterize extractive industry. The role of the independent in the development of California's crude oil resources was not a residual one. Moreover, these small businesses succeeded under geological conditions more favorable to majors than elsewhere in America.

Above all, the upstream oil business was based "a good deal on luck and good fortune," as the son of one Long Beach, California, independent reflected.[6] There was a high degree of uncertainty in extractive industry, owing to uncertainty in exploration. Managing risk accounted for much of the success in this segment of the business. Firms allocated capital according to minimum acceptable rates of return, which varied significantly, depending on costs of capital and risk preferences.[7]

Exploration for and production of crude oil became ever more competitive in California during the early decades of the twentieth century. The number of both majors and independents increased over time in all three of the state's oil-producing areas: the Coastal Region, the Los Angeles Basin, and the San Joaquin Valley (figure 1). As Ralph Lloyd's career shows, the role of the independent in California was not limited to minor fields that failed to attract the interest of majors, because the advantages of majors were not decisive in the search for oil.

"The small operator will always be a highly important factor in the American oil industry," insisted the editors of *California Oil World* in 1930.[8] And so they have been—and with the recent, "remarkable" transformation of the industry, they continue to be.[9] Independents have been especially prominent in the search for oil, as they have assumed higher risks and have accepted lower returns than majors. During the twentieth century, they drilled most of the exploratory wells nationally: 81 percent from 1946 to 1953 and 90 percent from 1968 to 1978, for instance.[10] In California's Coastal Region, where Ralph Lloyd primarily operated, independents accounted for 79 percent of the exploratory wells drilled from 1900 to 1940.[11] This contribution is widely recognized. Less often credited is the importance of independents in petroleum extraction. In the Coastal Region, they accounted for as much as 40 percent of annual production during this period. Yet even this metric understates the independents' contribution to the upstream segment of the oil industry. For only the operator of a well receives credit for its discovery and any subsequent output. Firms often spread the risks of drilling through various devices, including joint ventures, "bottom-hole money" (whereby other investors pay the operator to drill a well to a contracted depth), and "dry-hole money" (whereby other investors compensate the operator if the well fails to produce oil). Moreover, as we shall see, an operator who controlled the potentially productive acreage of field could play an integral role in its development without receiving official credit for it in published statistics.

In California, as elsewhere, independents found ample opportunity to search for and produce petroleum during the first four decades of the twentieth century. Charles A. Canfield, president of Associated Oil Company, the major with which Ralph Lloyd would become most closely involved, explained

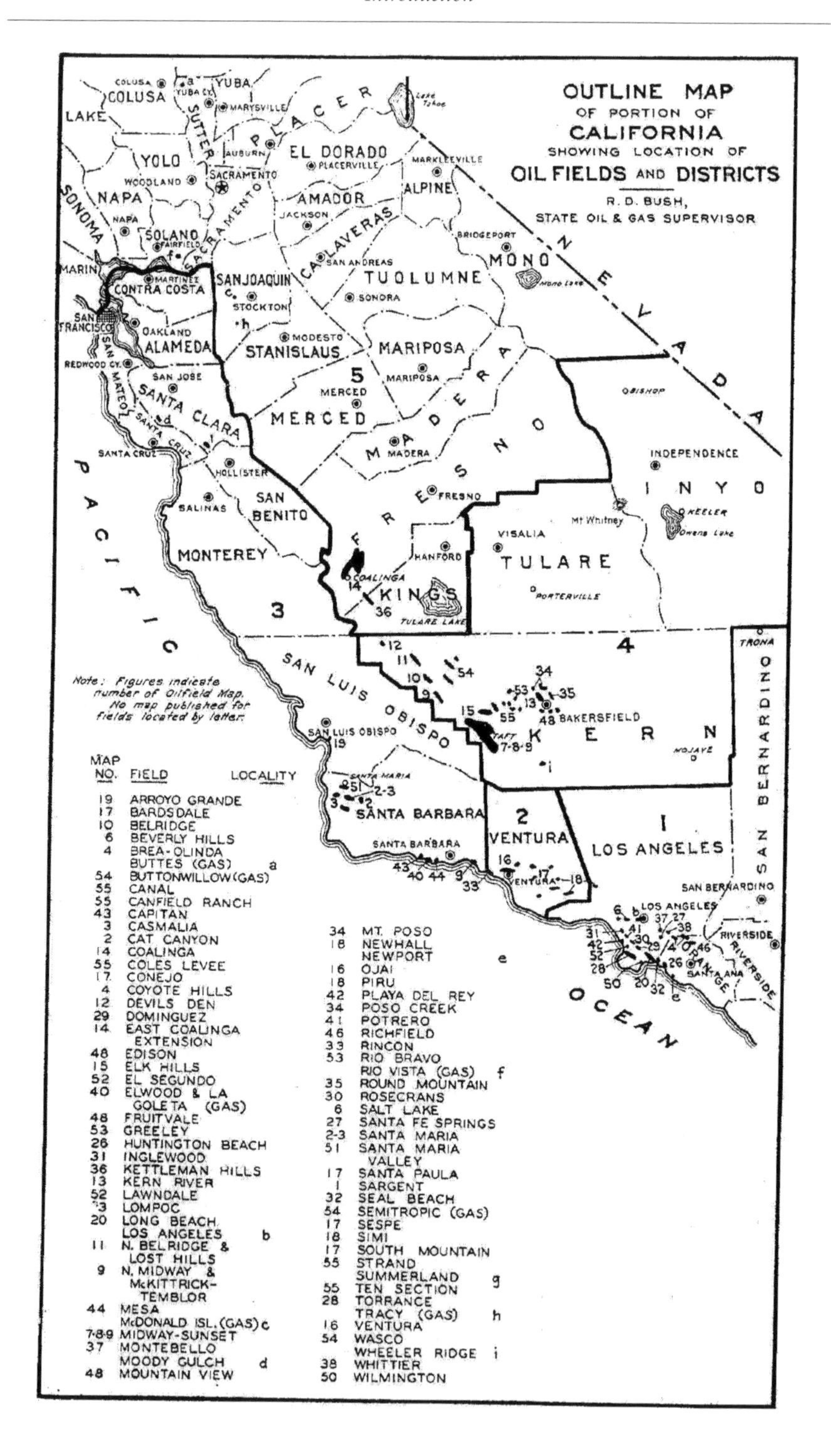
OUTLINE MAP
OF PORTION OF
CALIFORNIA
SHOWING LOCATION OF
OIL FIELDS AND DISTRICTS
R. D. BUSH,
STATE OIL & GAS SUPERVISOR
COLUSA
YUBA
LAKE
PLACER
EL DORADO
YOLO
SACRAMENTO
NAPA
SONOMA
SOLANO
AMADOR
ALPINE
CALAVERAS
MONO
NEVADA
MARIN
CONTRA COSTA
SAN JOAQUIN
TUOLUMNE
SAN FRANCISCO
ALAMEDA
STANISLAUS
MARIPOSA
MADERA
MERCED
SANTA CLARA
SANTA CRUZ
FRESNO
INYO
SAN BENITO
MONTEREY
KINGS
TULARE
PACIFIC
SAN LUIS OBISPO
KERN
SAN BERNARDINO
SANTA BARBARA
VENTURA
LOS ANGELES
ORANGE
RIVERSIDE
OCEAN
BAKERSFIELD
Note: Figures indicate number of Oilfield Map. No map published for fields located by letter.
MAP NO. FIELD LOCALITY
19 ARROYO GRANDE
17 BARDSDALE
10 BELRIDGE
6 BEVERLY HILLS
4 BREA-OLINDA
BUTTES (GAS) a
54 BUTTONWILLOW (GAS)
55 CANAL
55 CANFIELD RANCH
43 CAPITAN
3 CASMALIA
2 CAT CANYON
14 COALINGA
55 COLES LEVEE
17 CONEJO
4 COYOTE HILLS
12 DEVILS DEN
29 DOMINGUEZ
14 EAST COALINGA EXTENSION
48 EDISON
15 ELK HILLS
52 EL SEGUNDO
40 ELWOOD & LA GOLETA (GAS)
48 FRUITVALE
53 GREELEY
26 HUNTINGTON BEACH
31 INGLEWOOD
36 KETTLEMAN HILLS
13 KERN RIVER
52 LAWNDALE
3 LOMPOC
20 LONG BEACH
LOS ANGELES b
11 N. BELRIDGE & LOST HILLS
9 N. MIDWAY & McKITTRICK-TEMBLOR
44 MESA
McDONALD ISL. (GAS) c
7-8-9 MIDWAY-SUNSET
37 MONTEBELLO
MOODY GULCH d
48 MOUNTAIN VIEW
34 MT. POSO
18 NEWHALL
NEWPORT e
16 OJAI
18 PIRU
42 PLAYA DEL REY
34 POSO CREEK
41 POTRERO
46 RICHFIELD
33 RINCON
53 RIO BRAVO
RIO VISTA (GAS) f
35 ROUND MOUNTAIN
30 ROSECRANS
6 SALT LAKE
27 SANTA FE SPRINGS
2-3 SANTA MARIA
51 SANTA MARIA VALLEY
17 SANTA PAULA
1 SARGENT
32 SEAL BEACH
54 SEMITROPIC (GAS)
17 SESPE
18 SIMI
17 SOUTH MOUNTAIN
55 STRAND
SUMMERLAND g
55 TEN SECTION
28 TORRANCE
TRACY (GAS) h
16 VENTURA
54 WASCO
WHEELER RIDGE i
38 WHITTIER
50 WILMINGTON

in 1905 to the U.S. Bureau of Corporations why no firm could prevent independents from competing: "There are too many fields easy of access to market that can be developed by individuals every day. There are thousands of them in the state today which can be opened up . . . and you can't stop them. . . . Every man that has a piece of land with oil on it can drill a well, as they have been doing throughout the state."[12]

The availability of affordable acreage was crucial to the success of independents. California's relatively thicker and deeper oil-bearing structures supported comparatively fewer independents than those in Texas. As one independent who worked for an integrated oil company in both states explains, "You find six-to-seven-thousand-foot [deep] sections in a California reservoir that do not extend very far [horizontally], whereas in Texas you get twenty-foot sections that run for miles." [13] With more acreage and therefore more landowners involved in an oil-producing area, Texas's fields typically offered more leasing opportunities to independents than did the fields of California. As of 1940, Texas had almost three times as many independent producers than California, even though Texas extracted only slightly more than twice the amount of petroleum.[14] Even so, California's independents did not lack for opportunity.

Land ownership patterns shaped opportunities for entry on a field-by-field basis. Moreover, once established, leasing patterns generally persisted: the economic incentive associated with property consolidation diminished over time because reserves depleted in both physical and accounting terms. Three examples illustrate how leasing opportunities varied by land-ownership patterns in the Coastal Region from 1900 to 1940.

The Ventura Avenue field—the source of much of Ralph Lloyd's wealth, as we shall see—illustrates the legacy of ranchos—large land grants from the Spanish and Mexican eras. There were nineteen major ranchos in Ventura County. By 1870, most of them had passed into the hands of European Americans, who for the most part divided and sold them to ranchers or farmers. Yet these tracts typically remained much larger than property associated

Figure 1 (*opposite*). California's Oil Regions. The Los Angeles Basin corresponds to District 1, comprising fields in Los Angeles and Orange Counties. The Coastal Region corresponds to Districts 2 and 3, comprising fields in San Luis Obispo, Santa Barbara, and Ventura Counties. The San Joaquin Valley corresponds to Districts 4 and 5, comprising fields in Fresno, Kern, and Kings Counties. (*Annual Report of the State Oil and Gas Supervisor* 23 [April-May-June 1938]: 4.)

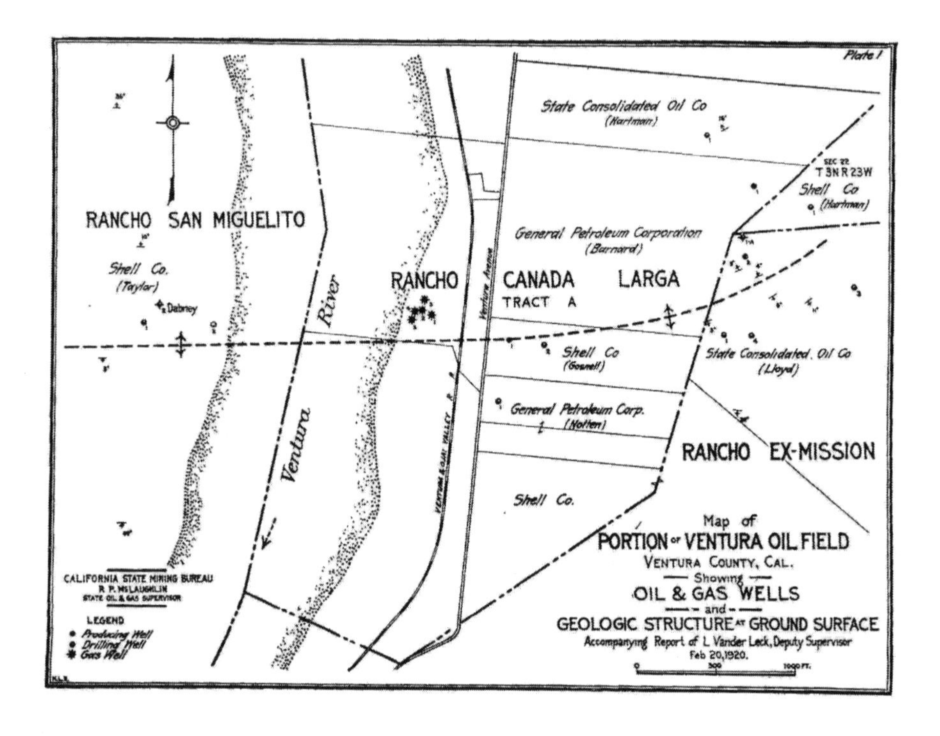

Figure 2. The initial proven area of the Ventura Avenue field in 1920, at the intersection of three former ranchos. (Lawrence Vander Leck, "Report on the Ventura Oil Field, Ventura County, California," *Annual Report of the State Oil and Gas Supervisor* 5 [February 1920]: 21.)

with U.S. government parcels. At Ventura Avenue, three ranchos intersected at the center of the field. Large holdings meant that half-a-dozen leases covered all the area Ralph Lloyd believed to be potentially producing and that few firms obtained acreage to explore the limits of the field once drilling confirmed the presence of substantial crude oil reserves (figure 2).[15]

In contrast, ranchers and farmers held much of the Santa Maria Valley, located in north Santa Barbara County, in smaller tracts. Prospective landowners acquired sections, or parts thereof, from the U.S. government, which throughout the nineteenth century remained eager to use public land sales to mobilize private investment on behalf of regional economic development.[16] In 1935, Union Oil Company of California, a leading California major, discovered a gigantic field near the city of Santa Maria.[17] Union controlled 3,500 acres in the vicinity of the discovery well, 2,000 acres of which proved to be productive over the next three years. Yet many independents executed leases with other landowners and joined Union in defining the limits of the field. As of January 1939, the proven area of the field encompassed 4,760 acres. Union remained the leading operator, accounting for about half of the field's daily output. At the same time, twenty-five other firms operated productive wells.[18]

Development at Elwood, which lay ten miles west of the city of Santa Barbara, featured drilling by independents both onshore and on state tidelands from piers that extended into the Pacific Ocean. In 1921, the State of California responded to the discovery of large pools of crude oil under its tidelands in the Los Angeles area with the Minerals Reservation Act. The law provided for the state's surveyor general to issue permits for tidelands exploration. When he stopped issuing permits in the interest of preserving the coastline, with the support of the governor and attorney general, operators who had been denied permits sued. In December 1928, the California Supreme Court decided the case of *Boone v. Kingsbury* in favor of the plaintiffs, forcing the surveyor general to issue permits that covered much of the south Santa Barbara County coast, including a dozen permits at Elwood. The lease that contained the discovery well covered land controlled by the heirs of Nicholas Den, to whom the governor of Alta California granted the Rancho Los dos Pueblos in 1842. Rio Grande Oil Company, a Los Angeles-based independent and the partner of Barnsdall Oil Company in a joint venture that held the lease, filed notices on four tidelands parcels to protect the investment, as permitted by the California Supreme Court's decision. The company failed to comply with the letter of state law, however, enabling others to post their own claims when the notices expired in twelve months. Several independents gained access to the field, enabling them to become leading Coastal Region producers. The opposition on the part of coastal communities and conservationists to the drilling frenzy at Elwood and in Los Angeles and Orange Counties spurred the state legislature to impose a temporary moratorium on additional leasing, giving legislators time to pass a more permanent measure, which the governor signed into law in May 1930. Nevertheless, public policy permitted a window of opportunity for independents to discover oil from tidelands drill sites.[19]

James E. Herley, a Long Beach operator of wells that his father had drilled in Los Angeles and Ventura Counties, characterized leasing as "do[ing] some pretty fancy footwork to get in there before somebody else got it all gobbled up."[20] Still, independents did not lack for ways of obtaining leases. Relationships with landowners provided local bankers, ranchers, professionals, merchants, and others with opportunities to obtain acreage.[21] Sometimes personal connections held the key to obtaining leases, even when the independent was not based locally. For example, Signal Oil and Gas Company secured a lease at Elwood because Samuel B. Mosher, its president, was a good friend of David Faries, a Los Angeles lawyer who obtained the rights to the tidelands leases once Rio Grande Oil Company's notices expired.[22] More often than not, independents from outside the region successfully approached landowners with proposals to drill in areas that they believed held potential. Moreover, the leasing pattern at Santa Maria Valley suggests that independents were able to break up the holdings of majors. For example, Arthur N. Macrate, a Long Beach independent who had drilled wells in half-a-dozen fields in the Los Angeles Basin before he turned his attention to the Coastal Region, leased two sections that lay adjacent to Union's properties in the most productive area of the field.[23]

Neither law nor politics prevented majors from leasing as much potentially productive land as they desired in these fields. But finding oil remained an uncertain business and capital budgets had to justify outlays in terms of expected returns on investment. Geology began to inform oil exploration in the 1890s.[24] Thirty years later oilmen remained skeptical of its utility. As one Mid-Continent oilman observed: "Practically none of the great discoveries have been made as a result of geological pioneering and advice. Even in California, the biggest pools which have been discovered were discovered in exactly the places where the geologists did not think that oil would be found." This record owed in large part, he argued, to professional conservatism that rendered geologists' advice useless: "They all play so safe . . . that they leave a fine opening for an alibi by saying 'only the drill will tell.'"[25]

Indeed, it was one thing to locate a possible oil-bearing structure; it was another to establish the presence of petroleum pools. One had to drill a well to prove a property. Companies might drill in the wrong direction or pass on leases that later proved to be productive. For example, at Kettleman North Dome, a San Joaquin Valley field discovered in 1928, Los Angeles–based Superior Oil Company discovered oil on a lease that two firms with producing wells in the field thought to be worthless. Three years later, the property accounted for more than 40 percent of the independent's statewide output.[26] The company's path to success demonstrates why nonintegrated operators thrived at the center of the upstream oil business.

During the 1920s, majors began to use so-called geophysical methods to improve the success rate of exploration. The interwar record of the new technology was mixed. Discoveries of many fields in Texas, for instance, could have been made without it: majors typically deployed geophysical methods to check acreage that already had presented surface indications.[27]

Even by the late 1950s, James McKie notes, neither geology nor geophysics "remotely approach[ed] certainty" in the search for oil.[28] Ultimately, the most important factor in the decision to drill a well was the interpretation of imperfect information. As Geoffrey Snow and Brian MacKenzie write, "Exploration decisions have been based in large part on intuitive, subconscious considerations that may be referred to as gut feelings, subjective judgments, or hunches."[29]

The California experience confirms McKie's observation that "the history of oil exploration provides many instances of erroneous evaluation of prospects by major companies, which have created opportunities for other major companies or independents to make the right intuitive judgments at the right time."[30] In one of several cases of a major missing the discovery of a large field during this period, Associated Oil Company drilled a dry well at Elwood in 1909. For almost two decades thereafter, geologists wrote off the area. Majors ignored it. In the summer of 1928, however, the aforementioned Rio Grande and Barnsdall companies drilled the discovery well on the advice of a young geologist from the University of California whose investigations persuaded him that the conventional wisdom was wrong. Moreover, Rio Grande filed its notices on the adjoining state tidelands when any company might have done so. The negative assessments of their exploration departments, rather than the legal regime, persuaded majors not to explore at Elwood. Instead, three majors placed their bets on the Capitan field, which lay a few miles up the coast. It turned out to be a far less lucrative play.[31]

California's numerous oil seeps, asphalt outcroppings and deposits, natural gas releases, and exposed geological formations offered "a specially tempting bait for the wildcatter," as one reporter observed—another factor that explains why independents were not confined to the periphery of the upstream segment.[32] Indeed, both the U.S. Geological Survey and the California State Mining Bureau documented these manifestations for the benefit of oil seekers. In a report on the Santa Maria oil district, for example, noted geologist Ralph Arnold concluded his extensive geological descriptions of the area with a list of tracts "that appear especially to invite testing with the drill."[33] Surface indications of oil as often as not attracted local interest before the majors' exploration departments could determine whether the property might be a worthwhile addition to their firms' portfolios.

California's majors invested more capital in accumulating properties than in drilling them, much as McKie found to be the case nationally after World War II.[34] From 1918 to 1923, for instance, Associated spent an annual average of $1.73 million in leasing or acquiring additional properties, compared to an average of $1.05 million annually on drilling wells.[35] Under Lyman Stewart, who had a "prodigious appetite for prospective oil properties," Union Oil Company accumulated hundreds of thousands of acres statewide, much of it in fee—that is, the company acquired both the surface and the mineral rights. As of 1919, Union controlled 286,417 acres statewide. In the Coastal Region, the accumulation of leases in the Santa Maria Valley provided the major with a "first mover" advantage in production. The company lost it, however, with the development of the Ventura Avenue and Elwood fields, showing that relative success in the upstream oil business was contingent on developing new reserves of petroleum to extract.[36]

During the early twentieth century, California's majors leased or held in fee increasing amounts of land: 503,800 acres as of 1919, rising to at least 1.2 million acres by 1940. Unsurprisingly, they also had higher ratios of unproven-to-proven lands than did independents, since majors historically have deferred possible returns on initial investment for relatively longer periods of time. As of 1940, for instance, the ratio for the three majors for which figures are available was more than thirteen-to-one (509,528 unproven acres to 37,996 proven acres). For five of the leading independents of the Coastal Region, the ratio was seven-to-one (111,268 unproven acres to 16,144 proven acres). These figures suggest that California's majors approached the upstream oil business in much the same way as their contemporary counterparts in Texas. Texas majors tied up large amounts of capital in leases without necessarily intending to test all their potentially oil-bearing properties with the drill. Rather, they "farmed out" acreage to independents willing to accept the risk associated with exploration. By this process, majors brought additional independents into the upstream oil business.[37]

As we shall see in the case of the Ventura Avenue field, lease terms added to the competitiveness of the upstream segment by exerting financial pressure on the lessee. A lease is a legal contract by which a private or public property owner authorizes the exploration for and extraction of petroleum on their property for a specific period of time in consideration of a royalty. Leases generally specify a period of time within which the lessee must drill a minimum number of test or development wells. Lessees of private land typically pay annual rental fees until they either surrender the lease or drill one or more wells. They often pay cash bonuses on a per acre basis to secure the right to drill. Also embedded in the lease are a number of implied covenants, the most potentially financially demanding of which requires the lessee to drill so-called offset wells at specified distances from producing wells on adjacent leases to prevent these wells from capturing an "excessive" share of crude oil in the pool. Leases on public lands are typically even more demanding. For instance, California's Minerals Reservation Act required operators of tidelands leases to begin drilling a test well within a year of the filing date and to sustain drilling thereafter until oil was found or forfeit the lease. Thus, operators could not hold such leases by making annual rental payments, as leases on private lands typically allow. All leases provide for the payment of a royalty to the lessor, expressed in terms of a percentage of production, from the conventional one-eighth royalty to one-half, as was the case with some Signal Hill leases during the frenzy set off by the discovery of the Los Angeles Basin field in 1921.[38]

Lease requirements could become an albatross around the necks of firms that pursued an aggressive land-acquisition policy, as both Associated and Shell Company of California, the Ventura Avenue field's largest operators, discovered during the Great Depression. From 1929 to 1931, the U.S. capital budget of the Royal Dutch/Shell Group, the parent of the California major, fell from $100 million to $16.6 million. The company surrendered $12 million in leases and farmed out acreage to avoid drilling expenses. Yet in 1931, the firm spent $6 million on drilling to satisfy lease requirements.[39] At the same time, satisfying the terms of its Ventura Avenue leases exhausted more than two-thirds of the budget of Associated's producing division in the early 1930s. Where the company's executives could save money, they did: retrenchment included not exploring an area of the field that the Lloyd Corporation would prove to be productive.[40]

The California upstream experience supports Donald Critchlow's argument that managerial choice within the specific cultural and institutional context of the firm is a critical factor in business outcomes.[41] As Amherst economics professor Willard L. Thorp put it at the time, "It is not their size, it is the ability of their leaders . . . that seems to be the outstanding cause for difference among business enterprises."[42] In the upstream oil business, one firm's skepticism of an area's potential created an opportunity for another to discover or extend a substantial field. As often as not, the opportunistic firm was an independent.

The development of California's crude oil reserves from the inception of the industry through 1920 facilitated the development of integrated firms, namely Associated Oil Company, General Petroleum Corporation (GP), Shell Company of California, Standard Oil Company of California (now Chevron), and Union Oil Company of California, each of which ranked among the nation's leading companies in terms of assets. As of 1911, however, the three integrated firms that had established themselves as industry leaders in California—Associated, Standard, and Union—controlled less than one-third of crude-oil production in a segment that included at least 460 other companies.[43] From 1911 to 1919, the majors, which now included Shell and GP, increased their share of production to 62 percent—a high-water mark for the period from 1900 to 1940. Yet the rationalization of America's oil industry by World War I did not constitute an end point in the evolution of the structure of the upstream segment. As the industry expanded—annual output almost tripled from 1920 to 1929 in response to a soaring, automobile-driven demand for gasoline—new entrants competed against and cooperated with existing firms to exploit California's petroleum reserves.[44]

After World War I the center of the state's oil industry shifted dramatically from the San Joaquin Valley to the Los Angeles Basin, where so-called town-lot drilling on private, urban leases by hundreds of operators characterized the development of the largest fields, most prominently Huntington Beach, Long Beach (Signal Hill), and Santa Fe Springs.[45] In these three fields, which together accounted for almost 70 percent of California's production in 1923, companies drilled one well for every two acres or less, in contrast to drilling an average one well for every eight to ten acres elsewhere.[46] As James E. Herley reflected: "At least in this area, it was the time of the small operator. There were a lot of opportunities."[47] Indeed, at 55 percent, independents accounted for a higher share of production in the Los Angeles Basin than anywhere in the state, as of 1940.[48] During the interwar period, when Ralph Lloyd hit his stride as a small businessman, independents increased their California market share from 38 percent to 46 percent.[49]

As of 1940, when the Lloyd Corporation was realizing success as an independent operator in the Ventura Avenue field, as we shall see, California's upstream industry was more fragmented and competitive than ever. The state's oil-and-gas division recognized more than fifty fields. Seven majors, which now included Richfield Oil Company and The Texas Company (Texaco), controlled more than 80 percent of the production in only seven of them, leaving just over half of the state's output to 1,226 independents—almost triple the number thirty years earlier.[50] In the Coastal Region, the number of independents with producing assets more than tripled from 1915 to 1940, rising from forty to 127. Independents held a 40 percent share of the Coastal Region's output, as of 1940—below the state average, owing to the weight of the Ventura Avenue field, but not a residual share by any definition.[51]

Yet, as this book explains, citing the share of output controlled by the Lloyd Corporation as independent operator would not explain in full Ralph Lloyd's contribution to the development of the Ventura Avenue field. His career shows how dynamic relations between independents and majors operated in California, which, in turn, illustrates how competition and cooperation shaped the industry during a sustained period of expansion.

During the first four decades of the twentieth century, the upstream segment of the California oil industry bucked the trend toward dominance in American business by large, multi-divisional enterprises. As small businesses, independents succeeded not by carving out a niche market with specialty product offerings, as was generally the case in distribution and manufacturing, but by exploiting their comparative advantages in exploration and production.[52]

An Oil-Urbanization Nexus

Ralph Lloyd invested millions of dollars of royalty and operating income in commercial real estate in an effort to convert a depleting asset—crude oil—

into assets that might produce sustained income streams. Beginning with the purchase of two lots on the corner of Union Avenue (now Martin Luther King Boulevard) and Wasco Street, in Portland, Oregon, Lloyd acquired hundreds of properties, including two large blocks of undeveloped property around Holladay Park on the city's East Side. Much of this book narrates and analyzes how Lloyd developed this district over three decades, culminating in the completion of the Lloyd Center, a regional shopping center, in 1960. In doing so, it shows how he competed successfully in a vast and fragmented sector that has received relatively little attention from scholars of small business.[53] Just as "every man that has a piece of land with oil on it can drill a well," as Charles A. Canfield observed, every person with a piece of commercial property might erect a building on it and lease space to tenants.

Ralph Lloyd's investment of his profits from petroleum extraction in commercial real estate established a direct link between oil and urban development on the Pacific Coast. Lloyd was not the first, and hardly the last, oilman to attempt to convert black gold into commercial real estate. Yet it is the rare study that discusses diversification into real estate from the oil and gas sector as a business strategy.[54] Discussion of investment of oil-generated wealth in commercial real estate remains largely ad hoc across biographies of successful independents.[55] As we shall see, Lloyd reinvested considerable earnings from petroleum extraction in drilling additional wells. Yet he also allocated a significant portion of his surplus capital to building projects hundreds of miles from the oil fields of Southern California. In examining the nexus of oil and real estate development through the lens of Lloyd's career, this book increases our understanding and appreciation of the extent to which the oil industry generally and an independent operator in particular could shape the built environment.

Martin V. Melosi conceives of "energy capitals" as physical regions, writing that "energy-led development has transformed many cities and regions physically, influencing metropolitan growth, shaping infrastructure, determining land use, changing patterns of energy consumption, and increasing pollution."[56] The career of Ralph Lloyd shows that the regional reach of Los Angeles, *the* "energy capital" of the 1920s, as realized both through the investment of capital accumulated from local petroleum extraction and the transfer of ideas generated by energy-led metropolitan development in the Los Angeles Basin, could be considerable.[57] Materialized as a gasoline-fueled automobile city with wide, asphalt-paved streets, the City of Angels provided Ralph Lloyd with both blueprint and imaginative possibility for Portland, where the oilman made his most substantial and sustained real estate investments.[58]

Richard Longstreth writes of the transformation of commercial space in Los Angeles between 1914 and 1941. Ralph Lloyd was one of the many speculators who drove it. Before plunging into the Portland real estate market in the mid-1920s, Lloyd acquired and developed property on the fringes of downtown and in Hollywood. The buildings that the oilman constructed in these so-called close-in districts comprised a portion of the "forgotten arterial landscape" that Longstreth has detailed and served as templates for vernacular projects that he would undertake in Portland. Signature landmarks, such as the Ambassador and Biltmore Hotels, also served as business and architectural models for Lloyd's plans for his property on Portland's East Side.[59]

An examination of Ralph Lloyd's career in commercial real estate also sheds light on contemporary relationships among the principal members of the building team, including owner, architect, structural engineer, and general contractor (or builder), particularly as they pertained to the delivery of vernacular projects, of which little, if anything, has been written. The prevailing method of project delivery in the commercial segment of the construction industry early in the twentieth century—indeed until quite recently—was known as "design-bid-build." Under this approach, the architect designed the project in consultation with the owner and the structural engineer. Builders who had no input in the design of the project then bid competitively on the architect's owner-approved plans and specifications, with the low bidder awarded the construction contract. The architect then managed the construction process as the owner's representative. The Lloyd Corporation used this approach, for instance, to build the Lloyd Center, which was one of the largest regional shopping centers in America when it opened in August 1960 on the East Side of Portland. As I have explored elsewhere, an alternative, increasingly widespread approach, known as "design-build," involves all principal members of the building team in the planning and design of the project.[60] In theory and often in practice, involving the builder in the early stages of the projects ensured that a project could be built within the owner's budget. Under this project delivery approach, construction contracts for private projects are typically negotiated.[61] During the early twentieth century, a number of engineering firms promoted design-build as a cost-effective means to deliver projects, particularly industrial buildings constructed from predesigned templates. As we shall see, Ralph Lloyd occasionally retained the services of such firms on this basis. Typically, however, he adopted a hybrid approach that may be called "design-negotiate-build." Under this configuration, Lloyd, as owner, retained an architect for traditional design services, but then provided a handpicked builder with the opportunity to suggest changes to the architect's plans and specifications. Once he was satisfied with the final design, after one or more iterations of revision, he then would negotiate the construction contract with his chosen builder. In considering the approaches that Ralph Lloyd used to deliver buildings, this book increases our understanding of how the commercial segment of the construction industry operated at the project level between 1920 and 1960.

As a "citizen outside the government," Ralph Lloyd became involved in Los Angeles's street program of the 1920s, which spent millions of dollars to widen, open, and extend arterial streets, along many of which he held property. He was a member or director of more than half-a-dozen improvement associations, corresponding with the geography of his commercial real estate portfolio. He was also a member of the Major Highways Committee of the Los Angeles Traffic Commission. Impressed with Los Angeles's commitment to solve its traffic problem and proud of the role he played in mobilizing municipal resources and expediting street work toward that end, Lloyd worked to persuade Portland to respond with comparable means and enthusiasm. And so, as we shall see, the Southern California oilman rarely acted in Portland without referring explicitly to his city-building experiences in Los Angeles.

At the same time, Ralph Lloyd acted as intermediary of the majors operating in the Ventura Avenue field who ultimately would wield substantial influence on the growth trajectory of Ventura, California, as an oil boomtown. Local business and civic leaders wanted their city to emulate nearby Santa Barbara as a recreational and tourist destination. For their part, the outside corporations sought to develop without restriction the oil field that lay adjacent to the city, thereby maximizing returns on their local capital investment. The Los Angeles– and San Francisco–based executives of Associated, GP, and Shell were pleased to see Ventura grow in terms of population, but showed far less interest in the quality-of-life aspects of boosters' plans. Concerned too with the maximization of crude oil production, Lloyd advocated on behalf of the outside firms whenever their interests conflicted with the vision of local leaders. He also enlisted boosters' support on local issues of concern to the major oil companies. In doing so, Lloyd largely succeeded in persuading Ventura's elites that their parochial interests aligned with those of cosmopolitan capitalists.

Once the Depression ended his initial bid to develop his East Side holdings as an extension of Portland's geographically confined downtown, Ralph Lloyd turned his attention again to Los Angeles. Beginning in 1931, he assembled a portfolio of properties along Wilshire Boulevard, including a block in Beverly Hills bounded in part by Rodeo Drive that today serves the highest end of the retail market. His belief in the potential of these properties and his efforts to develop them demonstrate the adage that there are always opportunities to be had in commercial real estate regardless of market conditions. Still, Lloyd, as developer, failed to contribute significantly to the commercial transformation of Wilshire Boulevard during the 1930s.[62] Ultimately, as we shall see, "regime uncertainty," associated in particular with federal tax policy, dampened Lloyd's enthusiasm for the private commercial real estate market and helped to persuade him that investing in testing the eastern limits of the Ventura Avenue field would yield greater profit.[63] At the same time, Lloyd was relatively risk adverse as a developer. As we shall see, other projects would proceed in the vicinity of his valuable properties even as he receded from the market.

Ironically, given his increasingly strident views on the growth of the federal state, Ralph Lloyd enjoyed rather more success in meeting the commercial real estate requirements of public agencies and completing projects underwritten with federal funds. Even as he railed against the New Deal, its revenue acts in particular, federal and state agencies provided him with opportunities for profitable real estate development. Moreover, by leasing abandoned space, these public agencies salvaged several of Lloyd's problem leases with private clients. Public capital investment enabled the Lloyd Corporation to bridge the two decades between the suspension of its building program in Portland during the Depression and development of the Lloyd Center. Moreover, the construction a cluster of office buildings for the Bonneville Power Administration set in motion the suburbanization of Portland's East Side and provided Ralph Lloyd with the confidence to pursue his development program for the city. As such, Lloyd's career illuminates the crucial role played by public capital in catalyzing the commercial real estate market in depression, war, and early postwar recovery.

The suburbanization of urban space on Portland's East Side culminated with the planning, design, and construction of the Lloyd Center. Gray Brechin visualizes Lewis Mumford's so-called Mega-Machine, "a constellation of five activities [that] has operated from the appearance of the first cities [to] give humanity its growing dominion over nature," as a pyramid. In this arrangement, mining (including oil) sits atop the apex, exerting "a seminal and dominant role" over mechanization, metallurgy, militarism, and moneymaking.[64] The regional shopping center as a market structure embodied the apotheosis of more than half a century of the development of the department store as an engine of consumerism.[65] Mass consumption may be visualized as a derivative activity forming a new base of the pyramid, shaping urban cultural, economic, and social trajectories in an increasingly post-industrial age. Mobilized by the Lloyd Center, mass consumption in Portland was linked directly and indirectly to petroleum extraction in Southern California.

For the whole of his career, substantial and sustained crude oil production fueled Ralph Lloyd's commercial real estate program, which did not become self-sustaining until after his death. His ambitions for Portland's East Side alone required him to invest continuously in oil exploration and production. And so, as we shall see, in the 1930s, the Lloyd Corporation would seek to extend the limits of the Ventura Avenue field as an independent operator when Associated and Shell showed no interest in doing so. But first we turn to the ranch lands that lay adjacent to the city of Ventura, which, a young Ralph Bramel Lloyd became convinced, sat atop a vast pool of petroleum.

Chapter 1

Developing the Ventura Avenue Field

The happiest people in the world are those with something to do and this is especially true when your interest in your work makes it a pleasure.

—Ralph B. Lloyd, 1932

The development of Ventura Avenue, which would prove to be America's twelfth largest oil field of the first half of the twentieth century, was a "life work" for Ralph Bramel Lloyd that underpinned his success as a small businessman.[1] As luck would have it, his father's speculation in real estate put him in position to orchestrate it.

Lewis Marshall Lloyd was born in 1835 in Lee County, Virginia. In 1854, his father, Absalom Lloyd, a farmer and Methodist preacher, moved the family to Andrew County, in the northwest corner of Missouri. Restless, impatient, and ambitious, Lloyd soon departed for St. Joseph, the seat of adjacent Buchanan County. The city had become a bustling staging post for westward migration, beginning with the California gold rush. There he read law with General Henry S. Tutt. On the eve of the Civil War, he was admitted to the Missouri bar. In a state of divided loyalties, Lloyd threw his lot with the Confederacy, enlisting in the Missouri State Guard under the command of Sterling Price, who served as brigadier general during the Mexican War and governor of the state from 1853 to 1857. Missouri remained under the control of the Union Army throughout most of the war. Lewis Lloyd left the army at the end of his three-year commitment. Under Reconstruction, however, Lloyd was disbarred, having aided and comforted the enemy. For a time, he taught school in the St. Joseph area before he returned to Andrew County, where he courted Sarah Elizabeth Bramel.[2]

Lewis Lloyd's fortunes rose when he secured a contract to supply U.S. Army fortifications in Colorado. With contract in hand, he persuaded Sarah Bramel to marry him. The couple moved near present-day Greeley, Colorado, where Lloyd cut hay for the U.S. Army. By the mid-1870s, Lloyd had relocated his growing family to Neosho, in southwestern Missouri, where, in October 1861, the remnants of the elected state government that supported the Confederacy had voted to secede from the Union. Lewis Lloyd resumed the practice of law with partner George Hubbert. On 28 February 1875, Ralph Bramel Lloyd was born. In all, his siblings included sisters Eleanor, Lorena, and Roberta, and brothers Lee and Warren. In 1878, Lewis Lloyd was elected to the Missouri Senate.[3]

In 1886, Lewis Lloyd came west in search of a cure for a variety of ailments. Taking advantage of the offer made by financier Jay Gould's Missouri Pacific Railroad of free passes to California to all current and former state senators, he traveled to Ventura, a small town located on the Pacific Ocean, some fifty miles northwest of downtown Los Angeles. The timing for successful real estate speculation hardly could have been more favorable. For Southern California was booming. The population of Ventura was doubling in response to petroleum and agricultural development and the extension of the Southern Pacific Railroad to Santa Barbara twenty-five miles further up the coast. Boosters would soon tout Ventura County as "a land flowing with milk and honey, oil and wine, and teeming with the richness of tropical and semi-tropical fruits."[4] Within days, Lloyd acquired a few thousand acres of the Rancho Ex-Mission (figure 2). He promptly sold most of them as town lots. With the acquisition of ranches in Hampton and Wheeler Canyons further inland—the latter in partnership with S. R. Thorpe—and another 480 acres in Simi Valley, located between Ventura and Los Angeles, Lewis Lloyd held some 4,500 acres of land. He returned to Neosho to retrieve his family.[5]

In 1887, Lloyd and Thorpe, together with N. Blackstock and W. E. Shepherd, formed the Ventura Land and Water Company (VL&W). Within a few years, Lloyd bought out his partners. By then, however, the regional boom had collapsed. In 1892, with the attraction of real estate speculation at a low ebb, Lloyd leased his Ventura ranch lands and relocated the family to Berkeley, where sons Lee and Warren matriculated at the University of California and son Ralph entered Berkeley High School.[6]

The death of Lee, the eldest son, two years later dealt a crushing blow to Lewis Lloyd. Disheartened and in fragile health, he nonetheless returned to Ventura in 1895 with his wife to take up cattle ranching. In that year, Warren Lloyd graduated with a law degree and left for Europe to continue his studies at the University of Berlin; Ralph graduated from high school and matriculated at the University of California (figure 3).[7]

By this time, both Lewis and Ralph Lloyd were convinced that oil lay beneath all the ranch lands that lay to the west and north of the city of Ventura. When Ralph was fourteen, Lewis Lloyd apparently barely escaped with his life when the flames of a wildfire that he was investigating on horseback reached a

Figure 3. Ralph Bramel Lloyd, University of California undergraduate, circa 1896. (The Huntington Library, San Marino, California, Photos drawer 1, box 4.)

spot where oil would be discovered some three decades later and erupted into a natural-gas-fueled inferno. So the story goes, flames engulfed the horse as Lewis leapt from the animal and rolled down an embankment to safety. However exaggerated it may have become in the telling, the incident was foundational to Ralph Lloyd's thinking about the potential mineral wealth that lay beneath the surface of the family ranch. At the University of California, Lloyd studied geology under Joseph Le Conte and Andrew Cowper Lawson, who were early proponents of anticlinal theory, which linked the presence of petroleum reserves to certain substrata structural folds. He concluded that the characteristics of the rugged terrain that surrounded the city of Ventura bore a striking similarity to the figures that his Berkeley professors were using to illustrate their lectures.[8]

Lewis Lloyd's increasingly poor health intersected with bad luck to cut short his son's academic career. In December 1896, he pleaded with son Ralph to return Ventura to manage the ranch. Lewis Lloyd had taken ill and was in no condition to keep the business viable during a drought that was afflicting the region. With the assistance of his father's top ranch hand, Teodoro Ortega, Ralph took a number of emergency measures to save the business, including killing calves and scouring the county on horseback in a desperate search for pasture. Drought conditions persisted through the winter of 1897–1898, exacerbating the difficulties with which Ralph Lloyd contended. For his part, an exhausted Lewis Lloyd gave up ranching in the summer of 1898 and decamped to Los Angeles. He left behind a family business with some $20,000 in mortgage debt.[9]

Ralph Lloyd carried on for five years before he too withdrew from the ranching business. Rainstorms that relieved the drought in the winter of 1898–1899 convinced Lloyd to borrow money to restock the ranch. Unfortunately, the animals that he purchased became infected with Texas fever—perennially a plague of the late nineteenth-century open-range cattle industry[10]—and only half the herd survived. His father's incapacity as a rancher may have enabled Ralph Lloyd to gain valuable experience that he eventually would apply with success to other business fields. But for the moment, with seemingly no means of overcoming the challenges of cattle ranching, Lloyd retreated. With the agreement of his father, he attempted to sell the land associated with the Rancho Ex-Mission, enticing potential buyers with its oil-bearing potential.[11]

In 1898, A. H. Koenig, a petroleum engineer based in Los Angeles, had become intrigued with the oil potential of VL&W's property around Ventura and had sent E. A. Rasor, who had some geological experience in the Fullerton field in the Los Angeles Basin, to map the anticline. Based on his findings,

Rasor counseled Lloyd not to sell the mineral rights to the property. Someday it would make him rich, he supposedly predicted.[12]

Rasor's favorable conclusions notwithstanding, Lewis and Ralph Lloyd were unable to persuade anyone to test the property with the drill. In 1900, for instance, Dr. A. H. Chandler, a friend of Lewis Lloyd, sent a petroleum engineer, D. W. Hudson, to investigate. Hudson's conclusions that "there is a vast bed containing oil under [the property]" and that one "can readily see that the location is pre-eminently an oil field" were not enough to convince Chandler to drill a well. In any case, parties that were keen to extract petroleum showed interest only in securing mineral rights, not in acquiring ranch land.[13]

Finally, in September 1903, Mariano Erburu, a native of Navarra, Spain, who had recently resigned his partnership in a local mercantile firm, acquired VL&W's Rancho Ex-Mission property for $20,000. Lewis Lloyd offered him the mineral rights for an additional $5,000. Erburu declined. He would later say that, at the time, he didn't have the money.[14]

With the execution of the sale, Ralph Lloyd turned his attention to the VL&W ranch in Simi Valley, aspiring to grow walnuts and black-eyed beans. In 1904, the year in which he married Lulu Nettie Hull, Ralph convinced his father, who preferred so-called dry farming, to install an irrigation system. Indirectly, Lloyd's turn to wet farming set him on a path to become a leading real estate developer in Portland, Oregon. Lewis Lloyd also agreed to acquire additional Simi Valley ranch land on which Ralph Lloyd would initially drill for oil in Ventura County. With his health at least temporarily restored, however, Lewis Lloyd also interjected himself into the daily affairs of the ranch. His constant presence on the property convinced Ralph Lloyd to yield the reins of management to him.[15]

In the process of installing several thousand feet of wood irrigation pipe that he purchased from the National Wood Pipe Company in Los Angeles, Ralph Lloyd had become acquainted with the men who managed the company, its president, William E. Hampton, in particular. The firm manufactured wooden pipes and other lumber products for mines, water works, and hydroelectric plants in Los Angeles and planned to open a factory in San Francisco. In February 1905, Lloyd interviewed with Hampton, who hired him with a view to install him as manager of the San Francisco facility when it opened. Until then, Hampton initiated what would now be called a management-trainee program for Lloyd at a salary of $60 per month.[16]

The earthquake that left San Francisco in ruins on 18 April 1906 put paid to William Hampton's plans to open a factory in that city. Ralph Lloyd continued to learn the business in Los Angeles. During this period, his daughters Eleanor and Edna Elizabeth were born. He also contracted typhoid fever, which damaged his intestines.[17] Thereafter, Lloyd would suffer from gastroenteritis, which on many an occasion debilitated him. Because he was not properly diagnosed, the condition would plague him until ultimately it would precipitate his demise.[18]

In July 1907, Hampton sent Lloyd to Olympia, Washington, to manage an underperforming plant. Three years earlier, Hampton had completed a $200,000 factory in the state capital for the Pacific Tank Company, a smaller concern based in Tacoma, Washington, which he controlled with two other partners. Now Hampton sent Lloyd to Olympia to determine why the plant was losing money. Lloyd reported that the factory was manufacturing products of poor quality that often had to be recalled. The problem was management. Impressed by the analysis, Hampton put Lloyd in charge of the plant at a salary of $200 per month. Within a year, the plant returned to profitability. As of July 1908, Hampton had increased Lloyd's salary to $385 per month. He also offered Lloyd the opportunity to buy stock in National Wood Pipe Company, which Lloyd exercised, accumulating a substantial block of shares.[19]

Ralph Lloyd's star within the company rose steadily. At the end of 1908, National Wood Pipe Company opened a sales office in Portland. Hampton put Lloyd in charge of it. And so Lloyd relocated to the city with his family, arriving in Portland on New Year's Day. The next day, the Lloyds moved into a house at Third and Multnomah Streets, on Portland's East Side. In the first quarter of 1909, Ralph Lloyd became the company's leading salesperson. Early in the second quarter, he secured enough business to keep the Olympia and Los Angeles factories operating at capacity for the remainder of the year.[20] At Hampton's behest, Lloyd stepped up efforts to secure a site for a new factory in Portland, which had been in the works for at least two years. On 4 July 1909, fire destroyed the Olympia factory, making the search a matter of urgency. Three weeks later, Lloyd acquired a site in Kenton, located in the northern part of Portland near the Columbia River. Lloyd supervised the construction of a $450,000 factory—"the largest and best equipped on the Pacific Coast"—which was completed in February 1910.[21]

For reasons of health and compensation, Ralph Lloyd cut short his career with the firm. Later in 1910, William Hampton combined the National Wood Pipe and Pacific Tank Companies into a new concern, the Pacific Tank and Pipe Company. It was a move that Ralph Lloyd apparently had been urging for some time. For National Wood Pipe Company, in which Lloyd held shares, paid no dividends. At the same time, Pacific Tank Company, in which only Hampton and two vice presidents held shares, paid generous dividends. Hampton offered Lloyd stock in the new concern, but on terms that the latter considered unfavorable. Lloyd sold his stock in National Wood Pipe Company. In September 1910, he tendered his resignation, effective 1 January 1911, citing overwork. Hampton promised to lighten Lloyd's responsibilities and make it worth his while financially, were he to agree to stay on. For the moment, Lloyd consented to remain in Hampton's employ. In February 1911, however, Lloyd told Hampton that repeated bouts with stomach distress prevented him from staying on past June. Before he left the company, Lloyd participated in discussions that aimed, but ultimately failed, to produce a merger of Pacific Tank and Pipe Company with three other regional wooden tank and pipe firms.[22]

Crucially for the trajectory of his career as a real estate developer, Ralph Lloyd fell in love with Portland during his two and a half years of residence. He and wife Lulu added two daughters, Ida and Lulu May, to their brood. Lloyd joined the white, largely Anglo-Saxon, and male Arlington Club, the bastion of financial and commercial leadership in the city. He acquired a second residence, adjacent to the family home on Multnomah Street. He was elected to the board of directors of the Bank of Kenton. In April 1910, Lloyd established an initial foothold in the local commercial real estate market with the purchase of two lots on the corner of Union Avenue (now Martin Luther King Boulevard) and Wasco Street, the site of the Adcox Auto and Aviation School. He contemplated demolishing the structure at the end of the tenant's lease and erecting either a family hotel or an apartment building in its place. In conjunction with Robert D. Grant of Los Angeles, Lloyd also attempted to negotiate the purchase of Holladay's Addition to the City of Portland (hereafter Holladay's Addition), which occupied much of the area immediately to the east of his Multnomah Street residence, from the Oregon Real Estate Company, of which Charles X. Larrabee was president. A successful banker, copper miner, and horse breeder in Montana, Larrabee had bought the platted, but undeveloped, property when he moved to Portland in 1877. Named for Ben Holladay, a shady railroad entrepreneur, who had acquired it in 1868, the land remained undeveloped. Were Lloyd to acquire the property, he instantly would have gained a significant foothold in the local real estate market. Ultimately, Larrabee declined to sell his holdings at this time to Lloyd. Once he secured his fortune in oil in California, Ralph Lloyd would return to Portland, purchase the property, and eventually become one of the city's leading commercial real estate developers.[23]

Exploring for Oil in Simi Valley

It is unclear what Ralph Lloyd may have had in mind for his career when he tendered his resignation to William Hampton in the fall of 1910. That he bought a modest two-story house at 1023 South Alvarado Street, just south of Tenth Street (now Olympic Boulevard), in Los Angeles, in December 1910, indicates an intention to settle in Southern California.[24] Crisis in the family business associated with his mother's terminal cancer persuaded him to resume managing the Simi Valley ranch, which would lead directly to his searching for oil in Ventura County.

Facing her mortality, Sarah Elizabeth Lloyd had drafted a will that provided for the distribution of her shares in VL&W to her five children, investing in them control of the company.[25] Since 1904, Lewis Lloyd had demonstrated his shortcomings as a grower; she wanted her children to have an opportunity to restore the ranch to profitability. As her husband and the largest shareholder in VL&W, Lewis Lloyd demanded that she change the will.[26]

At the same time, Warren Lloyd, who was practicing law in downtown Los Angeles with the firm of Lloyd, Cheney & Geibel, had accumulated an insurmountable debt load. He wanted to withdraw his 250 shares of stock in VL&W and pledge it as collateral for personal loans.[27] The company's stock, however, was allocated to a common pool. To improve VL&W's ability to pay off its mortgages and liquidate its outstanding obligations, family members had reached agreement, in March 1909, to hold the shares of the company as a block to prevent their use by any one stockholder as collateral or security on a loan or other debt instrument. The agreement would remain in force unless the shareholders annulled it by mutual consent.[28]

At his mother's request, Ralph Lloyd took time from negotiating the merger of regional tank and pipe companies to travel to Los Angeles. He persuaded his siblings to vote in favor of a motion to allow their brother Warren to withdraw his stock from the company. As we shall see, Ralph Lloyd's subsequent acquisition of his brother's shares would become a point of contention between family members once the vast pool of petroleum that lay beneath the family's former Ventura ranch was developed. Ralph Lloyd also convinced his mother to change her will to allow Lewis Lloyd to retain control of VL&W. She did so on the condition that her son return home and manage the company's ranches. On 26 June 1911, Sarah Lloyd died. Four days later, Ralph resigned his position with Pacific Tank and Pipe Company. On 9 July 1911, he returned to Los Angeles with his family.[29]

Ralph Lloyd's return to Southern California was well timed in terms of his entry into the oil business. For, in 1912, the promise of a local oil boom enticed his future drilling partners, Joseph B. Dabney and E. J. Miley, to Simi Valley. Oil booms of the late nineteenth century had established inland Ventura County and north Los Angeles County as the "cradle area of the California oil industry." During this period, entrepreneurs reinvested wealth generated in the oil fields of Pennsylvania in lands where seeps, asphalt outcroppings, natural gas releases, or exposed geological formations suggested the presence of large deposits. Capitalists from San Francisco became interested in these lands for their potential in supplying their city with illuminating fuel, eastern supplies of which the Civil War had interrupted. From the 1860s until the mid-1890s, most of California's exploration, and almost all its production, occurred between Ventura and north Los Angeles County. Indeed, by 1900, more than two dozen fields had been discovered. Yet output remained low by twentieth-century standards, peaking in 1888 at 690,333 barrels. Technology proved to be inadequate to meet the geological challenges that operators faced. With its frequent tilts and folds in its strata, the local geology was unlike anything that prospectors had found in Pennsylvania. By the turn of the century, the California oil industry had shifted its attention to the San Joaquin Valley, where both Dabney and Miley enjoyed success as independent operators.[30]

Born on a farm in Madison County, Iowa, on 21 January 1858, Joseph B. Dabney moved west because of the poor health of his widowed mother. After her death, in April 1889, Dabney worked as a reporter in Bozeman, Montana. Sometime after 1891, he and his brother Richard secured a federal contract to supply meat to the Army Corps of Engineers, which had undertaken a program of road construction into Yellowstone Park. Building storage sheds in conjunction with the contract inspired Dabney to speculate in timber. He settled in Aberdeen, Washington, where he accumulated a modest fortune dealing in timberlands and other real estate. At the turn of the century, Dabney was exploring for oil with Miley in the booming San Joaquin Valley.[31]

Born in St. Clair County, Illinois, on 22 October 1873, Emmor Jerome Miley was the brother of Lieutenant Colonel John David Miley, for whom a military installation at Point Lobos in San Francisco would be renamed in 1900. At seventeen, he moved to San Francisco, where his brother was stationed. After completing high school in 1895, he leased orchards in Solano County north of the city and shipped fruit until he turned his attention to the oil business.[32]

In 1900, both Dabney and Miley arrived in Kern County independently. Soon they formed a partnership to explore for oil. They leased land in the McKittrick field, which had been producing crude oil for two years, and drilled ten wells (figure 1). In May 1901, Dabney incorporated the Dabney Oil Company with a capitalization of $1 million. Miley sold his interest in the partnership. The two men then pursued separate business ventures until they joined Ralph Lloyd in searching for oil in Ventura County.[33]

E. J. Miley became a partner in the Silver Bow Oil Company, which held leases in the McKittrick and Midway fields (figure 1). Miley drilled one of the early wells in the Midway field. After the company ceased operating in 1903, in the midst of an oil sector slump, Miley mined copper in Nevada. When the financial reverberations of the 1906 San Francisco earthquake rocked Nevada's economy, Miley returned to San Francisco, where he worked in construction. In 1908, he resumed his search for oil in Kern County. Together with David J. Graham, Miley formed the State Oil Company and reentered the McKittrick field. Three years later, in March 1911, Graham, Miley, and A. M. Buley incorporated the State Consolidated Oil Company to absorb properties held individually by the partners. As lessors, Ralph Lloyd and Joseph Dabney would engage the firm to drill for oil in what would become the Ventura Avenue field.[34]

The early 1910s were successful ones for both Dabney and Miley. When they began drilling for oil in Ventura County, their companies were operating between them more than three dozen producing wells in the McKittrick and Midway fields. Dabney was becoming known in oil circles for an apparently uncanny ability to find oil, as he had drilled few dry holes (that is, holes did not encounter crude oil).[35] For his part, Miley was on this way to becoming "probably the best known individual oil man in California." Nothing that the two men experienced as independent operators in the San Joaquin Valley, however, prepared them for the geophysical conditions they would encounter in Ventura County.[36]

The arrival of Dabney and Miley in Ventura County was fortuitous for Ralph Lloyd, who remained stymied in his attempt to interest investors in the lands associated with the Ventura anticline. Early in 1912, Lloyd expected to see "some activity in actual development" after he persuaded an "English syndicate" to lease much of the area based on favorable reports from three geologists, including the increasingly renowned Ralph Arnold. The syndicate apparently failed to do anything else, according to Lloyd, owing to the outbreak of war in the Balkans, prompting several lessors to sue. Around the same time, a West Virginia investor allegedly offered Lewis Lloyd $40,000 for the mineral rights to the Erburu Ranch, but balked when the elder Lloyd demanded twice the price.[37]

In 1912, Joseph Dabney set in motion the series of events that would bring Ralph Lloyd into the oil business as an operator when he secured a lease at the western end of Simi Valley and spudded (began drilling) Simi No. 1. (A well is conventionally named for the lease on which it is drilled.) With a whiff of oil in the air, E. J. Miley and A. M. Buley arrived in the valley in search of property to lease. With Dabney drilling his well about a mile away from VL&W's ranch, Lloyd secured four leases on neighboring properties, comprising some 1,462 acres. In partnership with F. B. Ranger, an Oklahoma oilman, he spudded a test well that the two men quickly abandoned after encountering high gas pressure at shallow depths. Lacking capital to finance another well, Lloyd formed the Neosho Oil Company (named for his place of birth) as a vehicle to sell his leases.[38]

In the summer of 1913, Dabney, Lloyd, Miley, and Buley co-mingled their Simi Valley activities under the umbrella of the Hidalgo Oil Company. In the course of drilling their wells, Dabney and Lloyd had become acquainted. Dabney subsequently introduced Lloyd to Miley, who agreed to assume the leases held by the Neosho Oil Company. State Consolidated Oil Company agreed to drill a well on one of the leases. Lloyd also had leased land in nearby Tapo Canyon from the Patterson Ranch Company. Now he persuaded Miley and Buley to drill a well on this property. Dabney contributed bottom-hole money to offset the cost of drilling the well. That is, he agreed to pay State Consolidated Oil Company a sum of money if the well that Miley drilled was dry. Dabney, Lloyd, and State Consolidated Oil Company each agreed to receive one-third of the oil produced by the well. In July 1913, Dabney, Lloyd, and Miley incorporated the Hidalgo Oil Company as the vehicle for this arrangement, with each partner holding a one-third interest in the firm. The Patterson Ranch well, spudded on 20 August 1913, was dubbed Hidalgo No. 1.[39]

In all, E. J. Miley drilled five wells in Simi Valley, two of which struck oil. Based on Joseph Dabney's success with Simi No. 1, which was flowing high gravity crude oil at a depth of 1,538 feet at the end of August, Ralph Lloyd was initially optimistic that he and his partners had made a major discovery, reporting that he was "on top of one of the new oil fields, and have what the oil men say is the most valuable holdings in the field, and we have just struck 38 gravity oil worth $2.10 a barrel. . . . Everything indicates that I have a very valuable property, and that we have opened up one of the greatest high gravity oil fields in California."[40] In early December, Ralph Lloyd told his brother Warren that one of the Hidalgo wells that had been producing since August was "probably the best well ever struck in Ventura [County] . . . and triples the value of oil land . . . in that district."[41] As 1913 came to a close, Lloyd was giddy with enthusiasm over the prospects of the Hidalgo Oil Company, writing to the holder of his mortgage on the Adcox Auto and Aviation School building in Portland, "I have struck oil in California and have very bright prospects of making a fortune." Yet Lloyd's hopes were soon dashed.[42]

Ominously, even as the Hidalgo wells produced oil, they encountered high gas pressure at shallow depths. This pressure provided only an inkling of the conditions that the partners would confront at Ventura Avenue, but it was sufficient to cause alarm. For the latest oil field equipment was incapable of preventing wells subject to high gas pressure from blowing out.[43]

Cable-tool rigs, then widely in use, relied on drills with a sharpened, solid, cylindrical bit that worked vertically in the hole. The cable drill was suspended by a wire rope and activated by a walking beam. Unable to control natural gas pressure or prevent the sides of well holes from caving in the event of a blowout, the cable tools that worked satisfactorily in the fields of the Western Pennsylvania were no match for the freakish geology found below the surface of ranch lands in Ventura County. If the drilling crew were lucky, the blowout set off by piercing a natural gas pocket destroyed only the string of cable tools. If not, as Shell historian Kendall Beaton put it, "the tools, casing, rig, and all, would go vaulting into the air, a geyser of water and gas would blow for a few days, and then after patiently rebuilding the rig, they would start all over again." Contending with high gas pressures at shallow depths in Simi Valley, where at least one well blew out, gave the Hidalgo Oil Company partners a taste of things to come.[44]

Hidalgo Oil Company's early prospects for success fizzled within a year of spudding the first well. As 1914 unfolded, it was clear to the partners that they would have little to show for the $115,000 that they had invested in testing their leases. By August, Ralph Lloyd conceded to C. A. Havens, one of his lessors, that "oil developments in the Simi Valley have not been what we expected," as all of his wells had "proven very unsatisfactory."[45]

By this time, Dabney and Miley had abandoned the area. A surge of competitive drilling in the Midway field had compelled both men to protect their profitable investments in that field by drilling wells to offset those of their competitors on contiguous leases. Both men had resigned as officers and directors of Hidalgo Oil Company and sold their respective interests in both the Simi

Valley leases and the company to Lloyd, who in turn sold half of Hidalgo Oil Company for $20,000 to F. B. Chapin, a Toronto investor. Together Lloyd and Chapin attempted to interest oil baron E. L. Doheny in the company.[46]

As early as February 1914, Lloyd had approached Doheny, whose Petrol Company was developing a tract of land that lay adjacent to the portion of the Patterson Ranch that Hidalgo Oil Company had leased.[47] California's Coastal Region constituted a small part of Doheny's oil empire, but during 1916 he had acquired 6,000 acres near Ojai, in Ventura County, and had purchased the Petrol Company. Perhaps sensing an opportunity, in December 1916, Lloyd and Chapin offering to sell Doheny their company. Doheny demurred, pointing to the inferiority of one of Hidalgo's leases: "I am not sanguine of the results of drilling there, and would not wish to be compelled to drill a well on that property in the near future." Hidalgo Oil Company's corporate charter lapsed in March 1918. Ralph Lloyd's initial entry into the upstream oil business as an independent operator ended in failure.[48]

Return to Ventura

With his prospects of finding oil in Simi Valley sputtering, Ralph Lloyd turned his attention to the prospective oil lands that he had left behind in 1904, "determined to risk his capital and personal enterprise" in exploring them. As he reported retrospectively to his bankers in Los Angeles, in October 1913, Lloyd "[had taken] hold of development matters in the Ventura field believing that the opportune time had arrived to get control of a large area of valuable property at very favorable prices." After Joseph Dabney inspected properties that lay to the east of the Ventura River, the "Midway oil man" and Lloyd each agreed to take an undivided one-half interest in twenty-year leases on the Erburu (former Lloyd) Ranch and contiguous properties owned by the Davidson, Fraser, Gosnell, Hartman, Johnson, and McGonigle families. The leases provided for the payment of the standard one-eighth royalty and a monthly rental until the lessees commenced the drilling activity stipulated in the contracts.[49]

For operational purposes, Dabney and Lloyd split the Erburu Ranch into two units: a so-called Lloyd lease on the western one-third of the property, comprising 1,425 acres, which the two men targeted for immediate development, and a so-called VL&W lease on the remaining, more rugged portion of the property, comprising 2,150 acres. Reprising their Simi Valley roles, Lloyd and Dabney arranged for Miley to drill their wells. They did so by executing a sublease, which transferred their right to operate the Lloyd lease to State

Consolidated Oil Company. Under the terms of the contract, Lloyd and Dabney would receive a one-sixth royalty and 8,000 shares of State Consolidated Oil Company stock should Miley produce crude oil in commercial quantities (that is, a volume of oil that was sufficiently large to interest refiners in buying it). Dabney and Lloyd also assigned a one-third interest in the leases to E. J. Miley and A. M. Buley as individuals, so that each of the State Consolidated Oil Company partners owned a one-sixth interest in them.[50]

From the beginning, the drilling crews under Miley's direction were unable to contain the tremendous gas pressure and saltwater intrusion that they encountered. In March 1915, Miley spudded Lloyd No. 1 (figure 2). At a depth of 1,210 feet, saltwater flooded the well. Miley attempted to seal off the water by cementing in place the 12.5-inch steel casing in the well bore. Lloyd was optimistic that the procedure would work, reporting to his sister Roberta that "within the next thirty days the well will probably [be] opened up so that we can determine what we have struck."[51] At the same time, he recognized that "the gas pressure in the well is very great and we have to be careful bringing it in."[52] The cement job failed. With saltwater flowing over the top of the derrick, Miley nonetheless proceeded. At a depth of 2,219 feet, he again tried—and failed—to seal off the water by cementing the ten-inch casing—the second interval of steel pipe in the casing "string." At a depth of 2,290 feet, caving in the formation caused the 8.25-inch casing to "freeze." Miley persevered, drilling to a depth of 2,550 feet, at which point, on 8 January 1916, saltwater, gas, and a small amount of highly desirable light crude oil flowed from the well. Lawrence Vander Leck, deputy supervisor of the Department of Petroleum and Gas of the State Mining Bureau, would observe later that the well was allowing saltwater to "entirely destroy" the upper 600 feet of an 1,140-foot-thick light crude oil zone "for a considerable distance" around it.[53] For Ralph Lloyd, however, the showing of oil proved his long-held belief that a vast pool of petroleum lay beneath his family's property. Citing no less an authority than Union Oil Company's William W. Orcutt, one of several geologists who visited Ventura in the wake of the oil "strike," Lloyd assured his bankers that "there is no question but that we have rounded up an oil field," even as he conceded that he and his partners had no way of controlling its gas pressure.[54]

With the geologists of Associated Oil Company, Standard Oil Company of California, Union Oil Company, and others showing an interest in the Ventura anticline, Lloyd and Dabney decided to market their leases to a firm that was "big enough and strong enough" to handle the potentially lucrative field's extreme gas pressures, fractured geology, and water problems.[55] By this time, according to Lloyd, he, Dabney, Miley, and Buley together had leased "all the territory where the oil will be secured at a reasonable depth," encompassing nine square miles of land along the anticline, from the Taylor Ranch on the west side of the Ventura River, to the Joseph Sexton and William Sexton Ranches on the east. In March, Lloyd and his partners executed a lease with Louis Hartman, a son of Katherine Hartman, the owner of the Hartman property that Dabney and Lloyd already had leased. Louis Hartman's property lay adjacent to the Lloyd lease; the lessees agreed to begin drilling a well before the end of the year. Given their demonstrated incapacity to complete a well, however, their real interest lay in transferring risk by subleasing their operating rights granted to them by the lease (figure 2).[56]

Lloyd and Dabney agreed to confine the development of the field to a single, well-capitalized operator. As he explained to Louis Hartman, "In handling the Ventura field in this way, we believe that it is beneficial to every land owner in the district [because] the property can be uniformly and economically developed—no waste or bitter line contest should occur," which might be expected, were the field "divided up into many small holders."[57] By transferring all of their operating rights in their leases to one firm, Dabney and Lloyd would make possible the development of the Ventura Avenue field "slowly and in the most scientific and economical manner, thereby procuring the largest possible production per acre."[58] Unitizing field operations in this manner would prevent the competitive drilling that was giving the California oil industry a reputation as wasteful.[59]

In March 1916, Lloyd wrote to E. L. Doheny, reminding the "oil baron of the Southwest" that he had stated, "If you ever find in Ventura County a large area with the proper accumulative formation in an unbroken and uniform state, you will find something in oil worth while." Lloyd noted that he and Dabney had leased more acreage "than any one group of men can drill up." He flattered Doheny, telling him that he and his partner had decided that "it would be better to turn a portion of the territory to you rather than anyone else, and we have a motive in considering you first as we feel the aggressive policy that is shown in all your undertakings will be of mutual benefit." Doheny's reply, if any, is lost to history, but within a fortnight of his soliciting Doheny's interest in developing his leases, Lloyd reached an understanding with Shell Company of California, the firm he considered to be the most qualified "to solve the problems necessary to be solved in conquering this territory."[60]

By securing a lucrative agreement with Shell, Ralph Lloyd and his partners compensated for their ongoing failures in the field. In early April, B. H. Van der

Linden, Shell's head of exploration and production in California, visited Ventura at Ralph Lloyd's invitation. He was looking to expand Shell's production, which was confined to one field in the San Joaquin Valley. When Lloyd showed him outcroppings of oil layers along the east end of the anticline, Van der Linden became as convinced of the field's potential as Lloyd. Dabney, Lloyd, and Miley promptly drew up a memorandum of understanding that may have reached Shell's headquarters in San Francisco before Van der Linden had a chance to return to the city. In it, the partners asked for a jaw-dropping one-quarter royalty on ten properties that they would sublease to Shell. Shell would advance $275,000 on future royalty income and commit to drilling six wells in 1916. After negotiating for several weeks with Shell President William Meischke-Smith and Van der Linden, Lloyd and his partners reached an agreement with Shell that encompassed the Gosnell, (Katherine) Hartman, McGonigle, and Taylor leases, as well as the Lloyd and VL&W leases that encompassed the Erburu Ranch (figure 2). Signed on 12 June 1916, the contract provided for the standard one-eighth royalty to each lessor and an additional overriding royalty on each sublease to Lloyd and his partners. Shell reserved the option of buying out this overriding royalty at prices that ranged from $250 per acre to $750 per acre. The agreement also contained language governing Shell's acquisition of additional properties in the field that would become a source of bitter dispute between Lloyd and Shell's executive management. Finally, the partners would receive an additional $100,000 were State Consolidated Oil Company to produce commercial quantities of petroleum from the wells it was drilling on the Lloyd lease.[61]

For the moment, at least, State Consolidated Oil Company apparently had little chance of earning the production bonus. The ink was scarcely dry on the 12 June agreement when E. J. Miley restarted drilling Lloyd No. 1. Lloyd hoped that Miley could bring in the well within sixty days, even as he recognized that the gas pressure was making it "very hard to hold it down." In fact, caving ultimately caused the casing in the bore hole to collapse, leaving the partners with nothing to show for the $100,000 that they had sunk into the well.[62]

Miley realized a similar outcome with Lloyd No. 2, which he spudded in May 1916. Figure 4 depicts the well site in early September. Ralph Lloyd stands nearest the derrick, with Joseph Dabney, Miley, and the drilling team to his right. The neatly stacked casing in the foreground belies a measure of industrial control over nature that was not possible at the time, given a lack of effective blowout-prevention technology. Indeed, days earlier, Lloyd had reported that the well "came away from us last night."[63] Less than a week after this

Figure 4. State Consolidated Oil Company's operations on the Lloyd lease, September 1916. Ralph B. Lloyd stands nearest the derrick, to the left of his partners, Joseph B. Dabney and E. J. Miley (as they face the camera). (Photo by Schreder Photo. The Huntington Library, San Marino, California, Photos drawer 1, box 1.)

photograph was taken, Miley struck oil at 2,253 feet, at which point the well blew out, wrecking the derrick, rotary pumps, and bull wheels. "Part of the derrick went into the crater," as Lloyd described in vivid detail to Shell. "The force of the gas blast was thrown against the sides of the crater, which undermined the sides, causing them to cave in."[64] Still, Lloyd was unbowed. For however destructive its power, the high gas pressure demonstrated that the field would be a prolific one, once "we gain the experience necessary to control the situation."[65] Indeed, Lloyd No. 2 provided "further proof that our judgment is correct [namely, that Ventura] will prove to be the greatest high gravity oil territory on the acre basis ever opened up in California."[66]

To satisfy lease requirements, E. J. Miley began drilling Louis Hartman No. 1 scarcely a month after Lloyd No. 2 blew out. Again, Ralph Lloyd and his partners experienced no joy in the effort. With General Petroleum (GP) struggling with its first well on the adjacent Barnard property—which Lloyd and Dabney declined to lease because they determined that the steep plunge of the anticline on either side of the Ventura River narrowed the area of potential production (figure 2)—and Shell making little progress on its Gosnell, Hartman, and Taylor leases, Lloyd reported to his father that "we are all pretty well puzzled with the Ventura field. The great gas pressure, the large amount of water, and the looseness of the formation constitutes a series of difficulties that is testing the strength of the best of oil well material and trying the ability of the best posted men in the oil industry." Ralph Lloyd remained confident that, eventually, "we will all win out." At the same time, in asking for his father's cooperation in granting Shell relief from its drilling obligations on the untested VL&W lease, Lloyd beseeched him to accept what he only recently may have come to appreciate fully: "The oil will probably be quite deep" and therefore costly and difficult to reach.[67]

However elusive the successful completion of a well appeared to be, Ralph Lloyd and his partners soldiered on. Between May and July 1917, Miley re-drilled Lloyd No. 1 to a depth of 1,435 feet. With some 700 barrels of saltwater flowing out of the casing per day, Miley suspended operations. In September, Miley's men spudded Lloyd No. 3. Over the next two years, Miley's crews tried, without success, to complete both of these wells and Louis Hartman No. 1. When Deputy Supervisor Lawrence Vander Leck visited Ventura during November and December 1919, he found that all of Miley's drilling operations were compromised—stark evidence of the difficulties that the independent operator was confronting in the field.[68]

Ultimately, as independent operators, Ralph Lloyd, Joseph Dabney, E. J. Miley, and A. M. Buley spent some $750,000 in a futile effort to develop commercial production along the Ventura anticline. For his part, Dabney exhausted the capital that he had accumulated in the lumber and oil businesses. Recalled Lloyd: "We had nothing to show but craters blown out of the earth by gas, strings of wrecked casing deep in the earth, and a few puddles of oily saltwater. The gas and the saltwater seemed to overpower any efforts of man and machinery to get at the oil."[69] In this respect, their experience was typical.

Take Shell: Ralph Lloyd was confident that "the strongest [oil company] in the world" would succeed where State Consolidated Oil Company had failed. He assured A. L. Hobson, owner of land along the river that he and Dabney had chosen not to lease: "When they take hold of anything they go at it right."[70] Yet between August 1916, when Shell began drilling Taylor No. 1, and December 1921, the major operator invested $3 million in the field with little to show for it, demonstrating the limits of capital alone in successfully extracting petroleum from the Ventura anticline. Progress on each of four leases that crews tested was slow and demoralizing. Associated Oil Company Vice President A. C. McLaughlin, who had trained as a geologist at the University of Texas and had joined the Southern Pacific Railroad in 1907 in that capacity, observed that it took twice as long to drill a well in the Ventura Avenue field than anywhere in the Los Angeles Basin. Reaching drilling sites took valuable time, given the rugged and isolated terrain of the hills behind the city of Ventura. Shell spent considerable expense grading roads and, in the case of the McGonigle lease, building an aerial tramway from the nearest road so that workers and equipment could reach the drill site. Shell's early record in developing its Ventura Avenue leases demonstrated the risks inherent in the upstream oil business.[71]

In a scathing report on drilling methods employed at Ventura Avenue, filed in February 1920, Deputy Supervisor Lawrence Vander Leck of the Department of Petroleum and Gas of the State Mining Bureau spared no operators. In 1915, the state legislature had created the agency, vesting in it a trust responsibility for California's energy resources. By supervising the drilling, operation, maintenance, and abandonment of petroleum and natural gas wells, the department sought to "prevent damages to the petroleum and gas deposits of the state from infiltrating water and other causes." Vander Leck found that no operator in the field had dealt adequately with saltwater intrusion: 90 percent of fifty attempts to shut off water by pumping in cement under pressure had failed utterly; only two were "definite successes." As a result, for the month of December 1919, operators produced 4,103 barrels of crude oil and 100,000 barrels of saltwater from eight wells. He recommended that crews deploy rotary rigs exclusively.[72]

By drilling slowly and using heavy drilling fluids to "thoroughly mud each [oil] sand as entered," crews would be able to "completely kill the gas," in his estimation.[73]

At the time of Vander Leck's inspection, Shell was reassessing the value of its investment in the Ventura Avenue field. Having abandoned its test well on the McGonigle property in October 1919, it surrendered the sublease back to Lloyd and Dabney. In March 1922, Shell completed the first well in the field that enjoyed sustained, commercial quantities of production. Using rotary drilling equipment and heavy muds, as the State Mining Bureau prescribed, Gosnell No. 3 came in, flowing 1,040 barrels per day (figure 5). B. H. Van der Linden, who, according to Lloyd, "had strong convictions regarding the future of Ventura and . . . stood by [them] through a great many difficulties," felt vindicated. At his urging, Shell promptly bought out the overriding royalty that Lloyd and his partners held on the lease. But the parent organization remained skeptical. In 1921, Royal Dutch/Shell's assessors had fixed the value of Shell Company of California's interest in the Ventura Avenue field at $1 million—far less than the firm had invested in it to date. In 1924, Shell's executives in San Francisco considered selling the company's remaining interest in the field and investing the proceeds in the Los Angeles Basin, where the company was enjoying far greater success. They did not do so, but surrendered the Hartman sublease, in which they had invested $300,000, in October 1924. Shell's crews were also struggling at this time with saltwater intrusion in four wells that they were drilling on the Gosnell lease, where production was plummeting. Having retreated from the portion of the field that lay to the east of the Ventura River, Shell retained control of the portion of the field that lay to the west of the river by paying Ralph Lloyd and his partners $1.2 million for their overriding interest in their Taylor lease (and thus becoming the lessee of Alice Grubb, the owner of the Taylor Ranch). Shell's drilling teams proved the property the following year. (After World War II, the Taylor lease would become the field's most prolific producing area.) Even then, Shell's vice-president of production remained unenthusiastic: "We have spent a great deal of money in Ventura without benefiting thereby, and while very thankful that it now appears that we shall get our money out with some profit, the fact is that . . . we are still considerably in the 'red,' and therefore . . . desire to curtail expenditures in every line as much as possible." For his troubles, B. H. Van der Linden was recalled to The Hague. Soon thereafter he found himself in North Sumatra, where Royal Dutch/Shell was developing its first field in the Indonesian archipelago.[74]

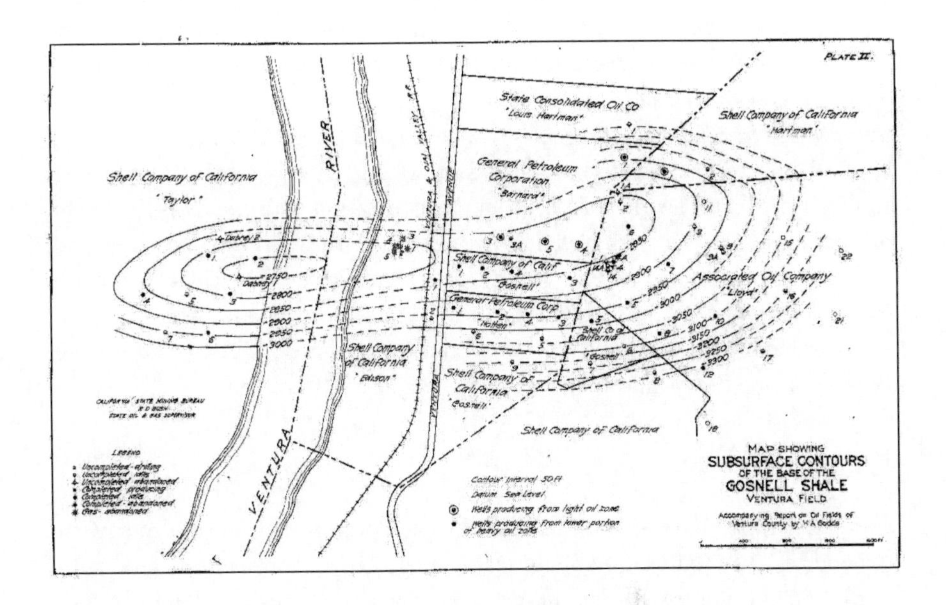

Figure 5. The Ventura Avenue field, as controlled by Associated, GP, and Shell in 1924. (Map accompanying H. A. Godde, "Oil Fields of Ventura County," *Annual Report of the State Oil and Gas Supervisor* 10 [November 1924]: 5–24.)

Ralph Lloyd believed that Shell's decade-long struggle in the Ventura Avenue field owed as much to management as to technology. He pointed to layers of bureaucracy in the organization that slowed decision-making. Moreover, the rotation of half-a-dozen managers through Ventura was evidence, in his mind, of ineffective personnel assessment. Only after general superintendent William C. McDuffie convinced Henri Detering, president of the Royal Dutch/Shell group, to give him direct authority over operations at Ventura, did Shell's performance improve, Lloyd concluded. For McDuffie grasped the significance of recent advances in drilling technology and had the managerial ability to get his drilling crews to deploy it effectively. Lloyd watched as McDuffie created a field-level organization with the requisite talent and resources to meet the challenges posed by the Ventura anticline and respond to the moves of competing operators. McDuffie relocated Shell's Ventura office from downtown to the field and brought the field's general superintendent to Ventura from San Francisco. These moves, according to Lloyd, enabled the company to reverse its record of failure.[75]

Break with Shell

When Lawrence Vander Leck inspected well conditions at Ventura Avenue at the end of 1919, Ralph Lloyd had been engaged in a protracted dispute with Shell for eight months. The company's response to his assertion that he had an exclusive right to secure leases for the company would so infuriate him that he would turn to Associated Oil Company to develop the Lloyd and VL&W leases that dominated the eastern portion of the field.

Based on his April 1916 memorandum of understanding with B. H. Van der Linden and conversations with Shell President William Meischke-Smith prior to the signing of the June 1916 agreement, Lloyd believed that he had Shell's assurances that the company would use him as its exclusive land agent. Should Shell develop an interest in any property within a defined band on either side of the Ventura anticline, Lloyd would approach the landowner. Were he able to lease the property, he would execute a sublease with Shell, thereby transferring his operating rights to the company. On this basis, Lloyd convinced landowners to refrain from leasing their property until Shell was either prepared to test it with the drill or indicated that it had no interest in it. The 1916 agreement did not explicitly define such an arrangement, but Lloyd believed that language in the contract supported his inferred understanding of the relationship.[76]

For his part, J. C. Van Eck, who replaced Meischke-Smith as Shell's president in May 1919, denied that Shell had entered into an exclusive principal-agent arrangement with Lloyd. In his view, the 1916 agreement defined a buyer-seller relationship under which Shell might acquire the right to explore and develop a property from Lloyd and his partners through a sublease, but also allowed the company's representatives to approach landowners for the purpose of leasing their property directly.[77]

Shell set off the protracted dispute by executing a lease with Southern California Edison. Ralph Lloyd was "surprised and in a way a little chagrined" when he learned of the transaction.[78] To date, Lloyd had worked on leasing all the potentially producing properties in the area of the field that had not been included in the 1916 agreement. Shell had shown an interest in none of them, including the Barnard property that GP subsequently leased and the Louis Hartman property on which State Consolidated Oil Company was drilling. When asked, William Meischke-Smith, whom Joseph Dabney once complimented for having "quickness of mind and keen intelligence," replied that did not "have the slightest recollection of ever having made an arrangement," as Lloyd claimed, and, moreover, never would have contemplated doing so.[79]

Meischke-Smith's rebuttal prompted Ralph Lloyd to change the way he dealt with the management of all the operators in the field. Fumed Lloyd: "If [Shell] wishes to adopt this [way of doing business] as a precedent for our future dealings, well and good, but I want them to set the precedent, for it will cut both ways." Further, he complained that he and his partners had agreed to so many concessions and modifications that "the original contract would not be recognized," and so why should Shell retreat behind its language now? Lloyd learned a lesson. In his future dealings with Shell and other operators, he would adhere to a strict interpretation of leases and other agreements. He would communicate through regular, written means to ensure that there would be no misunderstanding between parties. He also would monitor the actions of operators and correspond with their executive managers whenever he felt that their firm's field operations were failing to meet the letter of their agreements. In doing so, he would demonstrate the power of a well-placed lessor in the development of a gigantic oil field.[80]

In this context, Ralph Lloyd approached the management of Associated Oil Company. In doing so, he leveraged Joseph Dabney's relationship with L. J. King, the company's general superintendent. Employed by Associated since 1903, King's friendship with Dabney dated from the earliest days of the latter's arrival in the San Joaquin Valley. King called on Dabney whenever business took him to Los Angeles. When he learned of Shell's retreat from Ventura Avenue, King brought the decision to the attention of L. C. Decius, who headed Associated's leasing department. Based on a field investigation, Decius recommended that Associated lease the properties that Shell had relinquished.[81]

On 24 February 1920, that is, only two weeks after Shell rebuffed him, Ralph Lloyd and his partners secured another lucrative agreement, conveying their operating rights in the Lloyd lease to Associated. The major paid $250,000 up front on a sublease that provided for a one-fifth royalty until Associated recovered its $250,000. Then the company would pay a one-half royalty until the market value of production from the lease reached $500,000. Thereafter, Associated would pay a one-fifth royalty, payable in cash or in kind. This royalty was divided into the standard one-eighth royalty for VL&W and overriding royalties of 2.5 percent each for Lloyd, Dabney, and State Consolidated Oil Company, that is, E. J. Miley and A. M. Buley. Associated agreed to maintain two strings of tools on the lease continuously and complete two wells per year. The agreement would prove to be profitable for all concerned, enabling Lloyd to enter Portland's commercial real estate market and reenter the Ventura Avenue field as an independent operator and Associated to become the leading Coastal Region producer during the inter-war period.[82]

At the eleventh hour, B. H. Van der Linden had obtained permission from J. C. Van Eck to renegotiate the Lloyd and VL&W leases on behalf of Shell. Not surprisingly, from his position of strength, Lloyd declined Van der Linden's offer. After Lloyd and his partners concluded their agreement with Associated, Van der Linden was laudatory, stating that he would rather have Associated operating on the other side of Shell's fence lines than any other company. He "thought a great deal of [Associated Vice President A. C.] McLaughlin personally and his method of management," said Lloyd. Van der Linden's encouragement imbued in Lloyd confidence that the two majors could "cooperate to their mutual advantage" and to the benefit of his and Dabney's lessors.[83]

Recognizing that both sides were "bound to be closely associated in the Ventura matters for a period of time," Ralph Lloyd repaired his relationship with Shell. For their part, J. C. Van Eck and his executive managers knew that a failure to come to terms with the field's lessors "would eventually create resentment between all of us [and] would take the happiness of the entire project." Ironically, over the ensuing decades, Lloyd's relationship with Shell would be far more stable than his relationship with Associated.[84]

Rebuffing William Gibbs McAdoo

Despite adding Associated to the roster of the field's operators, Ralph Lloyd held to the principle of minimizing the operators at Ventura Avenue, as he demonstrated when William Gibbs McAdoo approached him shortly after the former U.S. treasury secretary arrived in Los Angeles in March 1922 to pursue his economic and political ambitions.

After serving for six years in the Wilson Administration, an exhausted McAdoo returned to private law practice in New York. In March 1919, he entered into partnership with Joseph P. Cotton and George S. Franklin. Cotton had served as consulting counsel to the Federal Reserve Board and two wartime agencies—the Emergency Fleet Corporation and the Food Administration. Franklin had acted as counsel to the War Finance Corporation and had advised the treasury secretary on the sale of Liberty Bonds. In November 1919, E. L. Doheny, who by 1907 had controlled half of Mexico's crude oil production, retained McAdoo to assist him in lobbying the Wilson Administration to protect, through armed intervention if necessary, his vast holdings from confiscation under Article 27 of the 1917 Constitution of Mexico, which vested all subsurface mineral rights in the state. McAdoo's relationship with Doheny would prove to be an important one for his economic prospects in California.[85]

Early in 1922, McAdoo resigned his New York law partnership and announced his intention to move his family to Southern California. Publicly, he was coy about his reasons for doing so, telling the *Los Angeles Times*, "I'm simply going to Los Angeles to make my home because Mrs. McAdoo and I like the climate and the country."[86] He was not necessarily more forthcoming in private. To his friend, George F. Milton, for instance, McAdoo explained that "we can get a larger satisfaction out of life out here. There is more available room and open-air life, and that we both like."[87] But McAdoo did not move to Los Angeles simply seeking respite from cold winters, as did many of the thousands of Midwesterners who fueled the city's population growth during the 1920s. For one, McAdoo was "consumed by political ambition," as historian Jordan A. Schwarz puts it.[88] Los Angeles would be a springboard to the Democratic nomination for the presidency that had narrowly eluded him in 1920. At the same time, McAdoo hoped to recover wealth foregone during his years of public service. Within days of arriving in town, he opened a law practice and secured the Bank of Italy as a client. Then, seeking to wield his legal expertise and reputation to strike it rich in the booming California oil industry, he rekindled his relationship with Doheny.[89]

In May 1922, McAdoo and his wife, Eleanor, the daughter of former President Wilson, visited Doheny on his sprawling 12,000-acre Simi Valley ranch, where more than half-a-dozen wells pumped oil from the reservoir that Ralph Lloyd and his partners had failed to tap successfully. As Doheny had just returned from a trip to Mexico City, where he had conferred with Mexican officials and executives from Standard Oil Company of New Jersey, Gulf Oil Company, The Texas Company, and Sinclair Oil Company on postrevolutionary conditions, McAdoo surely discussed matters pertaining to Doheny's Mexican oil properties. Indeed, Doheny soon put McAdoo on an annual retainer of $25,000 to continue to advise him on such matters. But the weekend also provided McAdoo with the opportunity hear the "call of the wild" that, he averred, had lured him to Southern California: a tour on horseback of both the Doheny estate and the adjoining Strathearn Ranch, a portion of which Ralph Lloyd had leased in 1913. Riding past Doheny's producing wells must have stimulated the imagination of the former treasury secretary, who was already interested in oil development in Oklahoma with his son, William Gibbs McAdoo Jr., and encouraged him to take up discussion of the California oil industry with the man who was invested in every producing district in the state.[90]

Citing the "wonderful developments" at Huntington Beach, Signal Hill, Santa Fe Springs, and other prolific fields in the Los Angeles Basin, where frenzied drilling was boosting output to fully 20 percent of the world's total (achieved in 1923) and helping to establish the City of Angels as the leading "energy capital" in America, if not worldwide, McAdoo took no time in acting "to get a small share of the good things that are going around." [91] In 1919, he had bid unsuccessfully to get in on the initial boom in the Santa Fe Springs field that lay some twenty miles southeast of downtown Los Angeles.[92] Now he jumped at the chance to invest in the newly organized Jameson Petroleum Company, which undertook a six-well drilling campaign in the field that McAdoo hoped would yield "a very substantial income" for shareholders.[93] (And it would: as fellow investor Dixon Williams, a Chicago industrialist, remarked in October 1925, "This has not been a bad investment!"[94]) In the fall of 1922, McAdoo and his son approached Ralph Lloyd about leasing properties on the eastern side of the Ventura River that Associated's managers to date had declined to lease.

Ralph Lloyd and Joseph Dabney concurred that the McAdoos' oil-field experience, which had been confined to the Mid-Continent Region (that is, Oklahoma and Texas), did not qualify them to test the "deep and difficult" Ventura Avenue field, which demanded "at least a year of educational work" even for a major, to understand how to drill it, as Shell had shown. After he discussed operations at Ventura Avenue with the McAdoos, Lloyd now felt "more strongly" than ever that it would be "best for everyone concerned to keep the number of operators down to a minimum." Indeed, in the wake of the McAdoos' bid to enter the Ventura Avenue field, Lloyd and Dabney hoped that Associated would sublease the remaining properties in their portfolio under the terms of the February 1920 agreement. Should Associated decide not to do so, the partners would approach not independents such as the McAdoos but "some concern that would be big enough and strong enough to handle the remaining property as one unit." Lloyd's advocacy of minimizing a field's operators anticipated California's turn to unitization as a means of conserving its petroleum resources under conditions of glut.[95]

In the event, Lloyd convinced A. C. McLaughlin to sublease the McGonigle property that Shell had relinquished, as well as the peripheral Johnson-Davidson-Fraser property in which Shell had never shown an interest. The McAdoos had expressed an interest in leasing both properties. Soon after Shell relinquished its Hartman sublease back to them, Lloyd and Dabney persuaded Associated to sublease that property too, which secured the company's dominance of the portion of the field that lay to the east of the Ventura River.[96] Given the size of the Taylor Ranch, the amount of acreage that each major controlled was relatively equal. But Associated had more lessors to please than Shell. The difficult geological and topographical conditions that its crews would encounter as they moved their drilling rigs to property that lay to the east and north of the Lloyd lease would intersect with the economic conditions associated with the Great Depression to make it difficult for Associated to satisfy these lessors during the period of overproduction nationally that developed in the late 1920s.

Breakthrough at Ventura Avenue

The Lloyd lease proved to be the field's first prolific oil-bearing property. Deploying a rotary rig, special muds, and a blowout prevention device developed by the Hughes Tool Company, Associated brought in Lloyd No. 5 in October 1922 (figure 5). Ralph Lloyd congratulated B. H. Van der Linden, for the well provided "additional proof of the correctness of your judgment in regard to this field." The well's initial output of 1,800 barrels per day, coupled with favorable reports from Associated's geologists, convinced top management to commit capital to an extensive development program.[97]

With the promise of the Lloyd lease increasingly evidenced as crews deepened Lloyd No. 5, Associated sought to renegotiate the October 1913 contract associated with the former Lloyd family ranch, covering the Lloyd and VL&W leases. The company's managers calculated that the eleven years that remained on the agreement provided Associated with insufficient time to exploit fully the properties' potential. Acting as the company's emissary, Ralph Lloyd brought the proposal to his family at a meeting of VL&W directors and stockholders, convened informally at a family picnic. On 8 August 1922, three days after the unexpected death of Warren Lloyd, VL&W agreed to modify the original lease that covered all the Erburu Ranch with two contracts that formalized Ralph Lloyd's and Joseph Dabney's division of surface property into two operational units. VL&W would continue to lease the land associated with the Lloyd lease to Lloyd and Dabney, but now it leased to Associated directly the land associated with the VL&W lease, which the major had not subleased from Lloyd and Dabney. Both contracts retained the royalty provisions of the October 1913 lease, with one exception: State Consolidated Oil Company was left out of the overriding royalty on wells drilled after 31 October 1933; Dabney and Lloyd would split it between them. Under both contracts, VL&W extended the right to drill twenty years, to 8 August 1942. In turn, Dabney and Lloyd extended the right of Associated, under the sublease, to drill on the Lloyd lease for the same period of time. The new contracts also stipulated that no well could be drilled within 250 feet of any producing well drilled prior to 31 October 1933—a clause that Associated would seek to renegotiate in the coming years, along with several clauses in the original lease. In return for leasing the 2,150 acres of the Erburu Ranch comprising the VL&W lease, Associated guaranteed VL&W a minimum royalty of $5,000 per month. Ralph Lloyd later would explain that he helped to arrange the two contracts to secure the financial health of the family business in the wake of his brother's death. Nevertheless, as we shall see, a decade later, the terms allegedly agreed to by family members on 8 August 1922 would become sources of controversy and litigation, with implications for the development of the portion of the field that lay east of the Ventura River.[98]

The breakthrough in production in the Ventura Avenue field came in late January and early February 1925 with the completion of a pair of wells on the

Lloyd lease at a depth of a mile apiece. Together, Lloyd No. 9-A, which flowed at a rate of 4,555 barrels per day of 30° API oil, and Lloyd No. 16, which flowed at a rate of 4,100 barrels per day of 28° API oil, doubled daily production in the field (figure 5). On behalf of his partners, Ralph Lloyd congratulated McLaughlin: "Everything indicates that we are going to have a great oil and gas field in Ventura, and we are glad to say that the more we see of you and your company the more satisfied we are that so large a part of our holdings in the field are in your hands." When the completion of Lloyd No. 101 later in the year promised to double the proven area of the field, Lloyd enthused: "I believe that the Associated Oil Company . . . has one of the greatest, if not the greatest, oil properties in America." A year later, Associated estimated the "ultimate value" of its Ventura Avenue property at $200 million.[99]

In the spring of 1926, Ralph Lloyd assured the citizens of Ventura of at least two decades of development and production from the field. Well completions that extended its proven area almost a mile to the east prompted him to proclaim to his sister Roberta that Associated held "one of the greatest properties upon the American continent." As well he might: the one-fifth royalty on the Lloyd lease alone was already pouring more than $180,000 per month, collectively, into the accounts of Lloyd and Dabney, State Consolidated Oil Company, and VL&W. E. L. Doheny, whose Petroleum Securities Company had taken a lease in March 1924 on the periphery of the field, concurred, telling a group of Los Angeles bankers and bond salesmen on a tour of Associated's leases that Ventura Avenue would become the greatest oil field in California.[100]

Substantial and sustained development of the field established Ventura as one of America's premier petroleum districts. By mid-1927, Associated was producing oil from the Lloyd lease from twenty-three wells at depths of one mile or more. Not long thereafter the firm's resident geologist estimated the field's ultimate oil recovery to be at least 250 million barrels.[101]

Associated's success motivated Shell and GP to develop their leases more intensively. In addition to enjoying success in developing its Gosnell and Taylor leases, Shell extended the proven area of the field further to the west. In 1929, production from the field reached an interwar peak of 21 million barrels and accounted for 85 percent of the output of the Coastal Region (figure 6).

In the meantime, Ralph Lloyd increased his leasehold interest in the field and therefore his potential profits from its development. By 1919, Warren Lloyd was again heavily in debt as a result of real estate speculation. Under an agreement that he proposed and drafted, Warren authorized his brother to redeem his VL&W stock, which he had pledged as security for various bank loans, and

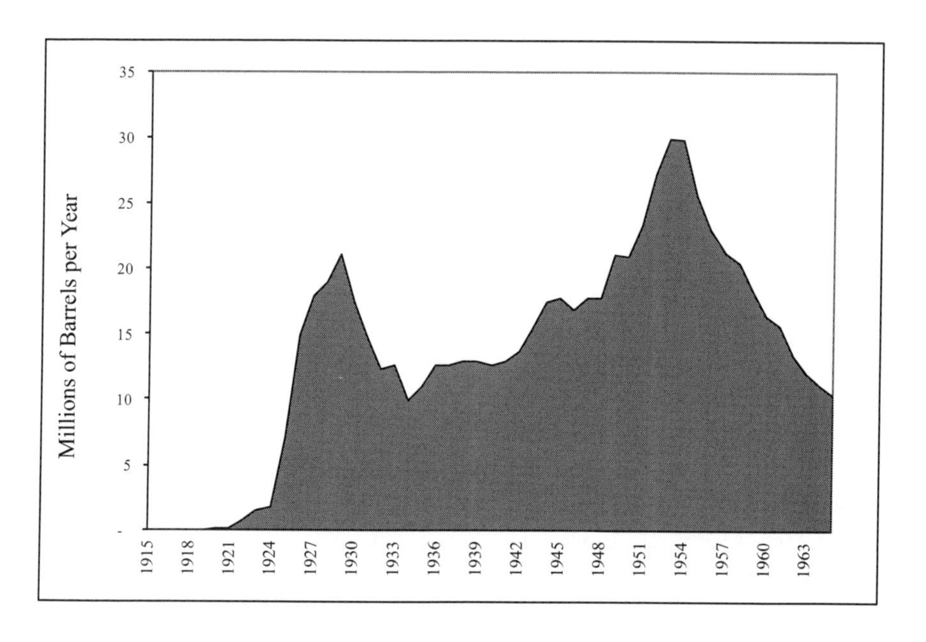

Figure 6. Crude oil production, Ventura Avenue field, 1915–1965. (Compiled data from "Crude Oil Production in California from the Industry's Inception to January 1, 1925," *Petroleum World* [February 1925]: 70-71; *Annual Report of the State Oil and Gas Supervisor* [various years].)

provided him with an option to purchase the shares. Ralph Lloyd agreed to pay off his brother's secured and unsecured debts as part of the transaction and exercised the purchase option. On 5 August 1922, Warren Lloyd succumbed unexpectedly to septic shock, ten days after nicking himself while shaving. At this point, Ralph Lloyd had acquired most of his brother's shares under the agreement. Together with the stock that he had received in the wake of his father's death just three months earlier, Ralph Lloyd owned 30 percent of VL&W.[102] Moreover, in June 1925, Lloyd paid Truman B. Gosnell $72,000 for his royalty interest in his lease. Gosnell was eager to sell, as Shell seemingly had no answer to the saltwater that was intruding into each of the wells that it was drilling. Lloyd also acquired A. M. Buley's overriding royalty interest in the properties subleased by Associated. Ralph Lloyd consolidated all of his holdings into the Lloyd Corporation, which he incorporated in October 1925.[103]

Portent for the Future: Lloyd's Relations with Associated and Shell

In his relations with the management of Associated and Shell, Ralph Lloyd saw himself as an entrepreneur, overcoming bureaucratic inertia and complexity by communicating directly with managers who were responsible for allocating resources among geographically dispersed projects. In Lloyd's view, a large corporation constituted "static power" that became "dynamic" only when its top management ensured the execution of business plans at the operational level. Believing that "the success or failure of any [oil] project usually rests with the individual," especially at the field level, Lloyd sought to play an ongoing and pivotal role in the development of the Ventura Avenue field.[104]

Ralph Lloyd's relations with Associated remained cordial while all of the major's properties in the field but the Lloyd lease remained unproven. As soon as he concluded the February 1920 agreement, Lloyd recommended a program of development to Vice President A. C. McLaughlin in the interest of managing the transition of drilling operations to Associated's crews on the Lloyd lease. From time to time, he reported on the progress of Shell and Associated to the managers of the other company so that the two operators might coordinate their activities. Moreover, Lloyd shared his thoughts on how to develop the field. He and his partners also modified their leases to grant relief on drilling requirements because they appreciated the difficult terrain and geology that operators encountered in the field. Communications during this period, however, were relatively infrequent. Associated was in position to develop the field "in the most advantageous way to get the maximum of production at the minimum of cost." Lloyd was confident that Associated had the ability and resources "to work out the theory and practice of development" for their "mutual benefit."[105]

Tensions would arise and Lloyd's communications would increase once Associated proved the properties adjacent to the Lloyd lease. In the meantime, the city of Ventura boomed.

Chapter 2

Local Elites, Outside Companies, and Ventura's Oil Boom

An oil field should not be and cannot be restricted and governed by the regulations that of necessity are applicable to city government.

—Ralph B. Lloyd, 1925

With the boom set off early in 1925 by Associated Oil Company's completion of two wells in the Ventura Avenue field, local boosters sought to achieve "the best kind of growth" for Ventura, as one editorial put it, rather than simply growth itself.[1] Business and civic leaders took pride in distinguishing Ventura's boom from its contemporary counterparts at Huntington Beach, Santa Fe Springs, and Long Beach (Signal Hill) in the Los Angeles Basin. They saw no reason why their city could not emulate Santa Barbara—a playground of wealthy capitalists and weekend destination of middle-class Southern Californians that lay twenty-five miles up the coast—as a recreational and tourist destination.[2] A consideration of their efforts toward this end supports the conclusion of urban scholars Rowan Miranda and Donald Rosdil that "the single-minded pursuit of development is hardly the only option that cities may exercise."[3]

At the same time, the oil industry influenced the outcomes of local initiatives by deploying financial resources and aligning themselves with key members of Ventura's growth coalition. Their paramount interest remained developing the extractive region that lay behind the city and west of the Ventura River without restriction. They were pleased to see Ventura grow quantitatively, but showed far less interest in the quality-of-life aspects of boosters' plans for the city.[4]

As the principal outside firms operating at Ventura Avenue, Associated, General Petroleum (GP), and Shell relied on Ralph Lloyd as their intermediary. Like the oil companies, Lloyd was above all concerned with maximizing petroleum production. Hence, as a "citizen outside the government," he lobbied on behalf of the industry. In doing so, Lloyd hoped to persuade local civic and business leaders that their interests mirrored those of the oil industry. At the same time, he took a personal interest in Ventura's development. After all, it was where he spent many of his early years. Because he shared local boosters' understanding of urban progress, Lloyd worked to secure financial support from the outside oil companies for some of their projects.

This chapter considers the resolution of these often competing, sometimes overlapping, views within the growth machine and civic boosterism frameworks of John R. Logan and Harvey Molotch, on one hand, and Kevin R. Cox and Andrew Mair, on the other. In this case, outside firms—that is, the oil companies—took a far greater interest in local issues than the "city as a growth machine" model of Logan and Molotch contemplates. In this model, outside capital cedes local control of the instruments of city governance to those whose interests are tied to "aggregate local growth or its specific distribution within the community," such as realtors and newspaper publishers. The firms in the extractive oil industry were also "locally dependent," to use the terminology of Cox and Mair. That is, Associated, GP, and Shell required a long-term local presence to realize the value of their enterprise. Unlike the utility companies and financial institutions cited by Cox and Mair, however, the value of their fixed assets did not depend on local population growth. Rather it depended on the degree to which they could deploy capital as they saw fit, that is, free from regulatory restriction. In fact, Ventura's boom created an environment that potentially threatened their interests. This chapter analyzes the circumstances that motivated the oil companies to become involved in local affairs and the mechanisms through which they exerted influence on behalf of their interests.[5]

Once local elites recognized that the Ventura Avenue field was capable of producing millions of barrels of crude oil annually for decades, they sought to use oil-generated wealth to implement a qualitative growth strategy. In charting a path for Ventura's growth, however, boosters encountered a complex interplay of interests. As a result, they garnered varying levels of industry and public support for their projects.

Anatomy of Ventura's Oil Boom

At the end of World War I, Ventura was an established, if relatively somnolent, community in a thriving agricultural region (figure 7). As such, it was unlike the rural boomtowns of the Mountain West, such as Rock Springs, Wyoming, and Rifle, Colorado, which owed their very existence to mineral extraction. A "farming aristocracy" embedded in the Santa Clara Valley to the east of Ventura dominated civic and community affairs. Yet local boosters had not taken the opportunity to promote Ventura as an agricultural processing center—a function that nearby Oxnard was happy to take on—or much else, for that matter. The breakthrough in oil production fundamentally altered both Ventura's socioeconomic underpinnings and the perspective of local elites, who now sought to establish their city as a recreational and tourist destination.[6]

Land use precedents established in the nineteenth century potentially constrained boosters' ability to chart a qualitative growth path for Ventura. Oil companies constructed a number of pipelines to connect their wells to tidewater, where the crude oil produced in Ventura County's hinterlands was loaded on tankers for shipment to San Francisco and Los Angeles (figure 8). (Water was the only economical means of transporting oil over long distances at the time.) In 1872, the City of Ventura constructed a wharf to facilitate the exportation of petroleum. Four years later, one of the predecessor firms of Standard Oil Company of California constructed a refinery near the waterfront. By 1890, numerous tanks stored crude oil on the beach. During the 1890s, the Union Oil Company constructed additional pipelines from its fields near its head offices in Santa Paula to the wharf for shipment to its new refinery on San Francisco Bay (figure 8). The industrial use of Ventura's waterfront posed a considerable problem for boosters who sought to attract tourists and beachgoers from the Los Angeles Basin and the San Joaquin Valley.[7]

The contrast with Santa Barbara, where opposition to the use of the oceanfront for oil development has been a recurring theme, was stark. Residents objected to unsightly derricks on the beach at the adjacent resort town of Summerland, where oil was discovered in the 1890s. The Santa Barbara Chamber of Commerce also opposed beachfront drilling. In August 1899, the *Santa Barbara Morning Press* went on record in opposition to industrial development for Santa Barbara generally and predicted that oil development would drive down property values and prompt well-to-do residents to leave the area. Explaining that "[i]t would be an unfortunate disaster if the beach near Santa Barbara's waterfront should be disfigured with the ugly derricks of oil wells," the newspaper urged people to resist any attempt to erect drilling rigs beyond Summerland. With no legal means of preventing entrepreneurs from drilling for oil on its state-owned beaches, property owners heeded the call to resistance, demolishing a derrick and part of the wharf of the adjacent upscale community of Montecito.[8]

Santa Barbara also prevented the building of infrastructure that would have dedicated its wharf for use as a shipping point for crude oil. In 1908, for

Figure 7 (*above*). Main Street, Ventura, circa 1920. (Reproduced by permission of the Museum of Ventura County.)

Figure 8 (*opposite*). Location of the City of Ventura to the Ventura Avenue and other fields of Ventura County on the eve of the boom. (Map accompanying H. A. Godde, "Oil Fields of Ventura County," *Annual Report of the State Oil and Gas Supervisor* 10 [November 1924]: 5–24.)

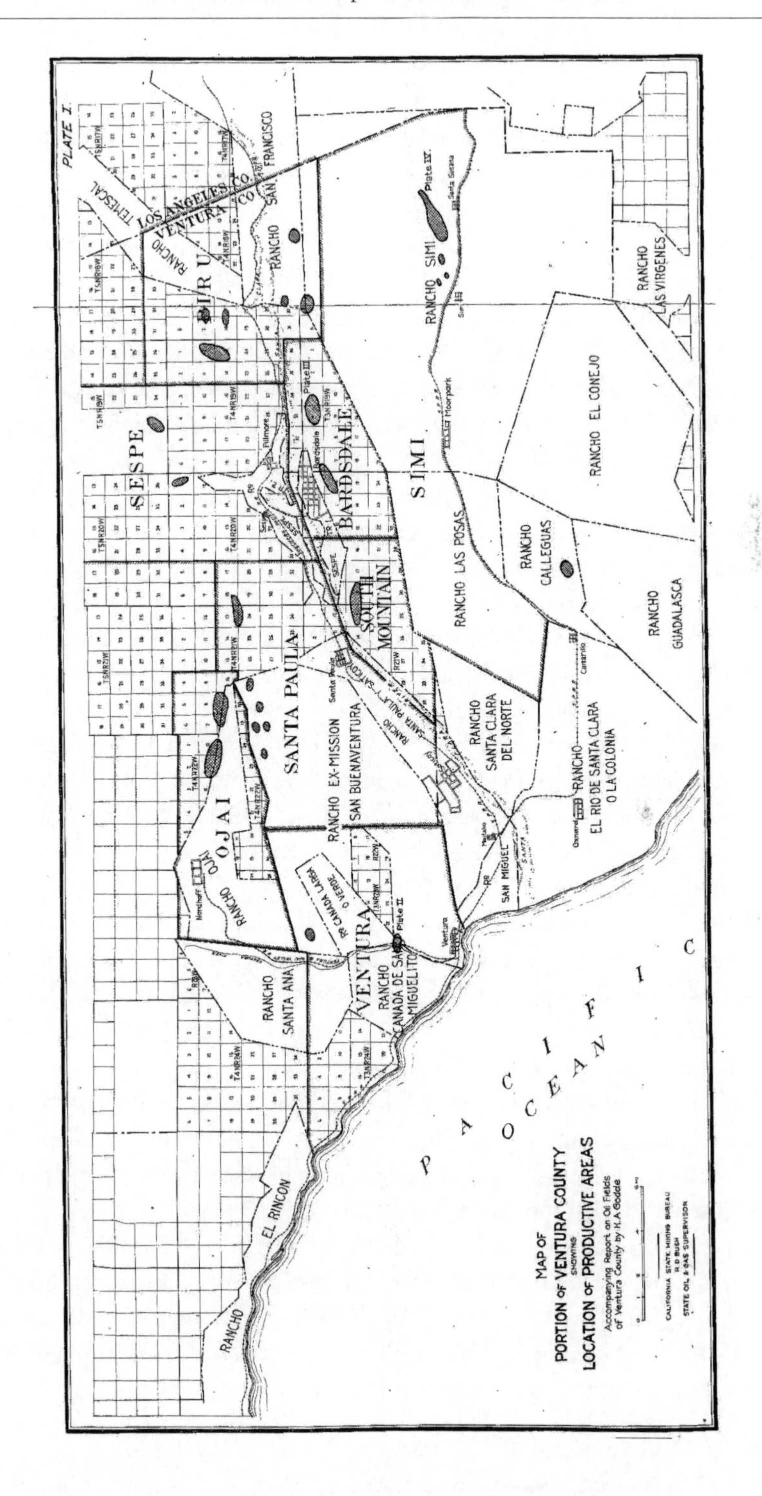
PLATE I.
MAP OF
PORTION OF VENTURA COUNTY
SHOWING
LOCATION OF PRODUCTIVE AREAS
Accompanying Report on Oil Fields of Ventura County by H. A. Godde
CALIFORNIA STATE MINING BUREAU
STATE OIL & GAS SUPERVISOR
LOS ANGELES CO.
VENTURA CO.
TEMESCAL
RANCHO TEMESCAL
PIRU
RANCHO SAN FRANCISCO
SESPE
BARDSDALE
SOUTH MOUNTAIN
SIMI
RANCHO SIMI
Plate IV.
Plate III
Fillmore
Bardsdale
Moorpark
RANCHO LAS VIRGENES
RANCHO EL CONEJO
RANCHO LAS POSAS
RANCHO CALLEGUAS
RANCHO GUADALASCA
SANTA PAULA
Santa Paula
RANCHO SANTA PAULA Y SATICOY
RANCHO EX-MISSION SAN BUENAVENTURA
RANCHO SANTA CLARA DEL NORTE
RANCHO EL RIO DE SANTA CLARA O LA COLONIA
Camarillo
Oxnard
SAN MIGUEL
OJAI
RANCHO OJAI
Nordhoff
CANADA LARGA O VERDE
VENTURA
Ventura
Plate II
RANCHO SANTA ANA
RANCHO CANADA DE SAN MIGUELITO
EL RINCON
RANCHO
PACIFIC
OCEAN

instance, the Santa Barbara Chamber of Commerce successfully opposed the building of a pipeline to the wharf—a position that enjoyed broad-based and sustained support within the community. As a result of efforts such as this one, Santa Barbara's wharf never served as a transshipment point of an oil hinterland.[9]

Ventura's boosters who sought to deploy revenues generated from petroleum extraction to reconstruct their community along the lines of Santa Barbara also faced very different circumstances with respect to oil industry interests. In Santa Barbara, non-oil-controlled sources of wealth prevented the oil industry from establishing a beachhead in local politics. The capitalists of the Gilded Age who made the city their residence possessed enormous wealth generated from industrial enterprise elsewhere. They preferred gardens, polo horses, and yachts to additional industrial activity and had the economic and political resources to implement a growth strategy that was compatible with their quality-of-life interests.[10] Nevertheless, the municipal code did not ban drilling for oil within city limits.[11] And so the discovery of oil in the Mesa neighborhood, overlooking the Pacific Ocean, in the 1920s threatened to dilute Santa Barbara's qualitative character. A few dozen firms snapped up leases. Soon the Mesa was covered with derricks. Ultimately, the field proved to be a minor one that did not attract the interest of majors such as Associated and Shell. Moreover, production was not controlled among operators, who drained the relatively small reservoir within a few years.[12] As a result, Santa Barbara's business and civic leaders avoided the need to confront an oil industry with a substantial stake in the outcome of local politics.[13]

By 1925, the consequences of intensive oil development for the cities of the Los Angeles Basin were widely known and deplored. Competitive drilling on small, "town lot" leases by hundreds of operators characterized the exploitation of fields at Inglewood, Long Beach, Santa Fe Springs, and elsewhere. Drilling wells and pumping oil under congested and chaotic conditions overwhelmed local infrastructure, fouled the air, polluted yards and streets, and—when wells and storage tanks exploded—cost lives. Ventura's residents were keen to avoid the direct, visible environmental consequences of urban oil extraction with which they were familiar.[14]

Ventura's oil boom created many of the social pressures that Roger M. Olien and Diana Davids Olien associated with contemporary oil booms in towns of the Permian Basin in West Texas.[15] With the breakthrough in production at Ventura Avenue, hundreds of workers arrived, in large part from the Los Angeles Basin: the number workers in the field increased from 260 in January

1925 to 1,200 by May 1926. Spurred by the influx of roustabouts, geologists, engineers, and other workers, and their families, Ventura's population grew 25 percent in 1925 and another 27 percent in 1926. By 1930, the city's population had almost tripled its 1920 level. Including adjacent, unincorporated areas, it neared the 19,000 mark.[16]

The surge in population strained the capacity of local infrastructure. The postmaster reported that his office was "literally swamped with . . . steadily mounting business," as receipts registered year-on-year gains of thousands of dollars. Requests for electrical connections during the first quarter of 1925 doubled over the first quarter of 1924; those for water, telephone, and natural gas connections of similar order of magnitude "swamped" the respective utilities. The most significant short-term negative impacts of the boom, however, were associated with a lack of housing and soaring real estate prices.[17]

The housing shortage remained acute for more than a year. "Antiquated" and "ridiculously high priced" accommodation in existing apartments, hotels, and rooms in homes exacerbated the inadequate supply of single-family houses. Many workers traveled fourteen miles or more from nearby towns to the field—a long and arduous commute at the time. Others shared space with fellow workers. One furniture dealer reported that he had "sent down a bed to a rooming house . . . where there [were] twenty-eight oil men living in a seven-room residence."[18] Demand for tents, camping equipment, and lumber for tent houses soared, as newcomers took up residence in the county park along the oceanfront. Observed Roy Pinkerton, who relocated from San Diego in June 1925 to found the *Ventura County Star*: "It is a commonplace that the tourist auto camps are so filled with permanent residents that tourists have to be turned away." Others squatted in makeshift houses or tents on city property. As the situation worsened in the summer of 1925, Pinkerton, who upon his arrival became one of the city's most visible boosters, declared that the construction of hundreds of "comfortable, pleasant homes where people can live happily" was Ventura's "greatest need." Pinkerton had cause for lament. Estimates of the significance of the oil field had persuaded him to establish a newspaper in town. Now he was operating it out of a garage, owing to a lack of "suitable business buildings for rent."[19]

Real estate speculation also posed an immediate concern. Realtors arrived from Los Angeles and Santa Barbara "all bent upon obtaining options on tracts of land."[20] Many residents set up real estate offices. Owners of farmland or orchards along Ventura Avenue, which ran parallel to the Ventura River, bordering the western end of the downtown, were particularly well situated to profit by selling out to speculators. For instance, one resident sold thirty-seven acres of inherited property to a group of investors, led by Joseph Argabrite, who, as president of the First National Bank and director of the Ventura Chamber of Commerce, was a key member of the growth coalition that coalesced around the boom. The group built twenty-five period-revival homes, selling them for as much as $3,000. Other homes along Ventura Avenue sold for as much as $5,000. At the same time, vacant, subdivided lots sold anywhere from $765 to $1,000. Along Main Street, commercial property "was changing hands rapidly, at big price increases."[21] Rents, too, were "unquestionably high, in some cases exorbitantly high." One landlord was "making 100 percent a year out of the rental of certain shacks which he bought at advantage some time back." "Fictitious real estate and farm values" characterized the real estate market. The *Ventura County Star* worried that the failure to bring them into line "with prices in other communities" would hurt "those least able to stand the pressure."[22]

Ventura's residents were less concerned about being swindled by disreputable promoters, even as many of those "who cleaned up at the expense of the public" at Long Beach, Santa Fe Springs, and other Los Angeles Basin fields arrived in "droves" in search of leases for purposes of selling stock in companies they would create ostensibly to drill for oil on those properties.[23] Owners of land "anywhere near" the burgeoning oil field were "besieged daily by men seeking options and leases."[24] Associated found it necessary to put up a fence across the road leading to its Lloyd lease to keep away the "hordes of people running over the ground . . . trying to get information upon which to base leasing."[25] Associated, GP, and Shell advised landowners to consult them before signing a lease, assuring them that they would profit more from royalties from production than from stock-selling schemes. Yet both promoters and reputable "oil men of vast capital," who also arrived in town in early 1925, found that—as Ralph Lloyd had orchestrated—three companies controlled most of the acreage that might form the basis for stock selling or investments in wells. The speculative fever soon withered.[26]

Despite the pressures produced by rapid population growth, the social fabric held together. Ventura remained a "pretty clean town," recalled Verne Patmore, an employee of an excavating and road paving firm that did much of its work in the oil fields. Richard Willett, whose parents grew apricots and lemons until E. L. Doheny's Petroleum Securities Company found oil on their property (on the periphery of the Ventura Avenue field), recalled that the oil workers had "rough edges" but were not necessarily "rowdy." Indeed, crime fell in the first year of the boom, from 644 civil and criminal actions in 1924 to 581 offenses in 1925. At the same time, there was lawlessness to the extent that defying Prohibition defined the term. Workers had little difficulty in "getting a drink if [they] wanted to drink," given the prevalence of speakeasies and "residential saloons." Moreover, poker parlors and saloons along Ventura Avenue allowed workers to gamble legally under local option laws. Nevertheless, Ventura did not become "a wide-open town," according to Patmore. Rather, "it was a nine o'clock town," as stores remained open until then to accommodate workers' schedules. After they closed, workers typically gathered on Main Street to "socialize."[27]

The assessments of Ralph Lloyd and others in the oil industry on the significance of Associated's breakthrough in production shaped local opinions of the boom. When Lloyd assured the city "of ten years' more" of oil development and at least another decade of production in March 1926, he also promised that there would be "no flood of oil produced to demoralize the market." Ventura thus had the "best prospects of any small city on the Pacific Coast." Lloyd predicted that the city's population would soon swell to 25,000.[28]

Local newspapers boosted Ventura's economic and urban growth prospects by reporting similar assessments, often under sensational headlines. Less than a month into the boom, the *Ventura Free Press* assured its readers that "oil men generally are well satisfied that this city lies on the border of an oil pool the extent of which cannot be estimated" and "do not hesitate to express themselves frankly as to the vast prospects for oil."[29] As a result, the *Free Press* proclaimed that "the spirit of optimism prevails everywhere," as Ventura's residents were "thoroughly live to the fact that the activities of the oil field may develop it into one of the greatest this state has ever known."[30] Three months into the boom, the newspaper proclaimed that "the oil industry here promises to endure for many years." [31] A year later the *Ventura Weekly Post* reminded readers that "oil experts" believed that Ventura Avenue was going to be "one of the most remarkable fields in the state." [32] In the interest of building readership and boosting Ventura's urban fortunes, editors played up oil development for all it was worth. Substantial, sustained production ensured that readers who believed them were not disappointed.

Newspaper editors and businessmen took their cues from industry observers to explain why Ventura would enjoy a rosy future. They agreed that the oil companies that controlled production would not "force a too rapid development, resulting in a mushroom growth for the city," as E. E. Wiker, the secretary of the Ventura Chamber of Commerce, put it. As a result, "fortunate"

Ventura was "assured of a continued, rapid, steady, substantial growth." The editors of the *Ventura Free Press* sounded a similar theme. Local growth would be based on demand, rather than "fly-by-night" business, owing to the leasing pattern in the field. Ventura would not become another Long Beach, as "people in other cities of Southern California" had predicted. Nor would it become a "second Florida," according to the city building inspector, referring to the contemporary real estate boom in that state. E. R. Hoover, head of a local real estate firm, observed: "Unlike other boom towns, there are no makeshift buildings going up, no curbside promoters working here, and all activity is designed with the idea of permanency."[33]

Ventura's residents joined its leading citizens in comparing their boom favorably to those of the Los Angeles Basin. According to a department store manager, his customers believed that oil development would produce "the very best kind" of growth: "conservative . . . and probably permanent in its value," rather than the "sky-rocket growth that will lead to a slump after a brief period of artificial inflation."[34] As a reporter concluded in florid terms, Ventura was a "booming," rather than "boom," town, with scant evidence "of the forced, impermanent, uncivilized growth that was apparent in Santa Fe Springs . . . It is not the unhealthy growth of a forced hot-house plant, but rather the starling upspring of a well-tended plant which has found a new source of strength in a potent fertilizer."[35] Ultimately, the city "encouraged oil to the fullest extent," as the editors of *California Oil World* noted. Local support of the oil companies, however, proved problematic for boosters seeking to chart a qualitative growth path for their city.[36]

The Local Growth Coalition

An emerging, if informal coalition led the movement to grow Ventura and at the same time increase the city's attractiveness to those seeking recreation and respite. The aforementioned banker Argabrite and newspaper editor Pinkerton, together with Olaf Austad, manager of the Ventura branch of the Bank of Italy, Sol N. Sheridan, a journalist and writer, Giovanni Ferro, a large property owner in both the city and the county, Charles Rea, Shell's superintendent at Ventura Avenue, and George Randall, proprietor of Darden & Randall, Buick's sales agency for Ventura County, numbered among its most prominent members.[37] All of them held positions of influence in Ventura's institutional life: Argabrite, Austad, and Sheridan within the Ventura Chamber of Commerce, which Austad considered critical to ensuring that Ventura grew in a "lastingly sound and healthy manner";[38] Ferro as city trustee (commissioner); and both Rea and Randall as mayor, from 1923 to 1927 and from 1927 to 1931, respectively. These men did not develop a master plan for Ventura (although there was a planning commission). Further, they differed in the extent to which they backed specific projects. Yet all of them worked to capture some of the wealth generated by petroleum extraction to fund projects that would establish Ventura as a recreational and tourist destination.

They soon learned, however, that their ability to implement their ideas depended on the extent to which they gained the support of the outside oil companies. The managers of these firms agreed with Ralph Lloyd that quantitative growth in terms of increasing population, jobs, and businesses was all well and good, if it resulted from the uncoordinated, cumulative private decisions of individuals, families, and firms. On matters that involved political economy, that is, municipal decisions on the use of funds for public projects, the oil companies and local leaders differed in their views. As the latter would learn to their dismay, the oil industry ultimately would constitute the dominant factor shaping Ventura's growth trajectory.

Ventura's business and civic leaders moved on a number of fronts to attract new residents, sustain high demand for real estate development, expand public infrastructure, and strengthen the institutional basis for future growth. They promoted Ventura in Los Angeles newspapers and urged residents to extol Ventura's advantages in letters to friends and relatives "in the east." They also called on residents to patronize one another's businesses (the "best way to get the ball rolling"), join the Chamber of Commerce, purchase homes to make the community "more stable and more desirable," and invest in local real estate ("no better investment in the world today"). Further, they established or reorganized mortgage companies to finance new construction, backed the provision of tourist and recreational amenities, and called on city planners and architects to emulate the Spanish Revival style recently adopted by Santa Barbara in the wake of a devastating earthquake. Finally, the boom afforded boosters the opportunity to promote two projects for which they had campaigned many years without success, owing to opposition from agricultural interests, namely, a harbor that could handle the largest cargo ships and anchor the development of the beach as recreational area and the Maricopa Road—a link to the San Joaquin Valley and the major highway between Northern and Southern California.[39]

In their promotional rhetoric, these local leaders explained why Ventura could grow both quantitatively and qualitatively. As "a natural crossroads and

trading center" with "the highways and railroads [that] are the natural arteries of commerce," Ventura had an opportunity to expand its population and economy. Moreover, sustained oil production reinforced Ventura's "steady" character—another theme to which their boosterism frequently returned. As the city building inspector put it two years into the boom, "Ventura is not only the fastest growing city in California, but it is one of the most stable." To be sure, realtors and investors flocked to Ventura because of the boom. Yet they arrived by the "hundreds," the *Free Press* opined, because the city was "known all over also as a steady business town." In addition, Ventura possessed enormous potential for tourism, recreation, and resort living. Like Santa Barbara and San Diego, it was "fortunate enough to be situated along the coast in Southern California" and therefore enjoyed "the ideal climate, neither too hot nor too cold." Ventura was "bound to advance because of the natural advantages of location." Isolated geographically from the coastline, the Ventura Avenue field would not obstruct the development of the city's potential as a resort town (figure 8). Enthused E. R. Hoover, the real estate agent: "No other city in California, or for that matter, in the United States, has the potential of Ventura."[40]

Construction associated with the boom put Ventura ahead of larger Southern California cities, such as Orange, Riverside, and Whittier, in terms of building permits. Residential buildings accommodated 550 families in 1925 alone, as Ventura began to address its housing shortage. Commercial buildings erected during the boom reflected the confidence of local investors in the permanence of Ventura's fortunes. They included a Masonic temple, a five-story hotel, an Elks Club lodge, a downtown theater, and the C. G. Greenlee Building, which housed chain grocery and drug stores. Along Ventura Avenue, commercial buildings joined residences in replacing the packinghouses and farms that had previously occupied the land. E. R. Hoover insisted that the new buildings were "above average" in quality. Moreover, there were

Figure 9 (*opposite top*). Ventura Avenue District, Ventura, 1927. The district in transition from agricultural to industrial, commercial, and residential use. (Reproduced by permission of the Museum of Ventura County.)

Figure 10 (*opposite bottom*). Main Street, Ventura, 1929. Four years into the boom, the central business district had assumed its contemporary scale and scope. (Reproduced by permission of the Museum of Ventura County.)

"no eye-sores or fire-traps" anywhere, since property owners had torn down all the old buildings.[41]

By 1930, the area around Ventura Avenue was materializing as a neighborhood of oil field workers and a cluster of oil service businesses (figure 9). Elsewhere, developers and landowners had subdivided city parcels and were building single-family homes and apartment buildings. Main Street was becoming a vibrant place (figure 10). Commercial businesses were proliferating. Well before the Great Depression halted the boom, construction of private buildings and public infrastructure alleviated the population pressures associated with it.[42]

The development of the Ventura Avenue field generated economic activity that satisfied many boosters' expectations for quantitative growth. For residents and newcomers, the economic opportunities offered by oil development outweighed "the attendant hardships of boom life," as the Oliens describe the West Texas experience.[43] Sustained and controlled petroleum extraction promised to minimize social dislocation over time, making it likely that residents would embrace the oil industry. And so they did, giving the oil companies a veto, in effect, over the projects favored by local boosters.

Boosters Bid for Qualitative Growth

Local boosters were unable to realize their expectations for qualitative growth, as the outcomes of the Maricopa Road and Ventura Harbor projects and bond issues for public infrastructure illustrate. Debates contesting both road and harbor show how the oil boom increased socioeconomic tensions among urban and rural interests. A lack of voter interest in certain bond issues and tepid support of the Ventura Chamber of Commerce by business owners suggests that the interests of boosters and many residents diverged. The failure of boosters to generate requisite support for their urban vision denied them the means to challenge the local influence of the oil industry, which ultimately was preponderant. As a consequence, Ventura developed one-dimensionally, that is, along the quantitative growth trajectory established by the initial boom.

For boosters determined to make Ventura "a real city," in the words of the *Ventura County Star*, both the harbor and road projects were vital. A new breakwater and expanded harbor would make Ventura "a port of entry second to none on the Pacific coast," concluded the *Ventura Free Press*. It would accommodate the oil, lumber, and fishing industries, on the one hand, and recreational boats and beachgoers, on the other. The completion of the Maricopa

Road would ensure "Ventura's future as AN IMPORTANT COAST CITY," the *County Star* proclaimed. In seeking financial support from the operators in the Ventura Avenue field, banker Olaf Austad and journalist Sol Sheridan privately emphasized the reduction in transportation costs between California's Coastal and San Joaquin Valley oil regions that would result from the road's completion. Publicly, the project's supporters stressed the year-round benefits of the valley's residents traveling to Ventura "to enjoy the refreshing surf," spend money, and perhaps even purchase a beach home.[44]

Proponents of the road project had to overcome the opposition of the five-member Ventura County Board of Supervisors, which historically supported agricultural interests such as the Limoneira Company. In the summer of 1925, the Ventura Chamber of Commerce organized the West Side Chamber of Commerce, with Austad as president and Sheridan as secretary, for the purpose of promoting the road. Their advocacy helped to convince the supervisors to form a road district with their Kern and Santa Barbara County counterparts to fund the $2 million project. Still, Supervisor H. L. Butcher of Santa Paula, whose City Council perennially backed growers' interests, objected: "There are not five percent of the farmer taxpayers of Ventura County in favor of this road." Funding the highway would divert resources from road repair and the expansion of the county hospital. "All this agitation," he lamented, "to please a few boosters who want to build roads, build roads, BUILD ROADS!"[45]

Agricultural interests remained unwavering in their opposition to the project. Indeed, litigation brought by the county's farm bureau under the provisions of the state law that authorized such joint organizations at the county level halted construction until the California Supreme Court ruled in favor of the road district.[46]

In October 1933, the Maricopa Road opened to great fanfare. Yet, as Olaf Austad and Sol Sheridan had intimated, it ultimately proved to be more beneficial to Ventura's oil services sector than to local tourism and recreation, reinforcing the city's character as an oil town. Firms such as the Ventura Tool Company and National Supply Company that had begun their corporate existences servicing operations at Ventura Avenue now were able to serve the oil fields of Kern County, collectively the largest in the state, and expand their operations. In so doing, they contributed to the long-term stability of the local economy.[47]

Ventura County's agricultural interests successfully opposed the Ventura Harbor project. Moreover, they championed the construction of a harbor at Hueneme that served their marketing requirements. Thus Ventura's boosters failed to implement a project that was critical to their achieving the type of growth that they envisioned.

The development of the Ventura Avenue field reinforced the industrial land-use precedents for the waterfront established in the nineteenth century. After waves destroyed the wharf during the winter of 1913–1914, the City of Ventura rebuilt it with the intention of serving the transportation requirements of the oil industry. In 1922, both Associated and Shell laid pipelines to the wharf. On the eve of the boom, Associated constructed the first underwater pipeline on the Pacific Coast after the company's executives rejected as cost prohibitive a proposal to extend the wharf to accommodate larger tanker vessels.[48]

With the destruction of the wharf, the Ventura Chamber of Commerce, under Sheridan's direction, promoted a breakwater and harbor to facilitate commerce and solicited the U.S. Army Corps of Engineers to study the idea. Not much more happened in this regard until Associated's breakthrough wells sparked the local boom. After a storm destroyed the wharf a second time, in February 1926, the Chamber of Commerce formed a committee, again headed by Sol Sheridan, to promote a harbor that could handle the largest tanker vessels and accommodate recreational activities.[49]

At the same time, agricultural interests, led by Richard Bard, sought to build a harbor near Hueneme, located about ten miles closer to Los Angeles than downtown Ventura. Thomas Bard, his father and one of the region's largest landowners of the nineteenth century, had constructed a 1,500-foot wharf at this location in 1871 so that local farmers might seek national and global markets. During the 1920s, lemon cultivation in the county expanded southward from the Santa Clara Valley to the alluvial Oxnard Plain, displacing sugar beets and lima beans. In 1922, the California Fruit Growers Exchange, marketers of the "Sunkist" brand of citrus products, erected two packing plants at Hueneme. To expand the ability of growers to market their products, Richard Bard planned to construct a deepwater port on land that he owned. He used his political influence to persuade the state legislature to charter the Ventura County Harbor District, whose governing commission was appointed by the County Board of Supervisors. The supervisors installed Bard and his allies as the majority on the commission, which soon announced its intention to accept Bard's plan for the harbor and issue $2 million in bonds to fund its construction.[50]

Ventura's growth coalition challenged the legitimacy of the harbor district through publicity, letter writing campaigns, and legal action. Ultimately, the

California Supreme Court declared the district to be unconstitutional. Yet Ventura's boosters were unable to prevent voters from approving, in April 1937, the formation of an Oxnard Harbor District whose $1.75 million bond issue funded the construction of a harbor at Hueneme.[51]

Ventura's business and civic leaders failed to realize their goal of a comparable harbor for Ventura. The outside oil companies lost interest in the project when interwar changes in the shipping industry made the smaller harbor planned for Ventura uneconomical. They supported the Chamber of Commerce's idea for a breakwater and contributed financially to early efforts to gain its approval. Yet they soon lost interest in that project as well. Ralph Lloyd, Associated, GP, and Shell initially opposed the Hueneme harbor project on the grounds that the proposed $2 million bond issue constituted "an unwarranted burden" on the majority of the county's taxpayers, including, of course, themselves. Yet the oil companies did not oppose the project once it proceeded under the authority of the Oxnard Harbor District.[52]

Beachfront areas that might have been reserved for recreation continued to be used by the oil industry. For instance, in 1926, City of Ventura's Board of Trustees leased municipal property on a bluff overlooking the ocean to Shell for $17,500 so that the company could build two 80,000-barrel storage tanks.[53] The placement of additional storage and processing facilities on the beachfront discouraged its development along the qualitative lines found elsewhere along the California coast.[54]

Santa Barbara's response to a contemporary proposal regarding the use of its waterfront for oil infrastructure highlights the considerable influence that the oil industry wielded at Ventura. In 1931, the Irvine Oil Company applied for a permit to build a pipeline from the Mesa field to the ocean. Civic leaders and newspaper editors railed against the request. Noting that tourism was Santa Barbara's principal source of revenue, E. W. Alexander, chairman of the board of the Santa Barbara Harbor Commission, warned that the City Council could "not afford to menace . . . the municipal beach with an oil pipeline." The *Santa Barbara Morning Press* declared: "It would be an absolute disregard for the interests of the community and the rights of the people to permit an oil company to lay a pipe line into the ocean that might damage our beaches irreparably." The editors concluded: "Santa Barbara can never become a manufacturing or industrial city. . . . Its success or failure [lies] in attracting tourists, visitors and those with money seeking a pleasant place in which to live . . . we cannot destroy or even endanger the things that are most valuable to us in attracting this 'trade.'" The City denied the permit.[55]

The failure of voters to approve bond measures for the expansion and improvement of Ventura's water distribution and fire protection systems, and of the business community to support the Chamber of Commerce, suggests that there was also a conflict of interest between ordinary citizens, which included many small business owners, and the city's elite. Boosters were dismayed when long-time residents and newcomers alike demonstrated a lack of interest in funding their vision for Ventura. The *Ventura County Star* called on its readers to approve the measures, if they wanted their city to take its place as one of California's important new urban centers. Indeed, the newspaper published a telegram from Ralph Lloyd in support of these "vital improvements necessary for the city's continued development." In the event, the water distribution and fire protection measures both failed to secure the two-thirds majority they required for passage. Roy Pinkerton's newspaper lamented: "Today Ventura remained the same old city with its antiquated and inadequate water system and fire department." The vote also perplexed the Chamber of Commerce's Olaf Austad, who speculated that citizens either "lack[ed] confidence in the present city administration or disagree[d] with the proposed plan of water development, or both." The business community also disappointed growth advocates when the Chamber of Commerce failed to meet its funding targets during the first two years of the boom. At the same time, voters approved a bond issue to fund the construction of a new high school, indicating that they might support projects for which there was a demonstrated need to ameliorate present conditions. Ordinary citizens may have simply believed that the price of achieving qualitative growth was too high. Or perhaps, as Austad observed, "In a fast growing city like VENTURA, people being [*sic*] prosperous and absorbed in their private affairs, are . . . more likely to neglect organized community effort."[56]

The response of Ventura's residents was not unlike that of their counterparts in many small-to-medium-sized towns of the East and Midwest. In such towns, residents were reluctant to fund public projects, even during boom times, preferring their civic leaders to keep taxes and spending to a minimum. In the context of this conservatism, major employers—such as the Ball Brothers Company, in Muncie, Indiana; the Ford Motor Company, in Dearborn, Michigan; and the General Electric Company, in Pittsfield, Massachusetts, and other small- to medium-sized cities of the Northeast—exerted hegemonic control over local development through civic engagement, charitable contributions, and support for educational and social welfare institutions. Indeed, this influence was all but welcomed, at least initially. Ventura's boom enabled business and civic leaders to coalesce around a qualitative growth trajectory for their city. Without recourse to personal sources of wealth and unable to persuade voters to approve key bond measures, however, Ventura's boosters needed the oil industry to support their goals.[57]

Debating the Annexation of the Avenue

On the issue of annexation of land that lay between the Ventura River and Ventura's downtown, the outside oil companies demonstrated the limits of local elites relying on external actors for support. Three times local leaders called for the incorporation of this district, named for Ventura Avenue, which bisected it roughly parallel to the Ventura River. On each occasion, Ralph Lloyd and the companies that he represented thwarted the effort, convinced that municipal jurisdiction would impair their ability to maximize the returns on their investments.

Annexation debates were key moments in the life of many communities in the Los Angeles Basin during the first half of the twentieth century.[58] Cities either embraced or rejected oil, depending on the values they wished to promote. Long Beach and Huntington Beach, for instance, controlled land that oil operators wanted to lease for drilling purposes and welcomed the revenues that they could capture from oil production. Other suburbs, most famously Signal Hill, incorporated to prevent other cities (in this case, Long Beach) from annexing oil properties. Still others, such as Torrance, established boundaries to exclude the industry. For its proponents in Ventura, annexation of the oil field or urban areas outside the city limits, or both, would enable the City of Ventura, rather than Ventura County, to capture the taxes associated with increased property values. It would also allow the City to regulate a rapidly urbanizing area currently outside of its control. But the outside oil companies, coordinated through Ralph Lloyd, stymied their efforts, demonstrating the ability of firms with their headquarters in downtown Los Angeles and San Francisco to influence decisions that affected the character of Ventura's growth. Lloyd persuaded local elites that Ventura would benefit economically more from unfettered, oil-driven development than from taxes on property encumbered by regulation.

A number of annexation proposals were proffered for the consideration of Ventura's residents. In late 1925, Giovanni Ferro called for annexation of the entire area along Ventura Avenue that lay west of downtown. He was concerned that the district was "being built up with buildings which would not conform with any building code, and without sanitation protection." He asserted that, "with a little work in street improvements, and the enforcing of city laws," the district "could be made . . . into one of the choicest . . . in town." By spring, the area under consideration for annexation had expanded to include the Rincón, a beach community several miles up the coast, and all the oil lands that lay between it and Ventura (figure 8). In 1928, the planning commission offered a proposal that included these areas, the hills that served as a backdrop to the central business district, and a portion of the oil field. Another initiative, floated in 1927, sought to annex certain sections of the Ventura Avenue district, but took care to exclude the field itself. According to Roy Pinkerton, annexation of the district had "long been desired" by the city's "far-sighted civic leaders." Local newspapers suggested that a majority of residents favored annexation. The Ventura Chamber of Commerce also thought it was a good idea. Indeed, when Ferro proposed annexation, Associated, GP, and Shell only could count on the support of Mayor Rea, who was also Shell's superintendent, among Ventura's business and civic leaders.[59]

Ironically, given Ventura boosters' disparaging of Long Beach's boom, Ferro and other civic leaders pointed to the latter as an example of the benefits of controlling oil lands within a city's boundaries. The Long Beach field, discovered at Signal Hill in June 1921, comprised some 1,100 acres in January 1925. Long Beach, whose population almost doubled from 1920 to 1924, owned 140 acres of the field, the royalties from which totaled $3.2 million from November 1921 through June 1925. The city used these revenues to fund water and park development and the construction of streets, fire and police stations, schools, and a public hospital. Moreover, between 1923 and 1925, Long Beach voters approved several bond issues for public projects, including $5 million for a breakwater. Long Beach's independent source of revenue enabled it to limit the influence of the oil industry over its municipal affairs and implement the type of projects that Ventura's growth advocates desired.[60]

For Ralph Lloyd and the operators at Ventura Avenue, Long Beach's experience with annexation sounded a cautionary note and provided them with a point of leverage. When the City of Long Beach approved a tax on each barrel of crude oil pumped within its limits, operators on leases held by private owners joined together in an independence movement. In 1924, they incorporated the two-mile-square City of Signal Hill, which at the time had almost as many drilling rigs in operation as people—about 2,000 each.[61]

The outside oil companies and the oil field's lessors adamantly opposed Ventura's various annexation proposals. Ralph Lloyd coordinated their efforts to derail them. First, he convinced the Ventura Chamber of Commerce to reverse its position. To do so, he arranged for the firms to contribute $1,250 to offset the initial expenses of the committees charged with promoting a breakwater for the harbor and the Maricopa Road, on the condition that the Chamber back the companies' position.[62] Lloyd then met with Ferro and other proponents to dissuade them from pursuing annexation and with local newspaper editors to explain the industry's stance.

Ralph Lloyd argued that annexation "in the long run would not be beneficial to either the city or the operating companies." Citing Long Beach and Huntington Beach as examples not to emulate, he insisted that any field as large as Ventura Avenue could not operate under municipal building codes and other regulations. At Huntington Beach, he explained, operators drilled far fewer wells than officials expected after the latter incorporated the field into the city's limits. As for Long Beach, Lloyd cited the "virtual war" that municipal officials were waging against operators who failed to comply with new pollution and fire prevention ordinances as additional proof of his argument.[63] These examples showed that "an oil field should not be and cannot be restricted and governed by the regulations that of necessity are applicable to city government," as Lloyd put it to the Bank of Italy's John A. Lagomarsino.[64] The "friction and dissatisfaction" that would ensue as a result of annexation would dissipate any financial benefits that might accrue to the city. To date, he explained to the Chamber of Commerce's Sol Sheridan, unencumbered oil development had facilitated the "wonderful growth of the city" and was responsible for its "healthy business condition" and the "cordial cooperation between the [city] on one hand and of the oil field on the other."[65] Should the City force the issue, Lloyd vowed that operators and lessors would "do what Signal Hill had done and incorporate a separate town or municipality."[66] George Legh-Jones, Shell's president, agreed: annexation would "be a grave mistake."[67]

Such pressure persuaded annexation's supporters to retreat. On each occasion, a majority of the City's Board of Trustees ultimately backed the oil industry. Retailers joined the Chamber of Commerce in supporting the oil companies, once the latter publicized their view on the issue. As Lloyd explained, they feared "the ultimate outcome if a breach is made between the city and the oil interests." Lloyd also persuaded editors that annexation was a bad idea. A letter from him, for instance, apparently convinced the *Ventura Daily Post* to recommend that the City drop the issue. The newspaper's editors explained that a "quarrel" with the oil companies was not in Ventura's interests.[68]

The local Chamber of Commerce was already convinced that, as a small city, Ventura could not afford to "do anything that will hinder these companies or any other industry concerned."[69] To ensure that the Chamber continued to rally local support for the oil industry, should the annexation movement revive, as it did on two occasions, Lloyd persuaded Ventura Avenue operators to contribute to the financial health of the organization, which perennially lacked adequate support from local business owners. In asking outside oil companies to fund the Chamber of Commerce, Ralph Lloyd explained that "a little consideration and cooperation along these lines" would "cultivate" the support of "our Ventura friends" on key issues and prove to be more effective "than an expensive program of resistance."[70] From 1925 to 1931, contributions from the Lloyd Corporation and the oil companies defrayed the organization's operating expenses and provided the margin of success in its fund drives associated with harbor and road promotion. Oil interests also funded Community Chest drives in 1926 and 1927. In 1930, the companies gave at least $10,000 in total toward the expansion of the Big Sister's Hospital when it reported a $40,000 cost overrun. Local editors noted these deeds prominently on the front pages of their newspapers.[71] As a result, Ralph Lloyd could report that the Ventura Chamber of Commerce would act in a manner that was consistently "friendly to the oil interests."[72]

Having "captured" the Chamber of Commerce institutionally, Associated, GP, and Shell could count on its support when their interests conflicted with those of local agriculture, as on the issue of property valuation. Agriculture, from the oil operators' point-of-view, was not paying its fair share of the county's taxes. As early as 1923, Associated Vice President A. C. McLaughlin complained about the assessed valuation of his company's properties. It was unfortunate, he wrote to Lloyd, that the county "should add so greatly to our difficulties as to hamper materially the development of the field." Should the assessments stand, he warned, it would be "many years before the field is fully developed." As of 1931, Ventura County extracted twice as much as Orange County and three times as much as Los Angeles County per barrel of crude oil produced. In that year, Will M. Reese, Ventura County's assessor, increased his assessment on the oil companies despite a drop in production (on which valuations were based). He did so by imposing a 25 percent surcharge on each barrel of production to account for oil kept in the ground because of the curtailment regime then in effect, as we shall see. When the supervisors, sitting as the board of equalization, denied Associated's petition for relief in a meeting that Lloyd characterized as "a little picture of Russia," missing only the firing squad, the Chamber of Commerce resolved to use its influence to ensure that the industry received fair treatment. Secretary John Wallace feared that, should the

Chamber fail, Ventura would be "handicapped in its efforts to promote industrial activities." Pressure from the Chamber helped to persuade Reese and the supervisors that Ventura County's assessments on oil leases were excessive.[73]

The Chamber of Commerce also supported the oil industry on the development of the tidelands along the Rancho el Rincón (figure 8). In doing so, it opposed William S. Kingsbury, California's surveyor general, who was leading the conservationist fight on behalf of landowners, real estate developers, and others who wanted to ban oil development on California's beaches. The Chamber argued that the lands at issue had lost all scenic and recreational value, owing to oil development on adjacent private properties. It proposed a county ordinance that would permit unrestricted development. The local oil workers' union, the Ventura County Building Trades Council, and the Merchants' Credit Association of Ventura followed the Chamber's lead with similar resolutions. The Chamber also sponsored local speakers to build support for the development of tidelands leases all along the Ventura County coast.[74]

Observed Ralph Lloyd in one of his annual pleas to oil company executives for their support of the organization: "The Chamber is the only organization in the County that the oil interests can count on for support." In the wake of the fight over tax assessments on petroleum properties, he reminded operators that the Chamber "is really the only organized friend of the oil industry in the county." Only when its members "got busy," Lloyd noted, did the "political element" take note of the industry's arguments. As the outcome of the harbor project demonstrated, however, the oil companies did not always reciprocate the Chamber's support.[75]

"To encourage the friendly elements of city" generally and to enable him "to be of some influence" on critical issues, Ralph Lloyd also encouraged the major operators to increase their funds in local banks, the managers of which ranked among the leaders of the community. With annexation on the table, Lloyd asked both Associated and Shell to deposit an additional $100,000 with the local branch of the Bank of Italy; for his part, he had already transferred $50,000 from his Los Angeles account. Lloyd kept a large balance in Ventura "to keep in touch with the local situation and in friendly cooperation with the local financiers." In Lloyd's view, such action would persuade Ventura's financial leaders that it was advisable "to leave things as they are than to cause enmity by forcing an issue between the city and the oil interests relative to the incorporation of the oil field into the city." Lloyd was "anxious to keep in a position to be of some influence, especially in the case another attempt is made to expand the city limits of Ventura to the oil fields."[76]

In this case, Lloyd failed to persuade the companies to act, illustrating the different level of interest in local growth issues on the part of integrated oil companies with investments in dozens of fields, on the one hand, and landowners and independent operators, who typically concentrated their investments in a single area, on the other. Associated considered Lloyd's recommendation. The record is unclear whether it ever reached a decision. For its part, Shell demurred, noting that it received similar requests from banks "almost daily." It was impractical to keep large deposits in banks all over the country, as it needed the funds for its operations. For the outside oil companies, placing additional funds on deposit in local banks was not necessary to maintain their influence on local politics.[77]

Operators built and maintained support among elites and residents for their unfettered presence in other ways. For instance, Shell drilled a water well that doubled the city's supply at the peak of the boom. On at least one occasion, an operator underwrote the cost of a public works project: in 1928, Associated contributed $4,000 to a street paving project. Ventura's residents were also well aware that they were lighting their homes and businesses with cheap natural gas produced from the Ventura Avenue field.[78]

Acting as a "citizen outside the government," Ralph Lloyd played an integral role in enabling outside oil interests to exert predominant influence over local decisions and, in so doing, shaping the character of Ventura's urbanization. Through his intermediation, operators and lessors used both "carrots" and "sticks" to persuade local business and civic leaders to set aside projects that might interfere with their ability to develop the Ventura Avenue field as they saw fit. As a result, Ventura relied more on industry largesse than direct taxation for its benefits from oil-related activity and the oil companies enjoyed substantial discretion in deploying resources to shape the character of local growth.

A Balance Sheet for Oil-Directed Growth

Ventura's oil boom ended with the Great Depression. In fact, depression and war combined to halt municipal growth for almost two decades. As of 1947, the city's assessed property value of $11.4 million was only 9.7 percent higher than it was in 1929. In almost two decades, Ventura's population grew by less than 5,000 people to 16,354.[79] Nevertheless, the transformation of Ventura into an industrial city during the second half of the 1920s remained a source of pride for its residents. For oil activity had spurred the creation of industries and had fueled population growth. The popular view of oil boomtowns as chaotic, or even anarchic, hellholes held no more for Ventura than it did for the towns of the Permian Basin in West Texas studied by the Oliens.[80]

And yet Ventura's boosters, who articulated a set of community values intended to drive the implementation of a growth strategy that focused on improving the quality of life, failed to persuade enough residents to support their projects financially. Without significant non-oil sources of wealth to tap, they turned to Associated, GP, and Shell for assistance. Because these firms had shown only moderate interest in qualitative growth, this dimension of the boosters' plans for Ventura largely withered.

The oil companies operating at Ventura Avenue wielded influence comparable to that of the resident industrialists who dominated the political economies of medium-sized towns in the East and Midwest in the early twentieth century. Compared to the Ford and the Ball families, who owned or controlled large amounts of real estate in their respective towns, Associated, GP, and Shell forged their influence more indirectly and with a smaller philanthropic commitment.[81] As was the case with large manufacturing firms operating in medium-sized cities during the Fordist era, however, the local influence of the oil industry was ultimately a function of the degree to which the economic well-being of the community depended on its presence. Ventura's interwar growth owed entirely to the orderly development of substantial petroleum reserves.

Cosmopolitan capital will become involved in municipal growth politics to establish a political environment that serves its interests if its local capital investments are significant and returns on investment depend on place. At the same time, asset values may depend on factors other than local growth or its distribution. With high sunk costs locally, Associated, GP, and Shell took an active interest in all manner of parochial affairs so that they would be able to exploit the petroleum pools that lay outside the city of Ventura unencumbered.

Ventura's experience illustrates the path-dependent effect of development during the initial stages of an oil boom. By asserting its influence over local growth politics, the oil industry limited boosters' ability to invest oil-generated wealth in tourism and recreation. Thus Ventura's growth trajectory diverged from the path imagined by boosters who sought to emulate coastal communities that used petroleum wealth to diversify their urban environments or tapped non–oil dependent resources to control their urban destinies. By "turning its back to the sea," Ventura's development as a place had more in common with inland California oil communities, such as Bakersfield and Fullerton, than with its coastal counterparts.[82]

Chapter 3

Making Portland a Wonderful City

Father always said that after he made some money in California he wanted to return to Portland.

—Eleanor Lloyd Dees, 1958

Subscribers to the *Sunday Oregonian* in the leafy neighborhoods around Holladay Park read of the "dawn of a new era" on the front page of the 1 March 1926 edition. They may have been jolted by the excited predictions of East Side businessmen of "business blocks reaching the skyline."[1] For Ralph Lloyd's acquisition of two immense tracts of undeveloped property "created quite a stir in the real estate market," noted realtor E. J. Lowe.[2] Local residents need not have worried about the erection of skyscrapers, for the time being. But Lloyd had ambitious plans that, if realized, would fundamentally alter the character of the place. In the near term, widening streets by lopping off portions of their front yards would be their greatest concern.

Ralph Lloyd found irresistible the potential returns from investing his mounting surplus of capital from petroleum extraction in commercial real estate. In August 1925, he had begun discussing the purchase of the property that he had attempted to acquire in 1911 from Charles X. Larrabee, which was now held in two blocks.[3] In late February 1926, the California oilman closed a deal for almost 600 unimproved lots held by the Oregon Real Estate Company, in which the heirs of Larrabee, who died in 1914, were interested, for $1.25 million. At the same time, he obtained a $500,000 option for 170 lots held by the Balfour-Guthrie Trust Company, which he exercised in September with a cash payment from a bank account flush with royalty income. The Balfour-Guthrie lots straddled Sullivan's Gulch, a ravine that extended from an inlet of the Willamette River, and lay between the Oregon Real Estate Company lots (figure 11). As we have seen, this was not Ralph Lloyd's first real estate acquisition in Portland, but it was certainly the largest. Lloyd's acquisition of a large portion of Portland's East Side illustrates the low barriers to entry and the opportunities available to small business in real estate. The size of the holdings provided the oilman with the opportunity to become one of Portland's leading builders.[4]

Real estate broker William M. Killingsworth dubbed Lloyd's acquisitions "Larrabee's white elephant."[5] Despite his "unbounded faith" in the East Side, Charles Larrabee did nothing to develop his holdings.[6] Notwithstanding his earlier failure to acquire it, Ralph Lloyd "felt that some day" he would buy Larrabee's undeveloped East Side property.[7] Many of those among Portland's business and civic elite may have seen "Larrabee's white elephant" as wasteland, but the oilman was convinced that Portland could become *the* trade and transportation hub of the Pacific Northwest with its development. Larrabee's portfolio offered room for the city to expand commerce beyond a downtown that was confined by the Willamette River and the West Hills, as would be required, in his view, if Portland hoped to fulfill its potential as a metropolitan center. As a booster and developer, however, Lloyd would discover that many East Side residents would remain ambivalent to, and some would vigorously oppose, the prospective commercialization of their residential district.[8]

For more than four years, Ralph Lloyd would devote his energies "as a citizen outside the government" and as an outside investor of oil-generated capital to literally preparing the ground for commercial development. His expectations and vision for Portland were informed by his formative real estate experiences in Los Angeles, which by the mid-1920s had emerged as *the* "energy capital" of America, if not the world. He and his representatives would work through the apparatus of municipal government, from the mayor's office and the City Council to the public works department to the planning commission, to ensure that the public works projects that Lloyd deemed to be essential to the realization of his plans were implemented. In large part, municipal officials would accommodate the Lloyd Corporation, approving its petitions to open, widen, extend, and vacate streets, establish setbacks, and adopt other measures, and then clearing the way for these projects to be constructed by either the City of Portland or the Lloyd Corporation, or both. In responding as they did, Portland's civic leaders demonstrated the influence that a small

Figure 11 (*folowing page*). Anglo-Pacific Realty Company map of Holladay's Addition, Portland, Oregon, circa 1914. Ralph Lloyd's acquisitions included blocks, half blocks, quarter blocks, and individual lots. Each block contained eight lots. The lots that Lloyd acquired are shaded. The map appeared in the 7 March 1926 edition of the *Oregonian* under the caption, "Two Million Dollar Purchase by Ralph B. Lloyd Stimulates Real Estate Activity." (The Huntington Library, San Marino, California, LCR drawer 2, box 7.)

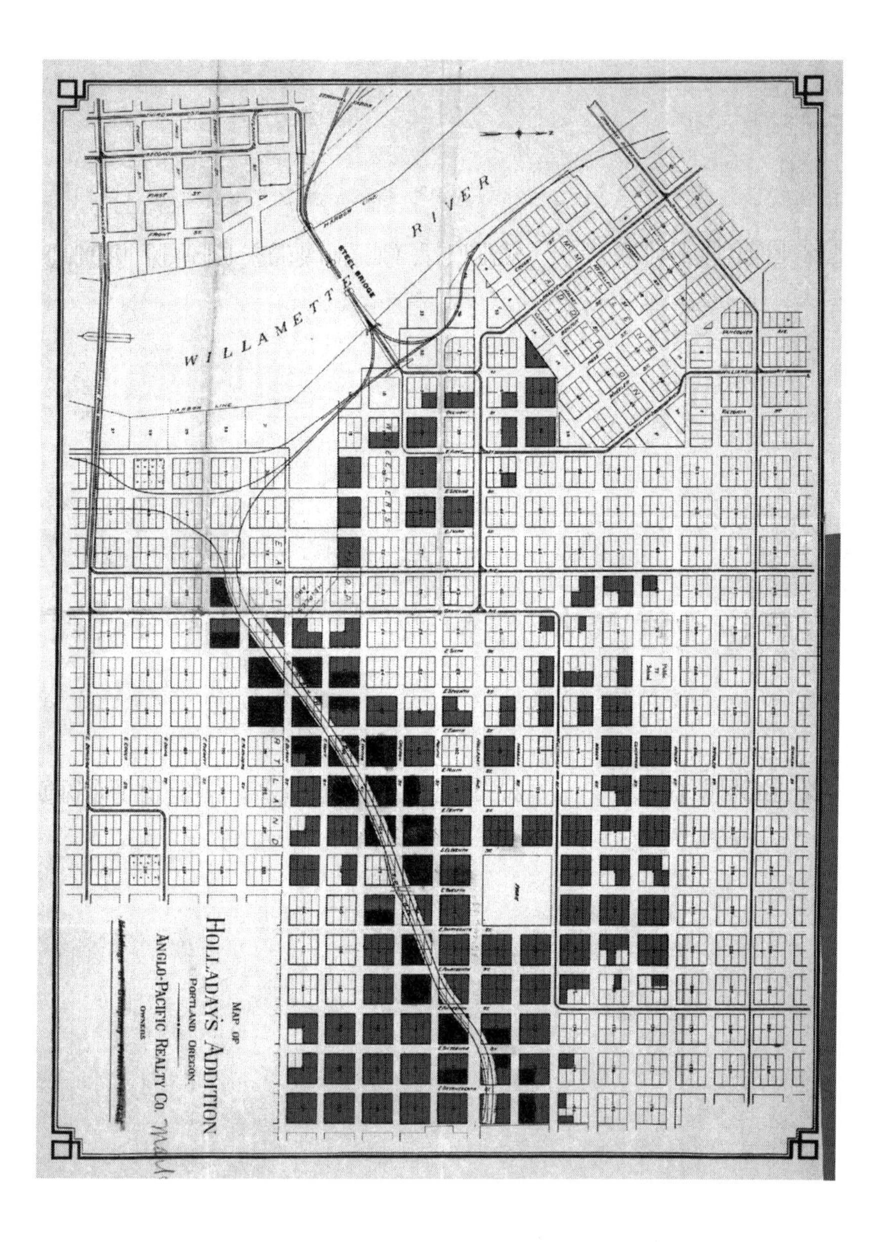
WILLAMETTE
RIVER
STEEL BRIDGE
HARBOR LINE
FIRST ST
FRONT ST
MAP OF
HOLLADAY'S ADDITION
PORTLAND OREGON
ANGLO-PACIFIC REALTY CO.
OWNERS

businessman with substantial capital might exert over local politics and real estate development. Ultimately, Ralph Lloyd would influence the development of Portland's East Side on a level commensurate with the scale of his holdings, as the appellation "the Lloyd District" in the modern city suggests. By any measure, as the proprietor of a small business with substantial capital, Lloyd operated in the center of the real estate business in a major metropolitan market. Reflective of the contingencies associated with commercial development, however, the Great Depression would set in before Lloyd would realize his initial plans. In this case, public infrastructure implementation in the context of residents' protests posed a greater obstacle to the developer than intracity market competition—though, as we shall see, competition for capital in a saturated hotel market would confound the oilman once he opted not to fund his program entirely out of retained earnings. Describing and explaining Lloyd's activities in Los Angeles and Portland during the second half of the 1920s in this chapter and the next sheds light above all on the risk and element of luck inherent in commercial real estate development.

Even though there were no productive oil wells within hundreds of miles of Portland, the materialization of the city's East Side became intertwined with the fortunes of the petroleum industry generally and those of Ralph Lloyd as a lessor in particular. Associated's breakthrough in the Ventura Avenue field provided Lloyd with substantial surplus capital that he chose to invest in Portland commercial real estate. As of January 1925, that is, at the onset of the boom, Lloyd was already worth almost $3 million and held practically no debt. Thus he was able to put down $300,000 on the Oregon Real Estate Company deal and retire a four-year, four percent note on the balance of the sale in one year and also pay cash for the Balfour-Guthrie lots.[9] Over the ensuing seven years, Lloyd would spend more than $5 million on real estate development, including more than $1 million on street improvements and other public works projects. He aimed to finance his program strictly out of cash flow, retained earnings, and short-term notes. In the late 1920s, this unleveraged conservatism, which characterized his approach to real estate development generally, would slow the pace of his initial program, ultimately compromising its viability. At the same time, it allowed him to hold property off the market until he was convinced that a project would pay off—a comparative advantage that any rival would envy. It would take more than a quarter of a century for Ralph Lloyd and his Lloyd Corporation to develop the area around Holladay Park on a scale commensurate with his vision—a vision inspired and shaped by his experiences in the Los Angeles real estate market.[10]

Cutting His Real Estate Development Teeth in Hollywood

As he devoted time and resources to growing walnuts and exploring for oil in Ventura County, Ralph Lloyd thought strategically about the commercial development of Portland's East Side. The area's population had increased nearly fourfold from 1900 to 1910, to about 120,000, and surged in the years running up to World War I. Lloyd anticipated that his relatively small holdings along Union Avenue would "become more valuable each year," especially with the construction of the Steel and Broadway Bridges and Union Avenue's designation as a key road in Edward H. Bennett's (City Beautiful) Plan for Portland. The Interstate Bridge, connecting Portland and Vancouver, Washington, promised to establish Union Avenue as the city's most important north-south thoroughfare east of the Willamette River and stimulated Lloyd to invest in Walnut Park, several miles north of Holladay's Addition. Certainly, there were no barriers to entry preventing him from doing so. All the properties that Lloyd added to his Portland portfolio between his return to Ventura County in 1911 and his closing of the aforementioned deals in 1926 lay in this part of the East Side. In acquiring these properties, he became acquainted with William M. Killingsworth.[11]

Killingsworth was a classic booster who promoted Portland for the singular purpose of capitalizing his speculative real estate investments. In 1885, he began accumulating land between the Willamette and Columbia Rivers. He focused his investments on Walnut Park, as indicated by the eponymous avenue that bisected the neighborhood in an east-west direction. In 1899, he joined other restless businessmen in creating the Portland Board of Trade as a more dynamic alternative to the sleepy Chamber of Commerce. He was also elected vice chairman of the Civic Improvement League. With his fellow boosters in these and other groups, Killingsworth shared an "unwavering confidence and faith in Portland becoming a City of great magnitude because of her matchless location [and] vast undeveloped resources." From their moment of first contact, Killingsworth sought to persuade Ralph Lloyd that Walnut Park was poised to become "the business center of Greater Portland" and that Lloyd could make it so, were he to buy the lots that Killingsworth had to offer. Killingsworth's solicitation provides an example of the opportunities that abounded in commercial real estate for the small businessman during the interwar period. In this case, Lloyd had the opportunity to develop an entire commercial district because of his relationship with the landowner. In time, the oilman would become Walnut Park's most significant developer.[12]

Ralph Lloyd agreed that Walnut Park was a potentially important commercial node but differed with Killingsworth on how long it might take to mature it into one. Between 1914, when he first contacted Killingsworth, and 1925, Lloyd limited his interest in Walnut Park to a few speculative lots. Frankly he found greater opportunities in Los Angeles. A subscriber to almost two dozen Pacific Coast newspapers and periodicals, Lloyd explained the geography of his investment strategy to Killingsworth in terms of regional urban rivalry, using language and arguments similar to those deployed by contemporary civic and business leaders in San Francisco who sought to meet the challenge of a rising Los Angeles.[13] Lloyd generally saw "the prosperity of the East gradually working to the West" and anticipated "an increasing movement of population to the West" after the war.[14] The cities of the Pacific Coast still had to compete with one another, however, to reap the benefits of this shift in demographic advantage. Portland faced intense competition from Los Angeles, San Francisco, Oakland, and Seattle—not just the latter, as Portland's boosters typically framed the terms of regional urban rivalry.[15] Until he felt that Portland was ready to rise to the challenge, Lloyd would focus on Southern California.

Ralph Lloyd used Los Angeles as his benchmark to discount Killingsworth's assertions regarding the speculative value of Walnut Park properties. The City of Angels, he argued, generally offered "the chance for quicker returns." As corroboration, he cited the skepticism of his oil industry colleagues, none of whom he could persuade to invest in Portland real estate. The financial calculus still favored Los Angeles when he added lots to his Walnut Park portfolio in 1923: "Our own city of Los Angeles is growing so rapidly and there are so many opportunities here to make attractive investments that I am hesitating in making further investments in Portland." Lloyd may have been quick to condemn the chaos that reigned in the oil fields of the Los Angeles Basin, but generally he applauded the organization and execution of city building throughout the metropolis because it provided him with ample opportunity to develop real estate.[16]

Still, Ralph Lloyd felt that he was young enough to justify his taking a chance on selected Walnut Park lots, "even if I have to wait [to] bring its reward" to the district.[17] Feeling that an investment in Portland would "win out 'OK'" in the end, in 1917, he acquired five lots along Killingsworth Avenue at the intersection of Union and Commercial Avenues (figure 12). Killingsworth was reluctant to let Lloyd "cherry pick" lots that he wished to sell as a block but relented

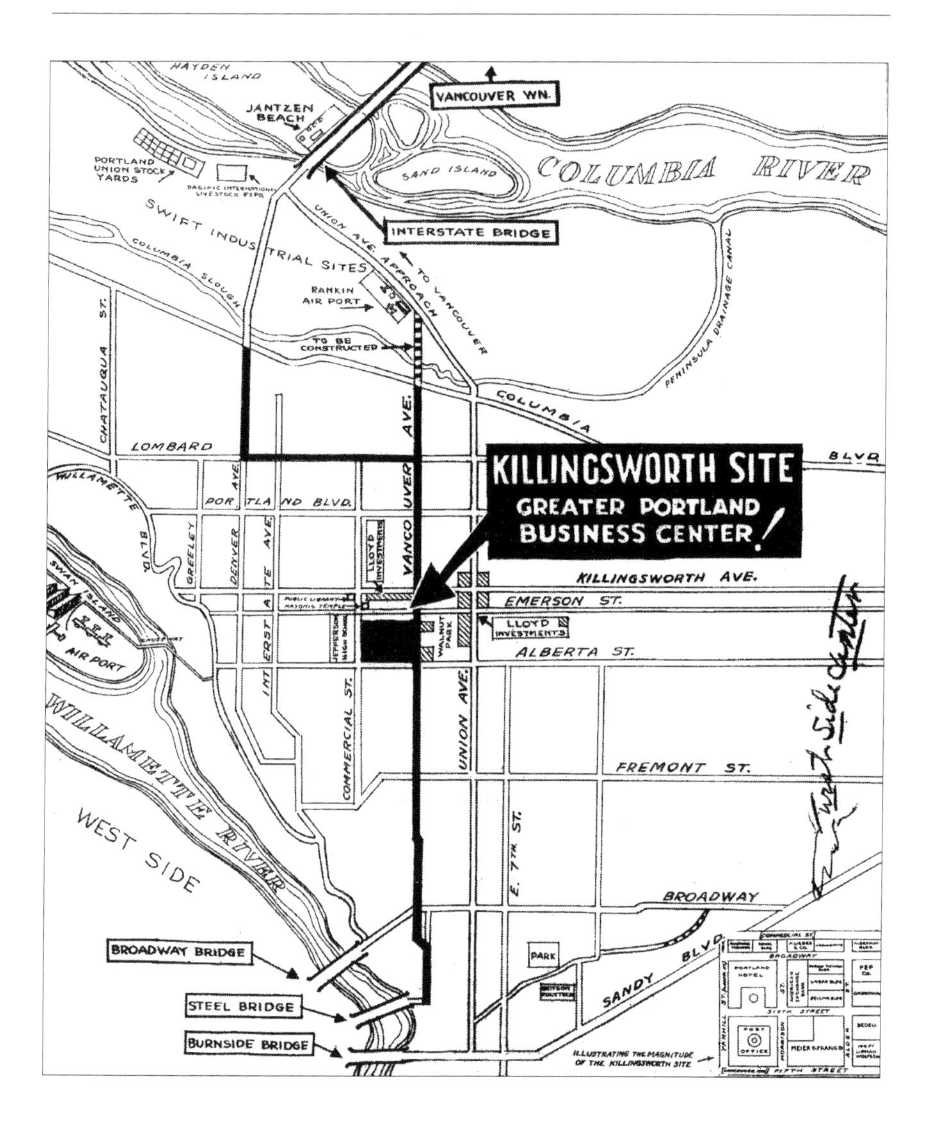

Figure 12. Killingsworth Site, Walnut Park, Portland, undated. Enclosed with a letter from William Killingsworth to Ralph Lloyd, 7 February 1929. (The Huntington Library, San Marino, California, LCL drawer 4, box 4.)

when the oilman promised to make additional acquisitions. When he had contacted Killingsworth in 1914, Lloyd had planned to develop these and his other Union Avenue properties with commercial or multi-unit residential buildings, but now he was purely speculating. He thought enough of Portland's future and had sufficient income from the oil business to resist brokers' offers to sell his small but growing East Side portfolio. Indeed, his success in the oil business provided him with the "staying power" to hold onto his portfolio regardless of market conditions—a key competitive advantage in the real estate business. For the next six years, however, Southern California oil and real estate development monopolized his attention and resources. Killingsworth would have to wait for Lloyd to return to Walnut Park.[18]

In the two decades after Ralph Lloyd returned to Southern California, Los Angeles took on its modern shape and character. During the 1910s, the city's population grew 85 percent. In the next decade, it grew an astounding 136 percent. In these two decades, its population increased from 319,198 to 1.24 million. By contrast, Portland grew much more slowly, increasing in population from 207,214 in 1910 to 301,815 in 1930. After World War I, the value of building permits issued by the City of Los Angeles soared, from $28.3 million in 1919 to an interwar peak of $200 million in 1923, and then averaged more than $137 million annually over the next four years. Lloyd seized the opportunity to profit in this real estate hypermarket.[19]

Ralph Lloyd was one of many investors who joined developers, appraisers, and real estate agents in transforming commercial space outside of downtown Los Angeles during the 1920s.[20] Beginning with his purchase, in April 1912, of two lots along Jefferson Street, four blocks east of the University of Southern California campus, Lloyd acquired at least a dozen properties on the fringes of downtown and in Hollywood. His holdings in the latter district included five lots of the "Hollywood Gateway Tract" at the southeast corner of Vermont Avenue and Hollywood Boulevard; the southwest corner of Los Feliz Boulevard and Hillhurst Avenue; and the northwest corner of Vermont and Willowbrook Avenues: all important nodes within the metropolitan grid. Lloyd's experiences with the Los Angeles real estate market shaped his expectations for Portland and the ways in which he dealt with its business and civic leaders.[21]

During the 1920s, Hollywood developed into one of the largest outlying commercial districts in America. Located northwest of downtown in a rectangular area bounded by Beverly Hills on the west, Beverly Boulevard on the south, Griffith Park and the Santa Monica Mountains on the north, and the upscale Los Feliz and Silver Lake neighborhoods on the east, Hollywood became a model of decentralization in the minds of Angelenos, owing to boosters' success in transforming the district into one of the fastest-growing areas of Southern California. By 1930, boosters were pointing with pride not only to the "skyscraper mile" that materialized along Hollywood Boulevard between Vermont and La Brea Avenues, but also to numerous, often elaborate buildings at strategic intersections that, together, offered the affluent residents of the area a wide array of goods and services. Constructed as one- or two-story structures with the needs of retail, office, and theater tenants in mind, these so-called taxpayer blocks were the "workhorses" that anchored commercial development in Hollywood and lesser known outlying districts of the region's metropolitan centers.[22]

Ralph Lloyd's plans for his lots on the southeast corner of Vermont Avenue and Hollywood Boulevard, however, were more in keeping with the "forgotten arterial landscape" of Los Angeles that architectural historian Richard Longstreth has richly detailed than with the more elaborate dreams and visions of many Hollywood promoters and business owners. In October 1919, Lloyd signed a five-year lease for the corner lots of the Hollywood Gateway Tract that included a gas and oil service station and made plans to erect a one-story building that replicated a modest taxpayer that he had constructed south of downtown. He had a commitment from a prospective tenant to lease the building for five years at $10,000 per year. As he was earning $2,000 per year on the two lots occupied by the automobile service station, he planned to build south of the corner. As this example suggests, a vast and variegated commercial real estate market accommodated entry on the part of developers large and small. Moreover, Lloyd executed the lease well before the Ventura Avenue field showed promise of being the lucrative play it would become, which indicates that his interest and participation in the real estate industry did not wholly depend on his success in the upstream segment of the oil industry.[23]

No lease materialized, however, before Lloyd executed a ninety-nine-year ground lease for the lots with Henry Laub and Frank L. Schaeffer in May 1921.[24] Now "located in the center of a rapidly developing business district," Lloyd's property lay across from Olive Hill, where several stores were planned, and catercorner from the site of a soon-to-be-constructed branch of the Security Trust and Savings Bank. Laub and Schaeffer were willing to defer their development plans until the lease on the service station expired.[25]

Two years later, Schaeffer sold his interest in the lease to the Capitol Company, the real estate arm of A. P. Giannini's Bancitaly Corporation. The purchase was in keeping with Giannini's intention to open some fifty branches of the Bank of Italy in Southern California. The company and Laub renegotiated the terms of the ground lease, agreeing to pay Lloyd a total rental of $694,000 and construct a building by December 1924. On the face of it, the deal constituted a significant breakthrough in Ralph Lloyd's career in commercial real estate development.[26]

Local business owners, however, were quick to criticize Lloyd and the Bank of Italy when the latter revealed its plan to construct a vernacular one-story building for a branch bank and two other tenants. In a petition, they complained that the proposed building "would set back the development of the district at least five years." Rather, the corner could "be made one of the very best suburban business centers in Los Angeles" if it were improved "with buildings of proper size and dignity to attract the highest type business houses," such as a 1,500-seat theater or a "fine store and office building." The planned extension of the Los Angeles Railway along Vermont Avenue to Hollywood Boulevard and proposed "improvements" to both Prospect and Vermont Avenues promised to establish Lloyd's corner as "one of the greatest transfer and junction points in Los Angeles." Beyond continuing to apply pressure, however, there was little that a group of business owners could do at the time to prevent an individual developer from proceeding as he planned. Ironically, the widening of Vermont Avenue, as proposed in the 1924 Major Traffic Street Plan for Los Angeles, derailed all plans to "improve" the property.[27]

Ralph Lloyd generally applauded Los Angeles's street program of the 1920s, which invested millions of dollars in projects to widen, open, and extend arterial streets, along many of which he held property. As a "citizen outside the government," Lloyd acted through his membership or directorship in more than half-a-dozen improvement associations, corresponding with the geography of his growing commercial real estate portfolio. He was also a member of the Major Highways Committee of the Los Angeles Traffic Commission, which funded the Major Traffic Street Plan produced by renowned urban planners Frederick Law Olmstead Jr., Harland Bartholomew, and Charles H. Cheney.[28] After voters approved both the plan, which proposed some 200 projects affecting several hundred miles of streets, and a $5 million bond issue to fund the first phase of construction, the Major Highways Committee secured additional funds and lobbied successfully for passage of four laws to facilitate the implementation of the plan. The legislation allowed the City of Los Angeles to expedite street work, fund construction and land acquisition, and overrule the protests of property owners. Through this committee, Lloyd became acquainted with engineer Harry Z. Osborne, whom the

City of Los Angeles had retained in 1921 as a consultant to investigate traffic problems associated with the boom in automobile ownership. It was a relationship that would benefit the developer when he turned his attention to Portland's East Side.[29]

His association with local booster groups notwithstanding, Ralph Lloyd put the concerns of his potential tenant ahead of his support for the widening of Vermont Avenue from eighty feet to one hundred feet from Wilshire Boulevard to Hollywood Boulevard under the 1924 plan. In this case, he privileged his private role as a developer over his more public role as a "citizen outside the government." He followed the lead of the Bank of Italy when it opposed the widening of Vermont Avenue, as it would threaten the branch building that the bank sought to erect. In response to pressure applied by both the Hollywood-Vermont Association and the Vermont-Beverly Association, Lloyd adopted a neutral stance on the matter. Explaining that he generally supported street improvements, he could not endorse this particular project. Notwithstanding Lloyd's support, bank officials evidently remained concerned. By the time that the Vermont Avenue project was completed five years later, the Bank of Italy had opened its planned branch several blocks north of Hollywood Boulevard. Lloyd's unsuccessful parley with the bank's managers illustrates the competition that developers faced in securing tenants for their buildings. It would not be the last deal to fall through for the developer. Nor would it be the last time that Lloyd would put a potential tenant's interests ahead of the concerns of boosters seeking to protect real estate values.[30]

Notwithstanding his failure to gain a foothold in the Hollywood real estate market, Lloyd was impressed generally with Los Angeles's commitment to accommodate the automobile and proud of the role he played in mobilizing municipal resources and expediting street work along several important thoroughfares toward that end.[31] With his income from petroleum extraction allowing him to think in terms of commercial real estate development on a grander scale, he resolved to persuade Portland to respond with comparable means and enthusiasm. Lloyd also thought so much of traffic engineer Osborne that he retained him to plan his Portland street improvements. Later, the oilman would recommend that the City of Portland retain him as a traffic consultant. As he saw it, under Osborne's supervision, "miles upon miles of streets [were] laid and improved and millions upon millions of dollars [were] expended," in a city that faced unprecedented problems associated with hyperactive, automobile-driven growth.[32]

In every respect, the development of areas outside of downtown Los Angeles provided Ralph Lloyd with a blueprint for the development of the East Side of Portland. As he would later observe: "It has been proven time and time again here in Los Angeles that those who anticipate improvements by making the district they wish improved attractive, have out distanced in new development many other business points by this policy."[33] Not only would capital generated from petroleum extraction in Southern California flow directly into the Portland real estate market, the materialization of Los Angeles as an "energy capital" would influence the character of the development program that such capital would fund.

Before he would consider constructing buildings in Portland, Ralph Lloyd would seek to ensure that the physical environments of the districts in which he hoped to operate were favorable to conducting a profitable real estate business.

Getting Started in Portland

Until he bought "Larrabee's white elephant," Ralph Lloyd concentrated his Portland real estate acquisitions along Union and Killingsworth Avenues (figure 12).[34] In 1923, he purchased lots on a third corner, overruling William Killingsworth, who recommended that Lloyd add lots to the corners that he had held for six years.[35] Lloyd explained that he did want "to get too much property at any one location, as it often works out that a person will stand in his own light and that of the immediate district surrounding him in the event he should be slow in improving his property."[36] In the wake of the breakthrough in the Ventura Avenue oil field, Lloyd reversed his stance, adding lots where Killingsworth Avenue intersected Commercial, Union, and Vancouver Avenues. They included properties that Killingsworth had recommended two years earlier. The reason: Lloyd was now prepared to "improve his property" with a two-story theater and store building of the type that developers had been erecting in districts west of downtown Los Angeles since end of the war. The building in turn would anchor the development of the commercial district as Killingsworth had long envisioned.[37]

Each of the three Walnut Park corners in Ralph Lloyd's portfolio might have supported a secondary business center, anchored by a building with stores on the ground floor and either offices or residential flats on the second. As Killingsworth noted, the area "greatly needed" such a building. By the spring of 1925, Lloyd decided to anchor the corner of Union and Killingsworth

Avenues with a mixed-use building that prominently featured a theater. Lloyd asked Killingsworth to investigate the rental prices that a theater on that corner might command, and to contact him should he hear of any businesses needing space to lease. Shortly thereafter, Lloyd announced plans to start construction on one or more buildings in Walnut Park in 1926.[38]

Imagine, then, William Killingsworth's disappointment with Ralph Lloyd's decision to redirect his surplus capital to property acquisition in Holladay's Addition and prioritize the development of that district. After reading the news of Lloyd's real estate transaction in the *Sunday Oregonian*, Killingsworth fired off a letter, reiterating his twenty-five-year conviction that Walnut Park was ideally located to host a substantial business center and his decade-long belief that Lloyd should be its catalyst. In his rebuttal, Lloyd ordered his investment priorities and hinted at his strategy for improving the East Side's thoroughfares in support of his plans. Holladay's Addition was "riper for development," he insisted. Walnut Park would be the next logical step. The ground was not yet prepared for the kind of development in Walnut Park that Killingsworth envisioned—and desired as both real estate agent and owner of a substantial property portfolio. North-south arterial streets, especially Union Avenue, would have to be widened to deliver the automobile traffic necessary for commercial success in Walnut Park.[39]

Ralph Lloyd envisioned the development of his holdings on Portland's East Side just as he had observed developers and investors expand commerce in Los Angeles beyond its central business district. He aimed to "break the lethargy that has been upon this particular portion of Portland for so many years." As was the case with Walnut Park, Lloyd feared to "stand in [his] own light and block advancing prices that would be [his] only avenue of profit." He therefore planned to develop a core of properties that might attract other investors to the district.[40] His estimates on what he would ultimately spend on his East Side development program ranged from $25 million to $40 million—huge sums for any small business.[41] Its centerpiece would be a $3 million hotel east of Holladay Park. Second in priority was the development that Killingsworth desired—a Walnut Park business center "of such importance as to be secondary only to the main business center on the West Side."[42] As for the balance of his Portland real estate portfolio, Lloyd would consider proposals from interested parties, but only if they intended to build on the property he sold them. In the wake of the breakthrough in production in the Ventura Avenue field, he was in no hurry, asserting, "I would sooner hold the land myself and believe I can afford to do that."[43] Ralph Lloyd had the resources to develop and maintain a position at the forefront of the local real estate market.

The Ambassador Hotel, which opened in Los Angeles on 1 January 1921, served as a model for Lloyd's Holladay Park plans in two respects, namely, its configuration and its location. Sprawling across twenty acres, the property featured gardens, a swimming pool, a golf course, and private bungalows. The main structure contained thirteen acres of floor space, with seven floors of guestrooms and restaurants, a theater, a bank, and high-end shops. Situated west of Westlake—now MacArthur—Park, to the west of downtown, the Ambassador quickly became "a hive of social, cultural, and business activity for not only local and regional but also national elites."[44] In anchoring the development of a close-in district that supported commerce and featured multi-unit housing, the Ambassador Hotel offered a blueprint that Lloyd explicitly sought to replicate in Portland.

So inspired, Ralph Lloyd envisioned a hotel that would "hold its own . . . with anything on the Pacific coast." It would be the convention and social center of the city, flanked by a theater and a multi-story, mixed-use building with shops and offices. A golf course would straddle a refurbished Sullivan's Gulch. The hotel would also serve a base for potential investors who "would stay long enough . . . to get interested in the city . . . fall in love with [it and] invest their money here," enabling Portland to "compete with every other city on the Pacific coast."[45]

With Los Angeles as his touchstone, Lloyd rallied Portlanders publicly and through private communications to meet their intercity competition and dissolve their intracity rivalries. It goes without saying that he did so to help ensure the success of his building program, as a surge in metropolitan resolve and unity would surely redound to his benefit. In some measure, too, he was steeling his resolve and boosting his self-belief in the face of criticism from business associates who thought—and would continue to believe—that both he and they could "do better with their money" invested in Los Angeles rather than in Portland. Conceding that both heart and head influenced his decision to purchase "Larrabee's white elephant," Lloyd sought to profit from shaping the urban trajectory of his second home.[46]

Thus Lloyd publicly aligned his program with Portland's oft-stated regional ambitions. To mobilize support for his particular projects, he called on Portland "to wake up . . . to meet the intensive competition that is being waged on both her northern and southern borders—especially her southern border," and thereby improve her position in both the region's and the nation's urban league tables.[47] There was no reason why she shouldn't be able to do so, as Portland and Los Angeles constituted "the two most substantial and desirable cities in which to be interested" on the coast.[48] Portland's agricultural hinterlands and natural resources provided it with a considerable advantage over Seattle, its traditional rival, he averred. Lloyd urged Portland's residents to play their role in realizing Portland's destiny to become one of America's great cities by urging their civic leaders to pursue needed school, street, and sewer projects.[49]

Demographics favored Portland, Lloyd insisted. Increasingly congested Southern California was "bound to throw a great many people to the Northwest." Indeed, "the overflow from this great city [Los Angeles] and Southern California [would] eventually make Portland and the Northwest." Regional opportunities were now comparable to those enjoyed by Los Angeles two decades earlier, according to the oilman. For competition was relatively muted and real estate prices were attractive. Lloyd therefore recommended that Portland's newspapers and its Chamber of Commerce advertise the city's advantages in both Southern California newspapers and national weeklies, such as the *Saturday Evening Post*, much as *Los Angeles Times* publisher Harry Chandler and other boosters were doing to "sell" Los Angeles's beaches, climate, and lifestyle to, as Mike Davis puts it, "the restless but affluent babbitry of the Middle West."[50]

Of course, Portland should take its cue from Los Angeles's entrepreneurs on how to get things done. As Lloyd saw it, this intangible factor would determine success or failure in the tournament of cities. As he put it, "I am quite hopeful that a little of the enterprising spirit that is so manifest in Southern California will take root in Portland. If it does, there is no question but that it will result in making Portland a wonderful city." He, for one, would do what he could do as an investor with a firm foothold in the real estate market to plant the seed. On this question, more than on any other, Portland's business elite would disappoint him.[51]

Ralph Lloyd appealed to downtown business interests to look upon the development of his holdings not as a threat, but rather as an opportunity for Portland to expand its central business district across the river. Urban historian Carl Abbott has called the Willamette River "the unifying factor" in Portland's development, but Lloyd was cognizant of the ways in which developers, investors, and other interested parties deployed the waterway as a dividing measure of real estate value. As he pointed out to the ever-eager William Killingsworth, for instance, commercial development in Walnut

Park would be "a slow developing project at best" because West Siders opposed the development of East Side business centers.[52]

The oilman used a variety of appeals to unify Portland's business interests. At a basic level, he sought to allay fears that his building program might threaten the central business district. When he visited Portland in September 1926 to exercise his option on the Balfour-Guthrie holdings, Lloyd announced his intention to cooperate with rather than compete against downtown proprietors. Commerce should be able to flourish on both sides of the river, he insisted. He called on West Siders to see his development plans as part of an enlargement of downtown to the benefit of all interested parties and implored them to work with him "in building up the great central metropolitan area."[53] In this context, he explained to Raymond B. Wilcox, president of the Chamber of Commerce, that he was "a booster for the City of Portland as a whole."[54] Lloyd also linked intracity unity to interregional urban rivalry, insisting that Portland needed business districts on both sides of the river if it hoped to rank among the great cities of the nation. Again, America's "energy capital" showed the way. The unity for which he was calling "is what made Los Angeles the city it is today and it will do the same for Portland."[55] Not inconsequentially, of course, such unity would also improve Lloyd's prospects for the profitable development of his East Side real estate portfolio.

On municipal finance, Ralph Lloyd's views aligned with those of local banking and business leaders. He concluded that Portland was spending too much money on public services even though it actually was lagging most other large cities on the provision of basic services and education. Lloyd warned Portland's residents as he was about to embark on his East Side development program that urban and regional rivalry applied to tax and fiscal policy too. State spending, by Lloyd's estimate, was already three times too high as it needed to be. Were Oregon to pass an income tax—a bill was pending—large investors like him would take their capital elsewhere. (As it turned out, two of the tax bills that worried Lloyd were defeated later in the year by large majorities.) An inheritance tax—rumored to be under consideration—would also deter investment, he warned. In a "Statement to the People of the State of Oregon," Lloyd indicated that he was prepared to withdraw from negotiations for additional East Side property until he was assured that such a law would never pass. Oregon was "on the eve of a big development." Rather than scare off capitalists with talk of new taxes, the state should court them by building infrastructure. Not coincidentally, Lloyd's appeal neatly dovetailed with his interests as a developer. Street improvements would be one area in which Portland would outspend most other cities of similar size during the interwar period.[56]

Portland's political economy made it possible for a developer with considerable capital to wield predominant influence in the local market. Power within municipal government was fragmented. There was a planning commission, but the City Council did not necessarily rely on it for advice. In the area of street improvements, for instance, the planning commission had issued a street plan in 1921. Yet the City Council was more likely than not to follow the 1924 plan of the city engineer, Olaf Laurgaard (who acted as a planning commissioner, too, from 1921 to 1934). For his part, Lloyd did not discount the utility of working with the planning commission or, for that matter, any institution of municipal government.[57]

A business community that equated economic growth with the public good was the decisive factor in land use decisions. The City's 1924 zoning ordinance was a case in point. Drafted by realty interests, it zoned a quarter of the city, including all streets with streetcar lines, for business and another 45 percent for multi-unit housing, incidental commercial use, or single-family homes. As elsewhere, including Los Angeles, the ordinance supplied far more space for commercial purposes than demand required. To be sure, the ordinance fueled speculation and ignored area, height, and density controls, but at the same time it encouraged multi-family housing and did not prohibit the commingling of houses and businesses along arterial streets—configurations that twenty-first-century urban planners applaud. As was the case in Los Angeles, Portland's zoning decisions afforded a small business such as the Lloyd Corporation with ample opportunity to participate in the local real estate market.[58]

For Ralph Lloyd, the downside of Portland devoting so much space to commercial use lay in the administrative bottlenecks that resulted from developers and business owners flooding the City Council with petitions for projects, thereby overwhelming the public works department, which was obligated to conduct an engineering study of, and document protests against, every petition. Ultimately, the excessive competition in the commercial segment of the market that resulted from the 1924 zoning ordinance ensured that the Lloyd Corporation would struggle to secure approval of infrastructure projects in a timely fashion.

But political fragmentation and the lack of formal planning did not translate into the haphazard planning and development of Lloyd's East Side holdings. Today we may think of Los Angeles as a sprawling and dysfunctional metropolis courting apocalyptic disaster, but Ralph Lloyd saw it as *the* model for constructing urban space. He drew on examples in Los Angeles, where developers were creating complete communities within the metropolitan region, to envision the materialization of the "close-in" East Side as a portfolio of commercial and multi-unit residential properties configured to complement Portland's downtown. There was an internal coherence to his approach, as was the case with many a development in Los Angeles. Lloyd would consult traffic engineers, business owners, and city officials both to determine his infrastructure needs and to integrate his plans with those of other districts and the city as a whole. He then would work through all of the responsible bodies of municipal government to implement his plans. To be sure, Lloyd equated his interests with the public good. Moreover, he often would be given the benefit of the doubt, given the relative scale of his enterprise and the traditional stance of Portlanders on the sanctity of private property. But in imagining the "close-in" East Side as an extension of the central business district and in thinking of Portland in regional terms, the California oilman would win near unanimous praise from local leaders.[59]

Preparing the Ground

Ralph Lloyd's modus operandi in Portland differed from the way in which he operated in Los Angeles. He opened an office and put staff on the ground, owing to the strains that frequent rail travel between Los Angeles and Portland promised to exert on his health and family. Indeed, for these reasons Lloyd almost ended negotiations for Charles Larrabee's former Holladay's Addition holdings late in 1925. Seeking someone "trained in the development work here in Los Angeles and with the spirit and initiative that has made this district, and one especially trained in the real estate business," Lloyd recruited C. W. Norton to represent him in Portland. Norton executed Lloyd's instructions, of course, but he also wove himself into the fabric of the Portland business community by playing an active role in its organizations, just as Lloyd had done in Los Angeles. He joined the East Side Business Men's Club. He organized the East Side Development Association and served as its president. Norton also served on the Portland Planning Commission from September 1926 to December 1929. For his part, Lloyd managed operations from Los Angeles, traveling to Portland whenever he felt that his presence was needed to advance his program.[60]

To literally prepare the ground for development, Lloyd worked through municipal government to win approval for ordinances on street improvements and setback lines on arterial streets. But he did not rely solely on public institutions for this advance work. As a partner in an informal public-private relationship, the Lloyd Corporation took responsibility for cleaning up and beautifying Sullivan's Gulch and constructing and lighting a boulevard along its northern edge. Lloyd cast these private efforts as public works initiatives—so-called Lloyd Boulevard would become a municipal thoroughfare upon its completion—and instructed Norton to cooperate with the City of Portland on the implementation of these projects.

Often speaking through Norton, Ralph Lloyd made it clear to local civic and business leaders alike that the realization of his East Side development program was contingent on the approval and timely implementation of his entire infrastructure program.[61] On one level, his statements to this effect constituted bluff or threat. On another, they underlay his approach to real estate development generally. Lloyd's experiences with, and observations of, commercial development in Los Angeles, where multiple districts dependent on the automobile for their success materialized rapidly during the 1920s, convinced him that the carrying capacity of key arterial streets was a critical factor in determining intracity competitiveness. His statements regarding the enlargement of Portland's downtown notwithstanding, Lloyd's intracity competition could only come from the West Side. In this regard, his tabula rasa on the East Side gave him a comparative advantage as a developer: ample room for automobile parking. But the Lloyd Corporation would capitalize on this advantage only if Portland's streets could deliver the requisite volume of traffic to its projects. Materializing "Larrabee's white elephant" as a commercial district thus depended on opening, widening, or otherwise reconfiguring the streets around Holladay Park and those that fed into it. With royalty income swelling his bank account, Lloyd was willing to contribute capital to this end, but he also understood that an infrastructure project often involved a lengthy approval process. Since he determined not to construct his hotel-centered commercial complex before the completion of a comprehensive street improvement program, Lloyd deployed such statements to align public action with his private interests. In this way, street improvements might become a more predictable element in his development schedule and therefore lower his risk.

Soon after closing the deal with Charles Larrabee's heirs, Lloyd hired Los Angeles traffic engineers Harry Z. Osborne and his partner Walter E. Jessup to design a street system in support of his plans and then assist the City's planning commissioners and engineer in integrating it into a revised major street traffic plan for Portland. Their brief included both reengineering the streets in the immediate vicinity of Holladay Park and designing projects that would increase "communication" between the East and West Sides of the city by expanding the capacity of arterial streets that fed the Broadway, Burnside, and Steel Bridges. The proposal that the Los Angeles traffic engineers developed also recommended that Portland establish a traffic commission and a major highways committee, as Los Angeles had done, and fund the engineering study required to revise Portland's 1921 traffic plan. Through this effort, Lloyd as a "citizen outside the government" jumpstarted Portland's dormant traffic planning function.[62]

With their proposed street map in hand, Osborne and Jessup traveled to Portland to meet with Olaf Laurgaard, who served as city engineer from 1917 to 1933. Laurgaard supported street improvements generally—he would later boast that "no other city engineer has widened as many streets as I have"—and quickly became an ardent backer of Ralph Lloyd's street program. After Laurgaard gave the street map his blessing, Lloyd directed Norton to take up the program with John C. Ainsworth. As president of both the United States National Bank and the Portland Planning Commission, Ainsworth would become Lloyd's most important local ally in matters of both business and municipal politics. A graduate of the University of California, Ainsworth was involved in transportation and manufacturing as well as banking. He had been a member of the City Beautiful Committee and Civic Improvement League that had funded and promoted the Bennett Plan for Portland. C. W. Norton found Ainsworth to be "very broad gauged [and] public spirited," and "very much interested" in Lloyd's East Side plans in general, and "ready to help us in every way possible."[63]

Ralph Lloyd's actions set in motion the revision of the 1921 traffic plan. As the City Council had never adopted it, it had not been updated. Ainsworth met with Norton and then charged A. C. McClure, the planning commission's secretary, with reviewing Osborne's proposed plan. McClure urged the planning commissioners to apply pressure on the City Council to adopt a revised traffic plan. McClure hoped that such a plan, if submitted by the planning commission, would become the official blueprint for an ambitious public works program.[64]

There was more than a hint of subterfuge in Lloyd's tactics as a developer. The Californian wanted to remain in the background so that his efforts, as executed by Norton, Osborne, Jessup, and others on his behalf, would garner no publicity until he was prepared to offer it. It was especially important to

Lloyd that he remained under the radar, so to speak, since at this point he had neither formulated his building program nor exercised his option to acquire the Balfour-Guthrie lots. It was initially his idea to give Ainsworth $1,000 of Osborne and Jessup's $19,000 fee to stage an invitation for the traffic engineers to come to Portland to consult City officials. This, he calculated, would save time and avoid making it appear as if a clique of Angelenos were forcing their recommendations down the throats of Portlanders. It is unclear to what extent this gambit was attempted. It was the case that Osborne and Jessup developed three versions of their proposed street map. In meeting with Ainsworth, C. W. Norton used a map that showed what streets Lloyd proposed to widen but did not highlight Lloyd's properties. The other maps showed Lloyd's initial acquisition of the almost 600 undeveloped lots from the Oregon Real Estate Company and the optioned Balfour-Guthrie lots in color. The color-coded maps allowed Norton to disclose Lloyd's plans to selected audiences until his boss exercised his option on the Balfour-Guthrie lots, after which time the developer planned to become more forthcoming publicly on his building program.[65]

Unfortunately for Ralph Lloyd, L. M. Lepper, secretary of the East Side Business Men's Club, gave away the game late in June. Invited by E. A. Clark, the club's president (and a vice president of Citizens Bank), Norton explained Lloyd's plans to the East Side business leaders. To illustrate his talk, Norton pinned up all three maps. An overly enthusiastic Lepper leaked news of the meeting to the press.[66]

Still, the evening had a silver lining. Ralph Lloyd gained an important ally in the East Side Business Men's Club. C. W. Norton's report of the meeting encouraged Lloyd, who wrote, "It looks as if they are going to be with us on any move we may make."[67] The group was already inclined to back Lloyd: Lepper had stated that Lloyd's East Side real estate transactions were "the best thing I have heard of in a long time."[68] Further, Clark had written, "We are fully determined to make use of your Los Angeles experience."[69] The group would faithfully carry water for Lloyd even if it meant adjusting their own priorities on specific projects. With Lloyd's well-publicized entrance into the local real estate market, the group's literature and correspondence began to read as if Lepper—the usual author—were either channeling Lloyd or lifting his words verbatim from his letters, as illustrated by the club's attack on a proposal to set a 60 percent threshold for property owners to meet, should they wish to overrule City Council decisions on setback lines: "Portland Citizenry as a whole seem to lack the FORWARD LOOKING SPIRIT—the

Los Angeles Spirit, if you please, as to this coming Growth of a Great City."[70] In securing East Side businessmen as allies, Lloyd established a platform from which to succeed as an outside investor in the local real estate market.

The incorporation of Ralph Lloyd's street program into the revised traffic plan shows how alliances with local groups such as the East Side Business Men's Club facilitated the Lloyd Corporation's access to key City officials and rationalized the process of working through municipal government to realize his plans. With advice from banker Ainsworth, C. W. Norton consulted both the City of Portland's attorney and City Engineer Laurgaard on submitting Osborne and Jessup's street map in support of a petition that would provide for the creation of setback lines on new structures. The measure would facilitate the widening of arterial streets. An East Side Business Men's Club committee of three, including Norton, then interviewed the city attorney on Portland's procedures governing setbacks and drew up the petition. Lepper attended the City Council meeting held on 21 July 1926, where he discussed the measure with Commissioner John M. Mann. Like all the men who served on the City Council during the 1920s, Mann lived east of the Willamette River and was eager to foster its development. Lepper and Mann presented the petition to the City Council, which referred it to the city attorney and Laurgaard, whose counsel the Lloyd Corporation previously had sought. In similar fashion, the Lloyd Corporation would submit dozens of petitions to the City Council over the next four years on setbacks and the widening, extending, and lighting of arterial thoroughfares and local streets in the vicinity of Holladay Park.[71]

In calling for the completion of projects that would enhance the physical environment surrounding his core holdings, Ralph Lloyd also sought the support of central business district owners. He pointed to the condition of the downtown waterfront and Sullivan's Gulch: together they constituted a "cancer in the heart of the city." Sounding the theme of public-private partnership, he promised to landscape the gulch using his money and urged West Siders to implement Olaf Laurgaard's 1923 waterfront improvement plan. When he exercised his option on the Balfour-Guthrie lots that straddled Sullivan's Gulch, he pointed to the public benefits of the deal, explaining that he added this property to his portfolio to keep the area free from both "objectionable industrial enterprise" and "poorer-class homes." Together, the Sullivan's Gulch and waterfront projects would help to "weld the great central metropolitan district of Portland into one unit," as befit a city of its size. As the projects allegedly would benefit the entire metropolis, he called on private interests on both sides of the Willamette River to "work in harmony" toward this end.[72]

As he considered the beautification of Sullivan's Gulch and the construction of Lloyd Boulevard to be "public improvement[s] for the good of the whole city," Ralph Lloyd consulted Ainsworth, who found these projects to be "so attractive" that they were certain "to have our city's close cooperation." After speaking with members of both the planning commission and the engineering department, Ainsworth assured Lloyd that he would receive "cordial cooperation in every way" from Portland officials.[73] Thus the California developer mobilized the apparatus of municipal government to facilitate his entry into the local real estate market.

All the effort involved in launching his East Side projects and attending to the development of the Ventura Avenue oil field, however, left Ralph Lloyd exhausted and suffering from a flare-up of his gastrointestinal condition at the end of 1926. Conceding that "I have been overworking for a number of years and it [is] beginning to tell on me," Lloyd planned to slow down for six months (or more, if necessary). For all of 1927 Lloyd aimed only to grade Lloyd Boulevard and materialize his holdings at the intersection of Killingsworth and Union Avenues with the construction of a theater, store, and office building.[74]

Designing the Walnut Park Building

Early in 1927, Ralph Lloyd refined his development vision for Walnut Park. He reiterated that he wanted to make a statement with his "first building of moment" in the city: a building that would be "an ornament to the district" and more appealing architecturally than any local building of its type. It would feature "one of the finest theaters in the city." Ground floor retail stores would front both Killingsworth and Union Avenues. The second story along Union Avenue would house medical offices. Lloyd expected to spend as much as $300,000 on the project, depending what his prospective tenant, Multnomah Theatre Corporation, a coalition of independent local theatre owners affiliated with the New York-based Universal Film Corporation, was willing to pay for a fifteen-year lease. In developing this project, Lloyd adopted a new commercial building delivery approach of which little is known regarding its application during the interwar period.[75]

Ralph Lloyd's use of Los Angeles as an urban development model for Portland extended to the architects he retained to design his signature buildings, beginning with this one. He looked first to two local architects, John Bennes and Lee Arden Thomas, to design the project. Bennes had designed the 1,500-seat Hollywood Theatre, which had opened in 1925 in a shopping district along Sandy Boulevard, a thoroughfare that bisected the East Side from the Morrison Bridge northeasterly to what is now the location of Portland International Airport. Thomas's Bagdad Theater was nearing completion in the Hawthorne District, which lay to the southeast of Sullivan's Gulch. Lloyd's tastes did not extend to the exotic themes that both architects incorporated into their designs, however. The Hollywood Theatre's elaborate terra cotta tower and rococo Art Deco façade and the Bagdad Theatre's terra-cotta-walled lobby that featured a mosaic of colorful frescos were simply too "jazzy." So Lloyd turned to Long Beach, California, architects Parker Wright and Francis Gentry.[76]

The architects, who are best known for their Greek Revival Masonic Temple, which was under construction in Long Beach when Lloyd contacted them, produced preliminary plans and a cost estimate that worked out to a rental of $1.25 per theater seat. Lloyd took this number to Universal Film executives, who rejected it because the theater would not be located in an established commercial district. The situation that Lloyd confronted was one that developers faced time and again, namely that the project, as designed by the architect without the input of the builder, did not "pencil," or meet their pro forma budget.[77] Indeed, Lloyd would later remark, more or less matter-of-factly, "We have always found that a building is never built for the estimate or contract price."[78]

To develop an alternate, more financially feasible, design, Lloyd turned to the Austin Company. Founded by Samuel Austin in 1878 in Cleveland, the Austin Company was perhaps the leader among a group of large engineering firms that integrated design and construction within well-capitalized, hierarchical organizations. Taking their cues from Henry Ford and other well-known innovators of techniques of mass production, construction engineers developed building templates that they marketed to industrial owners. By involving the builder in both the design and construction of a project, these large engineering concerns delivered buildings through an approach that would become popularized as "design-build."[79]

The Austin Company specialized in steel-framed structures whose elements could be mass-produced. Its Austin Method, introduced in 1901, offered a project delivery system that appealed to developers. As distilled in advertisements that appeared in newspapers, such as the *Los Angeles Times*, and national magazines, such as the *Saturday Evening Post*, the Austin Method guaranteed (1) a maximum price—that "the estimated cost of your building project will be the final cost to you"; (2) a delivery date; and (3) materials and workmanship of "the highest quality." Deploying the concept of "undivided responsibility" to promote its approach and professional expertise, the Austin Company urged potential clients to consult its engineers even before they were sure what they wanted to build or where they thought of locating the project.[80]

The Austin Company offered its services to commercial developers too. After World War I, it constructed multi-story office buildings, department stores, hotels, theaters, stores, and mixed-use structures. Little is known, however, about extent to which design-build was deployed to deliver commercial buildings during the interwar period. Ralph Lloyd's retention of the Austin Company for the redesign of his Walnut Park building thus provides valuable insight into the applicability and use of an early version of design-build to deliver a commercial project.[81]

The Austin Company designed a reinforced concrete, steel-trussed structure that would cost $100,000 less to produce than Wright's and Gentry's proposal. They achieved the savings in part by reducing the number of theater seats by 400, to 1,000. This reduction in seats also aligned the project with prospective demand. Thus the new design met Universal Film's leasing criteria. When he informed Wright and Gentry that he had retained the Austin Company to act as both designer and builder, Lloyd noted that the latter had promised to fix a maximum cost to construct the building on the basis of its estimate—a hallmark of design-build.[82]

Suggesting the novelty of the application of the Austin Method specifically or design-build more generally in a commercial project context, Ralph Lloyd expressed discomfort in using the company's engineers rather than an architect to design this relatively complex structure. Even as he replaced them as architects, Lloyd retained Wright and Gentry to review the Austin Company's design to ensure its safety. Further, he sought out the Portland office of Robert W. Hunt Company, a Chicago-based firm of consulting engineers that was the largest of its kind in the nation, to observe the concrete pours, check the strength of the building, and ensure that his project was "properly conceived and constructed."[83]

When, late in 1927, Ralph Lloyd organized the Union State Bank to support his local development efforts—evidence that the banking sector was another area open to the entry of small business—he also retained the Austin

Company to design a structure to house the financial institution, which, he believed, would "prove a factor in welding the east side and west side of the City of Portland into a harmonious whole." Owing to the need to redesign it, the mixed-use building was now set for completion in 1928, along with the bank, which would also be located in Walnut Park, on the northwest corner of Killingsworth and Union Avenues. Meanwhile, Lloyd retained the Austin Company to remodel the two-story apartment-and-store mixed-use buildings located on the opposite corners of the Walnut Park intersection, one of which would serve as temporary home for the bank.[84] The California oilman's apparent satisfaction with the Austin Company's design-build approach suggests that a method used initially to deliver standardized industrial structures may have been used to deliver numerous discrete commercial buildings before the sector ground to a virtual halt during the early 1930s.

Notwithstanding his experience with the Walnut Park mixed-use building, Ralph Lloyd did not dispense with the services of architects for the design of his buildings. In fact, the developer retained Southern California architects to design his most important buildings in Portland. Lloyd wanted to establish a local aesthetic standard and because in his mind Los Angeles was "in advance of Portland and any other city in the west in modern designs," the emerging "energy capital" would provide source material for his East Side buildings. Thus, for the hotel that he planned for Holladay Park, Lloyd would select two renowned Los Angeles firms, Walker & Eisen and Morgan, Walls & Clements, to compete for the commission.[85]

At the same time, for all but the most complex projects, Ralph Lloyd would involve "responsible" contractors with whom either he or the architect worked on a repeat basis in the design process. In Portland, for instance, he would use the same firm to design and construct many of his local projects. Architect Charles W. Ertz and his partner Thomas B. Burns, doing business as Ertz & Burns, Architects, and Ertz-Burns & Company, Building Contractors, would design and build auto and truck dealerships, auto service stations, a Coca-Cola bottling plant, and similar structures for the Lloyd Corporation.[86] Lloyd also used L. H. Hoffman, another prominent local contractor, to construct buildings designed by others. The contractor would review the architect's plans, specifications, and working drawings before agreeing to build the project and would make recommendations that often resulted in design revisions. Typically, Lloyd would bid the builder's contract on these projects, but he would restrict the bidding to selected firms. And, if he didn't like the bid, Lloyd would negotiate the price. This hybrid approach, which may be called "design-negotiate-build," offered Lloyd as developer a means of controlling smaller, vernacular projects and thus facilitated his entry into the lower end of the commercial segment of the real estate market. Indeed, even as he acted to launch his signature hotel project, Lloyd would establish himself in the Portland market with the completion of such less-visible buildings.[87]

While Ralph Lloyd recuperated during 1927, his man in Portland, C. W. Norton, worked through the Portland Planning Commission and the East Side Development Association to advance the Lloyd Corporation's street program. Crucially, Norton filed petitions for setback lines on the streets around Holladay Park, including those north of the park that were lined with elegant homes. Lloyd felt that these requests were entitled to favorable consideration from the City Council, given the amount of property that was involved and the number of substantial projects that depended on their passage. On the assumption that the Lloyd Corporation would embark on its development program in a timely manner, the City Council gave its unanimous approval to all of these petitions. With the requisite ordinances in place, a refreshed Ralph Lloyd turned his attention to the design of the hotel and plans for the area around it, as revealed to Portlanders in the *Sunday Oregonian* on New Year's Day, 1928.[88]

Chapter 4

False Start: Ralph Lloyd's East Side Dream Falls Short

The day is just breaking on the East Side and the future is all before us, with unlimited possibilities in expansion and development.

—Ralph B. Lloyd, 1928

On New Year's Day, 1928, the *Sunday Oregonian* announced that Ralph Lloyd's "metamorphosis" of the area around Holladay Park was well under way. The Lloyd Corporation had recently completed the roadbed for Lloyd Boulevard, which would not only provide "swift and convenient" access to its properties, but a "shortcut" across the Willamette River for residents who lived east and north of the park. Once the boulevard was completed, perhaps by the summer, "Los Angeles millionaire" Lloyd would move ahead with his substantial construction program. As the sketch that accompanied the article depicted, it included a $1.5 million hotel on the eastern border of the park, a theatre on Multnomah Street on the park's northern border, and a store and office building, half-a-dozen stories in height, along Holladay Avenue south of the park. In addition, the *Sunday Oregonian* reported, Lloyd would erect civic monuments and public restrooms on the four vacant blocks that he planned to append to the park and construct a nine-hole golf course.[1]

To his landscape engineer, Ralph Lloyd reiterated that his hotel would "hold its own . . . with anything on the Pacific coast." He envisioned that it would develop into the convention and social center of the city, as there was simply nothing else in town comparable to what he had in mind. He aimed to have the architect's plans and specifications for the hotel in the hands of his contractor by the end of the year.[2]

The developer's program for 1928 also included vernacular buildings delivered through the "design-negotiate-build" delivery method for the Adcox Auto and Aviation School and Henry Jenning & Son Furniture Company, a warehouse for International Harvester at the corner of First and Oregon Streets, and the mixed-use building in Walnut Park.[3] Implementing the program would benefit the entire city, Lloyd enthused to E. A. Clark, vice president of Portland's

Citizens Bank: "The day is just breaking on the East Side and the future is all before us, with unlimited possibilities in expansion and development." In his judgment, "the development of the near in East Side will add materially to the stability of the West Side. In fact, it will build up what the City of Portland must have—a metropolitan center that will include both the West Side and the near in East Side and a belt line joining the two with quick and easy transit."[4] Lloyd saw no reason why Portland's business and civic leaders should not support his efforts wholeheartedly.

Ralph Lloyd's activities as a "citizen outside the government" in pushing through the widening of Union Avenue suggest the extent of his influence on municipal politics. Yet it would prove to be insufficient to overcome intracity rivalry and deteriorating economic conditions in realizing the developer's grand vision for the East Side. Ultimately, the tsunami of the Great Depression would swamp Lloyd's plans for both Holladay Park and Walnut Park, demonstrating the extent to which external factors could be decisive in commercial projects. As the national economy stalled and then collapsed, the risk of the hotel project increased and alternative means to finance it evaporated. Lloyd became less and less able and willing to finance the hotel complex from retained earnings and so he would try to devise a financing scheme that would attract local capital. Ultimately, Portland's downtown business elite would disappoint Lloyd. To be sure, severe economic downturn would constitute reason enough for their collective reluctance to invest in Lloyd's project. But their ambitions for a downtown hotel also would factor into their lack of enthusiasm for it. Still, the oilman's failure to start construction of the hotel in a timely manner would owe in large part to his conservatism as a real estate developer.

Improving East Side Traffic Flow

The campaign to widen Portland's arterial streets in which the Lloyd Corporation had substantial interest came to a head in 1928. The outcome—a voter-approved bond issue much reduced in scope from one proposed just months earlier—shows how municipal fiscal constraints may limit the accomplishment of private ends through public means. However substantial they may have been, Ralph Lloyd's resources could not fund all the infrastructure projects in which he was interested. As he explained, the projects that he aimed to complete around Holladay Park alone constituted a "great financial burden."[5] Even if he had the means of doing so, Lloyd had no intention of underwriting the widening of miles of arterial thoroughfares that would benefit other property owners directly and the city generally. It goes without saying that a city's public works budget is never large enough to fund all the projects that municipal officials and private interests deem essential to urban growth. And so the Lloyd Corporation joined the fight among local interests for limited resources, as filtered through the apparatus of municipal government.

Widening Union Avenue remained Ralph Lloyd's highest priority, owing to the geography of his real estate holdings. In support of the project, he secured the backing of Commissioner Asbury L. Barbur, who saw it as a key part of a citywide effort to improve automobile access to the Broadway and Steel Bridges. Barbur, a former City of Portland auditor who served as commissioner of public works, initially gave his blessing to a petition to widen Union Avenue north of Broadway to that end.[6] Petitioners asked the City Council to include the project in a bond issue that was placed on the May 1926 ballot, but it rejected the request.[7]

On the recommendation of Ralph Lloyd and with the endorsement of Commissioner Barbur and City Engineer Olaf Laurgaard, the City Council subsequently approved setback lines along Union Avenue to provide for an 80-foot-wide street north of Broadway. Property owners, foremost among them the Lloyd Corporation, then petitioned the council to include the project on the November 1928 ballot. The City Council referred the petition to the planning commission, which was analyzing the merits of all the arterial projects before it.[8]

The report of the planning commission, presented to the City Council on 5 July 1928, deeply disappointed East Siders generally. Using the traffic plan published in December 1927 and various critiques of it as its guide, the commissioners relegated Union Avenue and other East Side projects to a list of "B" projects for further study. A list of "A" projects to be included in a $2.25 million bond issue privileged east-west streets, such as Broadway and Burnside, whose improvement would increase cross-town traffic, and feeder and outlet streets on the West Side associated with the Burnside, Steel, and Swan Island Bridges. East Side business owners called for a revision of the bond measure. On the motion of Commissioner John M. Mann, the City Council delegated to the public works department the task of writing a separate report that considered its previous support for East Side projects.[9]

City Engineer Laurgaard proposed to significantly increase the municipal resources devoted to East Side projects. For instance, he extended projects to widen Union Avenue and East Seventh Street from Broadway to Lloyd Boulevard and added the widening of Grand Avenue, which ran parallel to and between the two arterial thoroughfares. To do so, Laurgaard was prepared to alter the calculus of project funding. Since 1924, the City of Portland had split the cost with property owners on a 75/25 percent basis. To broaden the appeal of the projects, he asked owners to share half their cost. On 1 August, the City Council adopted Laugaard's report. Declared Commissioner Barbur: "If we are to be a modern city, we must do as other cities are doing in widening principal streets to provide for faster traffic."[10]

For his part, Ralph Lloyd weighed in from Los Angeles with a telegram to the mayor and the City Council that was read into the record at the 1 August meeting and with a private letter to Laurgaard. In both missives, he argued that Portland should consider widening arterial streets northward from Lloyd Boulevard, which was nearing completion, its highest infrastructure priority. At the same time, he counseled that equalizing traffic among major thoroughfares on both sides of the river would quicken the pace of metropolitan economic activity. Hence, he threw his support behind selected West Side projects too. Lloyd insisted that his prioritization of East Side projects would be "for the immediate good of the whole city," though a glance at a map of Lloyd's holdings suggests that there was also something in it for his development interests, the Sears store that the Lloyd Corporation was negotiating to build in particular.[11]

Ralph Lloyd's influence was insufficient to secure all the projects in which he was interested, however. The City Council ultimately determined that Portland did not have the resources to fully fund Laurgaard's ambitious program. It eliminated $5.5 million worth of projects, among them the East Seventh Street and West Sixth Street projects that Ralph Lloyd endorsed. It retained the Union Avenue and West Burnside Street projects that he favored above all, however, in an ordinance that called for a $3 million bond issue.[12]

As a developer with a substantial presence in the local real estate market, Ralph Lloyd adopted the posture of disinterested citizen and called on the City Council to reinstate the eliminated projects, in part to placate the East Side Business Men's Club, which favored the widening of East Seventh Street. As he explained to banker John C. Ainsworth: "These four projects would add more tone to the city than any similar widening projects in the entire [city]. While I may be called an East Side man, I see that a great city must be made by the building of both sides." The appeal did not alter the outcome, however. The prospective city builder could not overcome municipal officials' assessments of the taxing capacity of Portland's citizens.[13]

On the eve of the vote on the bond measure, Ralph Lloyd traveled to Portland with his daughters, Eleanor and Edna Elizabeth, both of whom had recently graduated from the southern branch of the University of California (now UCLA). "It is a combined business and pleasure trip," Lloyd explained to the *Oregonian* from his suite at the downtown Multnomah Hotel. "I thought I would bring them up here to show them what Portland is and where my holdings are." Surely, he would continue his conversations with Eric V. Hauser Sr., owner of the hotel and a successful builder who had completed projects for the Great Northern and Northern Pacific Railroads and was constructing a jetty at Coos Bay, Oregon, and a breakwater for the City of Long Beach. For Lloyd had been working for months on an agreement with Hauser whereby the latter would furnish and operate his Holladay Park hotel. The oilman also would see to it that his daughters would have "a good time, looking about the city and surrounding country." Lloyd took the opportunity to publicly support the bond measure: "They mean a tremendous lot to Portland. As a heavy taxpayer, I have been for the widening bonds ever since they were proposed. It is time to widen our streets here to meet modern conditions."[14]

The bond issue passed with 56 percent of the vote. Approval of the widening of Union Avenue suggests the extent to which a substantial investor who works through municipal government can achieve his goals. At the same time, the fiscal capacity of the City of Portland was the most important factor in shaping the "great arterial project" put before voters. For someone who was more familiar than local leaders with the extent to which Los Angeles accommodated the automobile, Portland's conservatism must have grated. Still, as Lloyd's plans for his hotel matured, the diminished arterial street improvement program posed less of an immediate obstacle to realizing them than residents in the neighborhoods around Holladay Park who sought to enjoin the Lloyd Corporation from widening their streets.[15]

Pushing for a Hotel Start

Ralph Lloyd hoped to begin construction of his hotel in 1929. To facilitate access to it and the other facilities that he planned to build around it, he determined that streets that either bordered or fed into Holladay Park, Multnomah Street in particular, had to be widened from sixty feet to eighty feet. He was anxious to get started. The City Council had already approved setback lines for the streets and had referred the Lloyd Corporation's petitions for their widening, paving, and lighting to the City's engineer. C. W. Norton, the Lloyd Corporation's local representative, expected Olaf Laurgaard to complete his engineering studies in the spring of 1929. The Lloyd Corporation would then proceed with hotel construction. Lloyd appreciated the fact that some residents might resist losing ten feet of their front yards. After all, while he had searched for oil in Ventura County, Irvington Addition and Laurelhurst, which lay to the north and east of Holladay Park, respectively, had matured as middle- to upper-middle-class residential neighborhoods. But neither Lloyd nor Norton expected that any actions taken by homeowners would delay the oilman's building program.[16]

Municipal officials happily ceded the lead on planning to Ralph Lloyd, who in their eyes was propelling Portland toward its destiny as one of the great metropolises on the Pacific Coast. Perhaps no official was more comfortable with this arrangement than Commissioner Stanhope S. Pier, who served on the City Council from 1923 to 1930. As C. W. Norton reported after consulting the commissioner on the proposed street improvements, Pier was "more than pleased to cooperate with us and frankly stated that when anybody comes to Portland doing as much as the Lloyd Corporation they should have the unanimous cooperation of any public body as well as the people."[17] Indeed, at the end of 1928, Lloyd was pleased with the performance of the City Council, remarking that it was "putting everything through for us as rapidly as possible."[18] For their part, both Laurgaard and the City's attorney remained "extremely desirous" to see the developer move forward with his street improvements and pledged to facilitate their implementation in any way they could.[19] The accommodation afforded the Lloyd Corporation by local officials reflected the size of Lloyd's real estate portfolio, the scale of his development plans, and the prominent position he enjoyed at the center of the market.

Despite these assurances, Ralph Lloyd was quick to threaten his own strike of capital when it appeared that the opposition of a few property owners might disrupt his project timetable. If he had to "fight for an opportunity to spend money in Portland," he warned, he would take his capital elsewhere, as "there are too many places where they are glad to have people that are willing and able to do something." In particular, were Multnomah Street not widened, he would drop his plans to build the hotel.[20] These statements were indicative of the level of opportunity available to Lloyd generally as a small businessman interested in commercial real estate. Further, they demonstrated his confidence in his ability to shape outcomes in local municipal politics as a "citizen outside the government."

Ralph Lloyd decided that "full disclosure" would be the best way to overcome residents' resistance.[21] And so he traveled to Portland, appearing before the City Council at its 3 April 1929 meeting, when its members took up Commissioner A. L. Barbur's report that addressed the street projects in which he was interested.[22] Before "the greatest crowd that ever jammed the City Council chambers," Lloyd reassured his supporters and tried to disarm his critics.[23] He explained that he "came here to present to the people of Portland some of my ideas as to the development of Portland as a great city. . . . I came here so that my motives and ideas might be known to all, and that, with equal frankness, I might get the motives and ideas of those who oppose these projects that we might meet on common ground and bring about a unity of action."[24]

Ralph Lloyd elaborated his plans for the hotel in the context of a promised, unprecedented $40 million development program for Portland's East Side. As he now conceived it, the hotel would be "the finest in the country." Designed "to make Portland a convention city," it would cost $3 million to build—more than double what the developer had suggested little more than a year earlier—and $750,000 to furnish. The structure, as sketched for readers of the *Oregonian*, would occupy a 200-by-400-foot footprint. In any urban context, the structure would be impressive in scale. On Portland's East Side it would loom as a colossus. Lloyd's plans for Holladay Park, if executed as planned, would create a place of respite and rejuvenation at the geographical center of a major American metropolis. He promised that the Lloyd Corporation would "provide a residential district that will be perpetually protected from the encroachment of industrial development." The hotel, if realized, would attract substantial investors who would stay in the hotel long enough "to fall in love with this city, as I did."[25] Before council members, Lloyd envisioned the suburbanization of urban space within sight of the central business district.

Ralph Lloyd drew on examples from Los Angeles to illustrate how his street projects would benefit the city. He then argued that a small minority should not be allowed to obstruct Portland's progress. Lest anyone ascribe selfish motives to the street improvements that he endorsed, Lloyd clarified: "I am not playing the game of the East Side, nor of the West Side. I am playing the game for the whole of Portland, for the street projects that are here contemplated will tie up all parts of the city closer together and develop every section of the city." The two thunderous standing ovations that he received surely must have convinced him that the hotel would materialize as he planned.[26]

Representatives of Portland's growth machine who attended the City Council meeting voiced their support for Lloyd. Businessmen who took to the microphone included Ivan C. Anderson, president of the Federated Community Clubs, architect Charles W. Ertz, C. C. Hall, president of the East Side

Commercial Club, real estate broker and investor William M. Killingsworth, W. H. Ross, president of the Portland Realty Board, F. L. Shull, president of the Portland Chamber of Commerce, and Robert H. Strong, principal in the Strong and MacNaughton Trust Company. Former mayor Allen Rushlight and former City Council member George Cellars endorsed Lloyd's proposed street improvements as well.[27]

In the days that followed, other business leaders registered their unequivocal approval of Lloyd's plans. The architectural firm of the late Albert E. Doyle, for one, was persuaded by the comparisons that the oilman made between Los Angeles and Portland: "Having watched the developments you mentioned in Los Angeles we appreciate the opportunities you see here and we are fortunate in having you with your vision and capital here." Indeed, Portland's most prestigious design firm, whose founder had planned Reed College and recently had directed the design and construction of the Pacific Building (1926) and the Public Service Building (1926–1927), Portland's highest structure, Doyle's firm offered its services to Lloyd in the design of his hotel. E. B. MacNaughton, Robert Strong's partner and vice president of the First National Bank, deployed the language of urban rivalry in a post-hearing letter to Mayor George Baker: "Too many of our citizens are laboring under an inferiority complex and look with envious eyes to the activity of Seattle and the cities of California, not realizing that in their own community they have a wonderful opportunity if only we all will put our hands towards making it come true." Others, such as the Chamber of Commerce's Shull, simply assured Lloyd that there was "no question" that most everyone in town backed his program.[28]

At its meetings of 3 and 10 April, the City Council adopted by ordinance A. L. Barbur's report that endorsed the street projects favored by the Lloyd Corporation and overruled the protests against it. On 17 April, it issued permits to allow the Lloyd Corporation to commence work, under the supervision of the city engineer, but at its own expense, on the portions of streets that affected no other property owners. With a timely start for the hotel seemingly assured, Lloyd moved ahead with other projects in Portland.[29]

A Sears Store for Portland's East Side

The opening on 16 May 1929 of a Sears, Roebuck & Company retail store at the intersection of Grand Avenue, East Glisan Street, and Lloyd Boulevard marked the completion of Ralph Lloyd's most successful project of the interwar period and solidified the small businessman's position at the forefront of Portland's real estate market. In this case, Lloyd as developer met the exacting demands of a client engaged in a sustained national building campaign.

To counter archrival Montgomery Ward's first-mover advantage into the retail sector, Sears expanded beyond its traditional mail-order business with the opening of a so-called Class A department store in Chicago in February 1925.[30] By the end of the year, Sears was operating eight retail establishments nationally. Only one Class A store opened in 1926. In December of that year, however, Sears announced two stores for Los Angeles, a 425,000-square-foot combination mail-order house and department store for Boyle Heights, a neighborhood east of downtown, and a 90,000-square-foot retail store at the intersection of Vermont and Slauson Avenues, west of the central business district. (In unsuccessfully pitching two of his properties as possible store locations, Lloyd became familiar with the company's executive managers.)[31] Both stores opened in July 1927. Now engaged in a fevered race with Montgomery Ward for the best locations, Sears opened another four Class A stores by the end of the year and twenty-one such outlets in 1928, including a third store in Los Angeles. In 1929, Sears would open twenty-four Class A stores, including one on property owned by the Lloyd Corporation on Portland's East Side.[32]

Above all, the soaring growth in automobile use as a means of transportation influenced the retail giant's thinking in siting stores. Traffic congestion and a lack of parking space ruled out downtown locations. Sears identified outlying areas that were only marginally in use or largely undeveloped, but at the same time provided a customer base sufficient to meet its sales targets. Lower land values associated with such sites allowed Sears to provide ample parking to customers. Lower overhead, rent, and taxes provided additional capital for stocking shelves with goods that would draw customers away from downtown emporia. At least during the 1920s, proximity to transit lines was a secondary factor in determining store locations. Thereafter, it would be hardly a consideration at all.[33]

In April 1928, C. W. Norton, Ralph Lloyd's representative in Portland, opened negotiations with Sears for a 70,000-square-foot, two-story, Class A retail store with basement to be located on a site along Grand Avenue. The Lloyd Corporation would be required to build the store per the plans and specifications of Nimmons, Carr & Wright, the architectural firm that designed the company's stores during this frenzied period of expansion. The location suited the needs of both the mail-order house and the Lloyd Corporation. It "is just the place for them, as they will appeal to the great middle class and we will want to hold the upper region [of our holdings] around the park for the more exclusive element." In other words, the throng of shoppers would not disturb the peace of the wealthy set enjoying the hotel and its amenities.[34]

The Lloyd Corporation negotiated with Sears as the City Council considered arterial street projects for the November ballot. Ralph Lloyd explained to Senior Vice President of Operations Lessing Rosenwald that the widening of key thoroughfares, if approved by voters, would complement the company's retail strategy by increasing and directing traffic to the proposed store location. In any case, extending Lloyd Boulevard one block to Union Avenue, which the Lloyd Corporation now planned to undertake, would feed traffic from Broadway, the major east-west thoroughfare, to the site. In all, a Sears store at the proposed location could attract as many as 385,000 shoppers annually, Lloyd estimated.[35]

Owing to "heavy purchases" that he planned to make in Los Angeles and the construction of the hotel, Ralph Lloyd preferred a ninety-nine-year ground lease to building a store to lease to the retailer. Under such an arrangement, Sears would design and build the store to its specifications and turn it over to the Lloyd Corporation at the end of the lease. Sears executives agreed to this arrangement. On 13 November, that is, just days after voters approved the bond measure to widen Union Avenue and other streets, Sears announced its plan to open a Class A store on the East Side. The next day, it executed the ground lease that Lloyd sought. Sears awarded the construction contract to L. H. Hoffman, the developer's preferred contractor, for a building to cost not less than $250,000. In fact, the Portland store would cost some $500,000 to construct.[36]

The three-story structure that Nimmons, Carr & Wright designed for Portland was comparable in size and style with other Class A stores that opened in the late 1920s (figure 13). Indeed, the Portland store was nearly identical to a store would open on Bladenburg Road in Washington, D.C., three months later, in August 1929. In their exterior design, architects George C. Nimmons and George Wallace Carr sought to evoke "a degree of stylishness without seeming pretentious or contrived," as Richard Longstreth puts it.[37] The *Oregonian* deemed the proposed structure to be "handsome" in design.[38] The effect was modern and would have met with Lloyd's approval. Following from the design of the mail-order houses, the Portland and District of Columbia stores, like their contemporaries, featured a prominent tower that that served as a landmark for customers arriving by automobile. In their horizontal massing, the two stores were less sculptural than stores designed only a year earlier. Still, notwithstanding Nimmons and Carr's incorporation of more subtle

Figure 13. Sears, Roebuck & Co. store, Grand Avenue, Lloyd Boulevard, and East Glisan Street, Portland, constructed 1928–1929; Nimmons, Carr & Wright, architect; L. H. Hoffman, builder; building altered. (Photo by Ackroyd Photography, 28 February 1949. The Huntington Library, San Marino, California, LCB drawer 1, box 1. Reproduced by permission of Thomas Robinson.)

surface treatments reflective of the Art Deco movement, the stores retained the civic character of their immediate predecessors.[39]

The opening of the store was front-page news. The *Oregonian* heralded "the dawn of a new era" on the East Side," for Sears, Roebuck & Company was the first retail "institution of such magnitude to choose a location east of the river." Mayor Baker, Raymond Wilcox, president of the Portland Chamber of Commerce, and East Side Commercial Club President C. C. Hall, joined Lessing Rosenwald, store manager L. R. Steelhammer, and regional managers A. M. Berry and H. F. Smith for the ceremonies. In the evening, Wilcox and Hall hosted a "royal reception," as Ralph Lloyd put it, for Rosenwald and his colleagues at the downtown Benson Hotel. Lloyd was pleased that "our friends" on both sides of the Willamette River were united in welcoming Sears. Bringing the retail giant to Portland established Lloyd as a bona fide developer intent on expanding the city's retail core beyond its downtown confines.[40]

Despite his intention of doing so, Ralph Lloyd was unable to attend the opening. He was glad to learn, however, that Sears "are very much pleased with their location and the future prospects of the store." Validating the company's national retail strategy, he noted that the store's parking lot was "constantly full" a week after it opened for business, demonstrating "the drawing power of parking area in a retail district." Two months later, banker E. B. MacNaughton reported that Sears was "doing a nice business by all appearances," if the automobiles that jammed its parking lot and filled the spaces available on nearby streets were indicative. Just before the stock market crashed at the end of October, manager Steelhammer reported that the store was experiencing a "steady and healthy" increase in business.[41]

Coming on the heels of the disappointment of having to suspend his mixed-use building project in Walnut Park, owing to the withdrawal of the Multnomah Theatre Corporation from its lease, the opening of the Sears store reaffirmed Lloyd's faith in Portland as a city that was favorable to enterprise and outside investors.[42] It would mark high water for the decade, however, as far as his commercial real estate endeavors on the East Side were concerned.

Running Out of Time

Ralph Lloyd had traveled to Portland in April 1929 to convince local business and civic leaders and residents of the value of his hotel and related projects. He did not expect his opponents to use the occasion against him. In his performance—and without a doubt, it was a performance—Lloyd presented an outcome that observers could use as a benchmark. Near unanimous approval of his program did not guarantee that all the street work that Lloyd deemed necessary would be completed in a timely manner, however. Opponents had twenty days to take action against decisions of the City Council. Hence, both City Attorney Frank S. Grant and City Engineer Olaf Laurgaard impressed upon the developer the need for the Lloyd Corporation to begin the work that the City Council had approved as quickly as possible. For doing so would make it rather more difficult for potential plaintiffs to demonstrate a lack of public necessity for the work, as they likely would charge. These municipal officials evidently appreciated more than Lloyd the importance of timeliness in achieving a hotel construction start.[43]

The legal action that Grant and Laurgaard for all intents and purposes predicted in making their recommendation arrived at the end of May in the form of an injunction filed in circuit court against the City of Portland and the Lloyd Corporation. Kate Thatcher, the plaintiff, owned two lots on East Eleventh Street, just north of Holladay Park. She charged the City with advancing a "speculative promotion scheme" that benefited a private interest without showing a public need for it. In the complaint, Thatcher acknowledged that Ralph Lloyd had made clear his intention to build a hotel and other structures on his vacant property, but observed that he was not obligated to do so. In so doing, she questioned the assumptions under which the City had endorsed Lloyd's development program. Her argument struck at the heart of Lloyd's premise that his projects benefited the public and not simply his pocketbook.[44]

Kate Thatcher's charge was similar to one of three that had been made and overruled at the 3 April City Council meeting, but now the matter was thrown into the courts, where its disposition would surely take longer. City Attorney Grant thought it was all a bluff and encouraged Lloyd to proceed with the authorized work. Portland Chamber of Commerce President Wilcox was sure that the matter would be resolved in reasonable amount of time. Nevertheless, the suit was a blow to Lloyd who began to act more and more conservatively as a developer. In his mind, he was in no better position than he was before he appeared before the City Council. In fact, he was incredulous that the small number of people who opposed him—he put the figure at five percent—could "block the wheels of progress." Lloyd found it particularly galling to have to go to court to secure the privilege of spending his money. Under the circumstances, he complained, it would be hard to interest outside investors in his project—and by this time he had decided that he would need such investors to build his hotel. No, he would not invest any more of his surplus capital in Portland until the suit was resolved.[45]

Luck historically has played an underrated role in the success or failure of small businesses, as Mansel G. Blackford has noted.[46] Yet luck is not necessarily blind. Indeed, Ernest Hemingway's counsel to his son, "You make your own luck," has become a cliché among business professors and management consultants.[47] In delaying the start of the hotel until all legal actions related to street improvements in which he was interested were resolved, Ralph Lloyd as developer failed to make his own luck.

Kate Thatcher wasn't the only resident who believed that the plans that Ralph Lloyd presented to the City Council constituted a purely speculative play for private profit. Property owners along Multnomah Street expressed similar concerns. Indeed, the longer that "Larrabee's white elephant" remained "unimproved," the more residents became persuaded that this was the case. The concerned property owners promised to drop their objections once the Lloyd Corporation started construction on the hotel. Architect Charles W. Ertz observed that building the hotel was the only way of convincing the "people among us with no vision [who] must be shown every step before they will believe."[48]

Olaf Laurgaard beseeched Lloyd to redouble his efforts to complete work on the portions of streets around Holladay Park that the Lloyd Corporation controlled. The Lloyd Corporation completed the street work to which Laurgaard alluded, but at the end of 1929 Ralph Lloyd still awaited municipal action on many other projects that had an impact on the value of his property in the vicinity of Holladay Park.[49]

In reply to architect Ertz, who hoped that "the little barking that may occur now and then" would not dampen his "desire and enthusiasm," Ralph Lloyd insisted that he could still wait until people "ripen to the different projects."[50] (What, if anything, the stock market crash portended for the economy was not clear. Indeed, there was consensus within the Los Angeles real estate community from which Lloyd took his cues that the woes of the stock market would encourage investors to punt on real estate in the coming year. In any case, Lloyd did not invest in equities.)[51] Still, it astonished Lloyd that people would file injunctions to stop his work. The California developer embedded his complaint in the context of urban competition. It was "a disagreeable feature" of Portland, he began, "that so many of its citizens [showed] such small caliber in their actions regarding public improvements," especially since in his view he was bearing their cost and they would receive much of the benefit.

Injunctions that delayed infrastructure projects were putting the city at a competitive disadvantage with its Pacific Coast rivals: "It is impossible to get outside people to come into Portland and undertake the management of a large hotel project or something of that nature unless they can actually see the arteries of communication opened up." It would be a "wonderful tonic," he asserted privately, to bring 10,000 Angelenos into town for a month or two. For doing so "would work miracles in Portland during the next ten years."[52]

Ralph Lloyd assured a concerned property owner who took the time to write to him that he was not about to abandon his plans. Nevertheless, he demonstrated that he was nearing the end of his rope, so to speak. Complained Lloyd, again using Los Angeles as a benchmark: "Unfortunately, some of my neighbors are, by injunction suits and otherwise, hampering and delaying my program in a way that is astonishing to a man who is accustomed to the encouragement that is given one for making improvements and capital expenditures in all districts of Southern California." As matters stood, his real estate program had been delayed at least a year, maybe two, he lamented.[53]

The Lloyd Corporation's man on the ground in Portland remained optimistic that the City of Portland soon would clear the way for the completion of the remaining street projects in which the developer was interested. Since passage of the November bond issue, C. W. Norton explained, the City engineer's office had been overwhelmed by legal appeals from property owners. City Engineer Laurgaard was working on the Lloyd Corporation's petitions, as Commissioner Barbur had reported in July 1929, but it was impossible "to obtain greater speed than has already been made" on them. In early 1930, however, both architect Ertz and Norton reported "a decided speeding up" of the arterial street program, owing to the resolution of various legal actions: only five of some eighty cases involving Union Avenue remained open, for instance. Unfortunately, from Lloyd's perspective, the City had assigned a lower priority to cases that involved streets such as those that bordered Holladay Park.[54]

For the prospective developer of a substantial area of Portland's East Side, this was an intolerable state of affairs. Ralph Lloyd fired off a letter to the City Council, detailing eighteen improvement projects in the vicinity of Holladay Park that had to be completed before he would move ahead with the construction of the hotel. They included street vacations that would allow the Lloyd Corporation to clear the street grid for parking, the golf course, and other appurtenances. In today's parlance, he asked the City of Portland to "fast-track" the approval of these projects as a block. Lloyd's position suggests that he had little inkling of the economic tsunami that loomed. At the same time, he remained attuned to Portland's politics from his office in downtown Los Angeles.[55]

Only a week earlier, the Citizen's Recall League had commenced a drive to collect signatures to recall the entire City Council, that is, Mayor George Baker and Commissioners A. L. Barbur, C. A. Bigelow, John Mann, and Stanhope Pier, on charges that they had wasted funds, mismanaged affairs, and betrayed the public trust. Olaf Laurgaard, too, was under fire. Plaintiffs seeking to stop the widening of West Burnside Street charged the City engineer with acting in his own self-interest in deciding which side of the street would bear the burden of the improvement. Ralph Lloyd found these developments "astonishing."[56] Nevertheless, mindful of the negative effect that the lack of progress on the hotel may have been having on the political careers of officials upon whose unwavering support his development program depended, Lloyd assured the City Council that he would complete excavation and foundation work in the next six months.[57]

At the same time, Ralph Lloyd was apparently prepared to withdraw from the Portland real estate market as easily as he had entered it. He vented to his local representative: "If the city is going to fight every project that anyone attempts that is a progressive development, the sooner we get out of the city the better it will be for all concerned." Lloyd had apparently miscalculated Portlanders' desire to meet their competition on the West Coast: "The majority of the people seem to have no imagination as to the possible developments of their city. This peculiar psychology will some day change but it is apparently a governing factor in the city at this time, causing development of any type to move slowly and response from development to be almost negligible." Again he deployed the language of urban rivalry: Portland risked losing out to Seattle, San Francisco, and Los Angeles insofar as the investments of "men of money and those controlling the industrial enterprises" were concerned. He was adamant that the hotel project would not go forward "unless we are permitted by the city to have the proper environment around this structure to justify the capital expenditure, and the things we are asking for are all essentials that must be granted us before we can proceed."[58]

C. W. Norton assured his boss that fully 99.9 percent of local residents would "do anything in their power to further the interest of Lloyd Corporation." The remaining tenth of one percent were "old timers that are just in the habit of crabbing everything and everybody." It was small comfort to Ralph Lloyd. A tenth of one percent apparently sufficed to block a street project in Portland.[59]

Exposed to the possible wrath of voters, Commissioner Barbur and his colleagues refused to take special action to accommodate the Lloyd Corporation. Procedurally, the public works department handled its projects on a case-by-case basis. Moreover, the projects retained their place in the queue. The department took them up as resources allowed. It would be a matter of too little, too late, as far as Ralph Lloyd was concerned. Ultimately, the local political economy of infrastructure and Lloyd's response to it were key factors in delaying the Lloyd Corporation's development of Holladay's Addition for some two decades.[60]

Financing and Operating the Hotel

Ralph Lloyd sought to distinguish his hotel project in a saturated market. Local occupancy rates were no more than 60 percent as early as 1925. Portland's situation reflected the national trend, which left occupancy rates below 70 percent after a boom in construction from 1920 to 1927. From that point, rates would plummet nationally, to a nadir of 50 percent in 1933, when 80 percent of hotels in America were in receivership. Nevertheless, even as Lloyd sought to finance his project, parallel efforts to develop a downtown hotel persisted, involving some of the business leaders whom Lloyd sought as investors in his project. Because Lloyd ultimately chose not to fund the hotel entirely from surplus capital, local real estate market conditions became an obstacle to his realization of the project.[61]

At the end of 1928, the Lloyd Corporation had reached agreement with Eric Hauser Sr., whereby the latter would furnish the hotel and operate it under a twenty-year lease. The parties would split the profits on a fifty-fifty basis. But then Hauser died unexpectedly.[62]

In the wake of the contractor's death, Lloyd mulled over financing options that included funding the project himself. On that basis, he sought to secure a construction loan. After the securities arm of the First National Trust and Savings Bank refused to originate one, Lloyd decided to proceed as investors had done in the case of the Los Angeles Biltmore—the largest hotel west of Chicago when it opened in 1923. To finance the $7 million project, investors, headed by Joseph F. Sartori, president of the Security Trust and Savings Bank, formed the Central Investment Corporation with a capitalization of $5 million, $4 million of which it issued as stock. Of this amount, the "most prominent men" of Los Angeles reportedly took as much as $2.25 million of the issue. The company then offered $1.75 million in common stock to the public.

To raise the remaining cost of the land and the building, the company issued $3 million in 6.5 percent bonds and targeted "leading citizens" as investors. Using the Biltmore plan as his model—yet another example of the influence that Los Angeles exerted through Ralph Lloyd on the Portland real estate market—the California developer would spend the better part of the next three years wooing potential investors among Portland's business community.[63]

Ralph Lloyd would try various formulas to attract local capital. But his argument would remain essentially the same: the success of a project of this character and scale in a city with the size and ambitions of Portland required the backing of the entire metropolis. Those who became involved with the Los Angeles Biltmore Hotel, he noted early on, were adamant that its successful financing depended on convincing prominent local business leaders to buy stock in the corporation. As Lloyd explained to E. B. MacNaughton, a prominent figure in the Portland real estate market, "A hotel is more or less a public utility [that] must have the support of the local people and interests as well as the general public outside of the city." With this mindset, Lloyd traveled to Portland to deliver his presentation at the 3 April 1929 City Council meeting.[64]

At the same time, the death of Eric Hauser left Lloyd bereft of an operator. And so, even as he sought investors for the project, Lloyd searched for a replacement in a deteriorating market. Before all was said and done, he would solicit the interest of both operators of independent hotels and the Bowman-Biltmore and United Hotels chains—the largest in America—in running his hotel. Once again the Los Angeles Biltmore Hotel provided a blueprint for the Holladay Park hotel project. In December 1921, the Los Angeles Biltmore Company was created to lease and operate the hotel owned by the Central Investment Corporation. In this case, the lessee and the operator owned the furnishings, too. In February 1929, the Bowman-Biltmore and United Hotels groups announced a merger that would have created an empire with more than one hundred properties. Reflecting the difficult conditions that plagued the industry, the deal fell through. At the end of 1929, Lloyd had not identified an operator for his hotel.[65]

Acting on the advice of Charles Ertz and others, Lloyd began to make a series of announcements regarding the design and construction of the hotel. In March 1930, he reported to the East Side Commercial Club that the competition among Morgan, Walls & Clements and Walker & Eisen was under way on a design that would compare favorably to any hotel "not only on the

Pacific Coast but in the United States as well." A month later, he traveled to Portland to confer with builders. At the end of the week-long visit, he announced that the structure would be twenty-four stories in height—that is, four stories higher than he had announced previously—and that the design would be finalized by July. He also reported on the work of the Lloyd Corporation to improve the grounds of the hotel and complete Lloyd Boulevard with sidewalks and a fence along the border with Sullivan's Gulch. Upon returning to Los Angeles, he assured Ertz that the competing architects had nearly completed their models of the hotel.[66]

Ralph Lloyd's visit accomplished its public relations objective. It proved the Californian's faith in Portland, the *Oregonian* opined: "When men from outside come and see Portland, they show their faith by making large investments. When they come again, their faith is confirmed and they make further investments." At a time of economic uncertainty, Lloyd's faith in Portland should suffice to give the city confidence in itself, according to the newspaper's editors. Candy manufacturer and East Side commercial property owner Edwin A. Hollinshead, a frequent correspondent of Lloyd's on Portland's municipal and economic affairs, averred that Lloyd had succeeded in creating "a 'success complex' in the minds of the Portland people." Indeed, Hollinshead enthused: just as William Wrigley Jr. had accomplished with the construction of his iconic Wrigley Building across the Chicago River from the Loop—the Windy City's central business district—Lloyd would show with his hotel that business could establish a profitable beachhead beyond Portland's downtown.[67]

Ralph Lloyd built on the momentum that his visit generated by keeping Portland's business elite apprised of progress on the project. On 10 May 1930, he sent banker Ainsworth twenty-four pages of photos and architects' notes and asked for comment from a "conservative and cool headed business man." A few days later, he let MacNaughton know that the architects had completed their models of the hotel. Ahead of a planned stopover in Portland during an upcoming family vacation to Canada, Lloyd assured Ainsworth that excavation of the site would begin in August. In reply, the banker encouraged Lloyd to start excavation "as early as possible in order to avoid the inconvenience occasioned by Fall rains."[68]

Upon his arrival in Portland on 20 June with his family, Ralph Lloyd continued his public relations offensive, releasing photos of models of the hotel that the architectural firms vying for his custom had developed. He assured the city's leaders and residents alike that he would soon approve final plans for "what will be the most beautiful hotel on the western coast." He promised to stop again in Portland on his return from the Canadian Rockies, at which time he hoped to announce the winning design.[69]

In the event, not until early August did Lloyd select Morgan, Walls & Clements as his architect. He did so during a visit to Portland that featured a groundbreaking ceremony for the hotel: "the big news of the week," according to the *Oregonian*. Not only had excavation of the site commenced, both the landscaping of the grounds and construction of the golf course would begin at once so that both projects might be finished in a years' time. Lloyd hoped that the hotel would be completed in time to accommodate visitors who traveled along the Pacific Coast in conjunction with the 1932 Olympiad in Los Angeles. The announcement was a masterstroke of public relations. According to the *Oregonian*, it "was joyously received by the city as a whole," not least for the jobs and purchases of materials that it promised. Certainly, concern about the economy was in the air, to borrow Alfred D. Marshall's phrase. Yet there was little indication that Portlanders (or, for that matter, Angelenos) expected the economy to spiral downward into deep depression in the summer of 1930. Indeed, together with an announcement of the expansion of the Lerner chain of women's apparel stores into Portland, Lloyd's "big news" promised "a future that augers well" for the city and its hinterlands.[70]

Within scarcely more than a month, however, dwindling royalties from crude oil production intersected with deteriorating national economic conditions to raise considerable doubt regarding the viability of the hotel project. On 8 September, Lloyd and his architects met with a committee from the National Association of Building Owners and Managers at the Multnomah Hotel, where the model of the hotel was on display, and submitted the plans and specifications for the committee's review. Remarkably, given the amount of time that it had taken for the project to materialize to this point, Lloyd conceded that his architect's plans represented merely his ideal vision for the hotel. Now, almost two years after he initially had hoped to open the property, he was seeking practical advice. By invitation, the committee included many of the men whom Lloyd hoped to interest in the operation of the hotel, among them Sidney Sterling of the Los Angeles Biltmore Hotel and Edwin H. Lee of the Baker hotel chain. Also on hand were banker Ainsworth, architect Ertz, contractor Hoffman, Robert H. Strong (who, along with his business partner E. B. MacNaughton, was championing a downtown hotel and had persuaded their firm, the Commonwealth Trust and Title Company, to acquire a downtown site for a new corporate headquarters), and the owners and managers of the downtown Benson and Multnomah Hotels. The committee weighed in on the elaborate design and scale of the building with a view to determining its feasibility as an enterprise.[71]

After the meeting, but before the committee issued its findings, Lloyd suggested for the first time publicly that the project might be delayed. "It may be necessary to modify the plans in some respect to make the hotel an economically sound proposition," he conceded. He could make no commitment on the letting of the construction contract until the building owners and managers weighed in. If they recommended extensive changes, it would take time to revise the architect's plans and specifications. Privately, the conference was a wake-up call that put front and center in Lloyd's mind the need to secure financing for the hotel if the project were to move forward.[72]

As late as July 1930, Ralph Lloyd had vowed to fund the construction of his hotel out of cash flow. With production from the Ventura Avenue field running almost 20 percent below the peak established a year earlier and the benchmark, Signal Hill, price for a barrel of crude oil softening, as we shall see in the next chapter, Lloyd was feeling squeezed financially. In addition, a number of projects in Los Angeles and elsewhere in Portland were making demands on his financial resources. Lloyd already had decided not to acquire additional real estate so that he might focus on completing work in progress. Before he returned to Los Angeles, he pitched a financing plan to Ainsworth—who would serve as his intermediary—and other bankers and potential investors. As was the case with the Los Angeles Biltmore Hotel, the proposal included the formation of an operating company, the Holladay Housing and Heating Corporation, which would place a six percent, $500,000 preferred stock issue with some two dozen local business leaders and issue $500,000 in common stock. The corporation would use the funds from the preferred stock issue to furnish the property. The Lloyd Corporation and the operating company would back the issue. Either the Lloyd Corporation or the Holladay Housing and Heating Corporation would own the hotel. Together with the operator, the two companies would own the common stock. Lloyd would select men who subscribed to the preferred stock issue to serve on the board of directors. In the context of the gathering economic storm and a competing proposal for a downtown hotel, Lloyd would struggle to generate interest among Portland's business elite in his proposal.[73]

In offering his plan, Ralph Lloyd insisted that money was not the issue. Rather, the execution of his proposal held the key to securing an operator. He had in mind the operators of the Los Angeles Biltmore Hotel—an alliance at which he had been hinting since May. The financial commitments and directorships that he was seeking would increase his chances of interesting the Bowman-Biltmore group in operating his hotel because they would demonstrate that Portland welcomed outside capital and management. The lack of local investment in the project was the main, if not the only, issue, that needed to be resolved, Lloyd averred.[74]

The response of Portland's business elite to the proposal was tepid, at best. Because Ralph Lloyd had insisted to date that the Lloyd Corporation was in position to build the hotel at its expense, the prospectus that banker Ainsworth circulated at Lloyd's behest generated myriad questions about the preferred stock issue, but no commitments. The proposal to form a new operating company elicited surprise among those whom Ainsworth had approached. Had not Lloyd already lined up the Biltmore group as his operator? For his part, William Hemphill, a broker with Commonwealth Securities, told Ainsworth that Lloyd would have to offer more than six percent on the preferred stock issue if he hoped to place it with investors. A ten-year note issue might be marketable, if construction of the hotel structure were underway. Hemphill added that, if local support was what Lloyd sought, he would obtain it by selecting a board of directors from among the men who received the prospectus.[75]

At the same time, many of the "leading citizens" of Portland whom Lloyd targeted were warming to the idea of a new, 600-room downtown hotel. Ironically, the initial ideas of Robert Strong, the project's champion, on how to finance the project were not dissimilar to those expressed in Lloyd's proposal. Eric V. Hauser Jr., taking up as a builder where his late father had left off, noted as much when he counseled Lloyd to drop his financing proposal. To generate local support and to put to rest all talk of a competing downtown hotel, Hauser urged Lloyd to begin construction of the hotel by the end of the year.[76]

Nevertheless, in late October 1930, Lloyd pressed Ainsworth to approach his potential investors and relay his answers to their questions. In the meantime, he promised to secure an operator and let the construction contract. That same month, Columbus Marion "Dad" Joiner's wildcat well, Daisy Bradford No. 3, struck oil in East Texas.[77] The discovery of the largest oil field in the world in an already glutted market would have a greater impact on the outcome of Lloyd's hotel project than any macroeconomic effect of the Great Depression.

End Game

Rumors of a lease with the Bowman-Biltmore hotel group, the source of which was Ralph Lloyd himself, and the letting of the construction contract generated belief among Portland's business and civic leaders that the Holladay Park hotel would open in 1931. Together with plans for a downtown hotel and the Commonwealth Building, a twenty-story office building, the realization of Lloyd's hotel project, which appeared to be close at hand, convinced local observers that Portland would weather the economic storm better than its urban competitors. In the event, none of these projects materialized, sinking the optimism that they had created.

John C. Ainsworth and his colleagues were pleased to learn late in 1930 of the interest of the western affiliates of the Bowman-Biltmore hotel chain in operating Lloyd's hotel, as it suggested that the California developer would finally construct the project. By reporting the development of a possible link of his project with the national hotel chain, Ralph Lloyd surely hoped to increase the interest of the local business community in injecting their capital into the venture. He conceded as much when he cautioned Ainsworth that Lee Phillips, an investor in the Los Angeles Biltmore Company, had advised him that his company would not commit to the operation of the Holladay Park hotel until Lloyd set a completion date. Moreover, Los Angeles Biltmore Company would not invest in a Portland counterpart. The value of the tie-in lay in the association with the brand. Both Lloyd and the Biltmore people expected Portland's business leaders to buy stock in the operating company—something none of them seemed inclined to do. In a final move to attract the interest of outside capital, Lloyd let the construction contract.[78]

The move coincided with the initiation of discussions on the part of Robert Strong and his colleagues regarding an operating agreement for a downtown hotel on the site of the old Portland Hotel with Frank A. Dudley, president of the United Hotels chain. Dudley, who had established the company in 1910 with the construction of the Ten Eyck Hotel in Albany, New York, and was now overseeing the operation of some of America's finest hotel properties, thought that the proposed site of the hotel was the best one that Portland had to offer. The West Sider's financing plan intrigued Dudley, who calculated that the project would "pencil" if two floors of retail space could be leased. Suggesting what he must have thought about Lloyd's project as well, Ainsworth advised Dudley that it would be difficult to finance a local hotel under present conditions.

Nevertheless, Dudley promised to travel to Portland to discuss an operating agreement if Robert Strong and his partners obtained financing for their project.[79]

When he learned of the interest of the West Side's business elite in a hotel on the site of the old Portland Hotel, Ralph Lloyd meekly noted that "it would be a good location and probably a good thing for the entire city for it to be built." C. W. Norton pointed out the obvious: such a project, if realized, "would prolong our whole scheme of development around Holladay Park." Now Lloyd was in direct competition for capital with the men whose funds he was soliciting for his own project.[80]

Ralph Lloyd also moved ahead with the construction contract to support municipal officials whose careers were now under threat, owing to their support of various street projects. For instance, at the City Council meeting held on 10 December 1930, property owners protested the widening of Grand Avenue to eighty feet from Holladay Avenue to Multnomah Street: a project in which the Lloyd Corporation was especially interested. Many of the protesters had signed petitions in favor of the project, but now they objected to the size of their assessments. After meeting with property owners, Commissioner A. L. Barbur recommended that the City Council delay the widening of Grand Avenue six months to give the project's opponents an opportunity to put a bond measure on the ballot to offset its cost.[81]

On 10 January 1931, Ralph Lloyd awarded the construction contract jointly to W. S. Dinwiddie of San Francisco and local builder L. H. Hoffman. Established in 1911, the Dinwiddie firm was renowned for its construction of the celebrated Russ Building—San Francisco's tallest building from its completion in 1927 until 1964. Both Dinwiddie and Hoffman were expanding the A. E. Doyle–designed Meier & Frank department store, Portland's largest emporium. The contractors agreed to split the fee, as they had done on that $920,000 project. The installation of footings would begin in two or three weeks, promised Lloyd, just as soon as Morgan, Walls & Clements released the working drawings that they were revising in response to the recommendations of the National Association of Building Owners and Managers committee for significant alterations to the first two levels. In fifteen months, Portland finally would have the hotel that the California developer had been promising for nearly five years.[82]

The announcement generated the local "buzz" on which Lloyd was counting to sway the minds of the men he was targeting as investors. C. W. Norton lauded Lloyd for selecting the Dinwiddie and Hoffman firms as contractors and reported that hotel was "becoming the talk of the town." Banker E. A. Clark enthused that the announcement "has created a mighty optimistic feeling all over the city." Louis T. Merwin, vice president of Northwestern Electric Company, a predecessor of the modern Pacific Power and Light Company, let Lloyd know of "the general feeling of pleasure and satisfaction" generated by the announcement, adding that many of us "stood firm in our conviction that in due time your announcement would be forthcoming," even as many others "began to be very skeptical about the whole proposition."[83]

Ralph Lloyd was encouraged by the general response to the announcement, but he became dismayed when local business leaders showed no interest in his operating company prospectus. Almost immediately, Lloyd began a retreat from the hotel project that soon turned into a rout. A week after he made the announcement, Lloyd reminded Ainsworth that, while he was prepared "to make some sacrifices" to realize it, the project still depended on local investors committing $500,000 in capital to an effort that was going to cost some $5 million. Still, he promised to deliver a new model of the hotel within ninety days. Ainsworth promised to display it in the lobby of his bank, where some 3,000 people would have the chance to view it daily.[84]

Ralph Lloyd was also dismayed to find that he could buy cement, steel, and other building materials most anywhere along the Pacific Coast and ship them to Portland more cheaply than he could purchase them locally. As he told C. C. Hall, president of the East Side Commercial Club, he was now in the market for $100,000 in materials and wanted to buy them in Oregon, but he refused to pay the prices on offer. He could only shake his head. His colleagues in Los Angeles thought he was foolish to devote so much of his energy and resources to a project that was taking seemingly forever to materialize. Now he had more reason to believe that Portland was hostile to outside investors.[85]

In fact, according to local historian E. Kimbark MacColl, Portland's building suppliers were colluding on the Lloyd Corporation's bids for materials. Nevertheless, in a few weeks' time, it was becoming clear that the local prices of materials were becoming incidental, at best, to the completion of the hotel. At the end of June, the architect's plans and specifications for the hotel remained incomplete and the model that Lloyd had promised to deliver to Ainsworth was two months overdue. By then, the crude oil flowing from more than 1,000 wells in East Texas was flooding markets, driving down the spot price of a barrel of crude oil in that state toward ten cents.[86] In California, the collapse in prices was costing the Lloyd Corporation and VL&W together some $3,000 per day in royalty income.[87] With the pressures of his business affairs straining his health, Lloyd confessed to Robert Strong—whose Commonwealth Trust and Title Company recently had abandoned its plan to erect a headquarters building on the Sixth Street site he and his partners had purchased less than a year earlier—that he was "somewhat at sea as to know what is best to do."[88]

In the eleventh hour, Lloyd met Frank Dudley in Portland and again in San Francisco to discuss the possible inclusion of his Holladay Park hotel in a reorganization scheme that the United Hotels president was contemplating. Dudley sought to split United Hotels Company of America into five regional holding companies. For the Pacific Coast, he had "a definite plan in mind," as C. W. Norton put it. Dudley consulted Lloyd on including Portland's Benson Hotel and San Francisco's Sir Francis Drake Hotel in a Pacific Coast group and broached the idea of folding in the Bowman-Biltmore group's western properties. With many of his properties in distress, Dudley reversed the role of investor and supplicant, asking Lloyd to invest in the reorganization plan. But Lloyd had no interest in participating in the proposal.[89]

With the oil industry in chaos and with neither the Bowman-Biltmore group nor United Hotels in financial position to back it, and with Los Angeles offering him a relatively more enticing commercial real estate market, as we shall see, Ralph Lloyd effectively terminated the hotel project at the end of August 1931. He reported to Ainsworth that he would not start construction "until I can definitely see that matters are on the mend and that we are again on the way for a new cycle of business and financial activity."[90]

In Ralph Lloyd's view, only Portland could save the hotel. Were the city's leading capitalists to participate in the project, the developer promised to supply the junior financing needed to secure a first mortgage at a low rate of interest. Indeed, with no possibility of attracting outside interests, the hotel project would fail unless local capital rescued it: "It is up to us and Portland to put the hotel project over." Specifically, he offered to contribute $1 million and land to the project. Characterizing the hotel as a civic or semi-civic project might persuade local bankers and businessmen to open their pocketbooks.[91] Rather than take the proposal to his colleagues, as he had done previously, Ainsworth threw cold water on it as soon as he received it. There was little chance that the local business community "could be of much assistance," given business conditions.[92]

With the hotel industry generally in a "most deplorable condition," Lloyd concluded that building the hotel with his funds alone "would be absolutely foolish."[93] And so, three months after Ainsworth reported the lack of interest of the local business community in his latest gambit, Lloyd "temporarily

Figure 14. Lloyd Corporation properties, East Side, Portland, circa 1935. Lloyd Boulevard connects the Sears store at the bottom of the photograph and the golf course, completed in 1932, which straddles Sullivan's Gulch. The excavated hotel site lies east of Holladay Park. Photo undated. (Photo by News Pictures, Ltd. The Huntington Library, San Marino, California, LCB drawer 1, box 1.)

suspend[ed]" the project, having only the excavated site to show for his efforts (figure 14).[94]

With his political career on the line, A. L. Barbur asked Lloyd to explain to him and his fellow City Council members why construction of the hotel had stalled. None of the recall petitions circulated in the spring of 1930 had gained the signatures needed to subject any commissioner to a vote. Yet C. A. Bigelow had resigned in August and Stanhope Pier had lost his seat in the November election. Barber reminded Lloyd that the City Council had approved his petitions to widen streets over the objections of property owners only because they had believed that he would carry out his plans. Canceling the project would expose the mayor and the four commissioners to charges of being "too liberal" on public improvements.[95]

Ralph Lloyd empathized with the precarious political position of Barbur and his colleagues. He insisted that he was doing everything in his power to sustain the project. Nevertheless, construction of the hotel would have to await the return of more prosperous times. Lloyd's reply must have provided no comfort to Barbur, who lost his seat in the November 1932 election, or to John C. Mann, who was accused of larceny and recalled by voters in May 1932 after fifteen years of service and who failed to regain his seat in November.[96]

Ralph Lloyd's influence as a "citizen outside the government" on Portland's growth path validates E. Kimbark MacColl's observation that private developers took advantage of a weak formal municipal planning function to shape outcomes in their favor. Indeed, Lloyd's actions were more broadly in keeping with the steps that developers, owners, and proprietors of commercial establishments had taken nationally at least since the turn of the century to ensure the success of their real estate ventures in an urban context.[97] At the same time, the relatively large scale of Lloyd's real estate holdings and the resources generated from petroleum extraction he was willing to devote to developing them encouraged municipal government to engage in metropolitan-wide planning. For instance, Lloyd was the impetus behind the planning commission's revision of its traffic plan for Portland. His self-interest was always obvious. Still, his support of specific public works projects in support of his development plans, inspired by what he observed and experienced in Los Angeles, was coherent and well conceived at the municipal level, given the imperative of automobile accommodation. Local civic and business leaders lent the projects in which Lloyd was interested widespread support because they appreciated their potential multiplier effect on metropolitan growth. It is improbable that so many self-interested individuals were simply duped by an outsider's argument regarding regional urban competition. They perceived an opportunity to leverage the Californian's promised investment in the development of the East Side to the benefit of their city. Ralph Lloyd's vision for his holdings around Holladay Park—pitched as a high-class hotel "with suburban surroundings" in the center of a major regional metropolis[98]—in fact foreshadowed post–World War II development across America in terms of density and land use. In three decades, as we shall see, this suburban vision would unfold as the Lloyd Center, an office and retail complex that would feature one of the few regional shopping centers to materialize in an urban setting in early postwar America.

Municipal approval of public works projects literally helped to prepare the ground on which the Lloyd Corporation's projects eventually would materialize. Ralph Lloyd's experiences in Los Angeles did not prepare him for the resource constraints that he faced in Portland, however. The small businessman could appreciate the fact that streets could not be widened all at once, given the availability of public funds and the need to conduct engineering studies. He had far less patience with Portland's seemingly inadequate toolkit for dealing with protests that delayed the implementation of approved projects. It mattered little that, as traffic planner Harland Bartholomew observed, "no city of Portland's size did so much in the way of street openings and widenings in the period 1917–1931."[99] Delays in real estate development easily can be fatal and fighting various legal battles pushed back Lloyd's schedule into the Great Depression. One might ask, as one resident did at the City Council meeting on 3 April 1929 at which Ralph Lloyd presented his plans, why it was necessary for the developer to widen, pave, and light all the streets around Holladay Park before he built his "nice hotel." Given an overbuilt local hotel sector and unresolved intracity commercial rivalry, tapping Portland investors for the external financing that Lloyd ultimately sought for his hotel was always going to be difficult. The conservatism he displayed in conditioning the hotel's construction on the completion of a suite of public works projects ensured that capital from other investors would fail to materialize before East Texas crude oil flooded the market and the national economy collapsed. In a time of depression, public sector clients and federal housing programs would lower his risk as a developer to a point at which he would become comfortable enough to "green light" building projects. On this basis, Ralph Lloyd would begin to transform the East Side of Portland.

Chapter 5

The Lloyd Corporation Becomes an Independent Operator

Drilling can not be stopped—it is difficult even to regulate it.
—*California Oil World*, 1930

Tensions between Ventura Avenue's lessors and Associated Oil Company, on the one hand, and between various lessors, on the other, arose once the California major demonstrated its ability to tap the vast pool of petroleum that lay beneath its leased properties. As of 1933, Ventura Avenue was the largest producing field in California for the company, accounting for 60 percent of its total output.[1] In the context of statewide efforts to stabilize the petroleum market through natural resource conservation and voluntary curtailment of operations, Ralph Lloyd acted foremost on behalf of lessors, each of whom demanded that Associated maximize development of their property in the minimum amount of time. Lloyd generally pressed Associated to drill "diligently and in good faith" to satisfy all of its lessors, even if it meant exceeding the express provisions of individual leases.[2] After all, as future U.S. Supreme Court Justice Willis Van Devanter had written in *Brewster v. Lanyon Zinc Company*, a landmark case, "Whatever, in the circumstances, would be reasonably expected of operators of ordinary prudence, having regard to the interests of both lessor and lessee, is what is required."[3] Associated would be unable to head off legal action on the part of its common lessors who charged the major with failing to perform its implied duties under the prudent operator standard first fully articulated by Van Devanter. Ultimately, Associated's conduct would persuade Ralph Lloyd to reenter the Ventura Avenue field as an independent operator.

During the late 1920s and early 1930s, Ralph Lloyd stood on firm financial ground. Despite investing hundreds of thousands of dollars in real estate development and experiencing a drop in royalty income as a result of plummeting prices and restricted production, the Lloyd Corporation netted $4.07 million in profits from 1927 to 1934. VL&W, in which Ralph Lloyd held a 30 percent interest, earned $8.3 million over the same period. As of September 1934, when Lloyd decided to determine the eastern limits of the Ventura Avenue field, the Lloyd

Corporation had paid-in capital of $5 million, a surplus of $467,175, and only $438,679 in debt.

Still, Ralph Lloyd's real estate program was far from self-sustaining. And opportunities to profit in the private commercial real estate market during the 1930s would prove to be rather elusive for Lloyd, as we shall see. Above all, Lloyd reentered the Ventura Avenue field to accumulate capital that would enable him to realize his vision for his Portland portfolio. Associated's poor performance in the Ventura Avenue field relative to Shell convinced him that he could not rely on royalty income alone to realize his ambitions. As an independent operator, the Lloyd Corporation would do what Associated failed to do and extend the proven limits of the Ventura Avenue field eastward.[4]

Before then, Ralph Lloyd advocated directly for the price of Ventura Avenue royalty oil in an unstable market. In this ongoing effort, he had support of the other lessors, even if gaining that support required dogged effort on his part to keep them acting as one in negotiations with the refiners who purchased their royalty oil.

As crude oil prices collapsed in the early 1930s, Lloyd lobbied for price support, both directly and through production control, as president of the Oil Producers Sales Agency of California and as a member of other boards and organizations. In this public-private role, however, he based his recommendations for statewide action on his experiences with Ventura Avenue field operations. In so doing, he acted in the spirit of associationalism advocated by Commerce Secretary Herbert Hoover. Like many a supporter of Hoover as president, however, Lloyd became deeply troubled by the New Deal. Indeed, federal tax policy of the mid-1930s contributed to his decision to drill wells in the field as an independent.

All the while, litigation threatened to derail the development of the field east of the Ventura River. As holders of the most productive lease in the field, Ralph Lloyd and Joseph Dabney became the target, directly and indirectly, of lessors, including members of the Lloyd family, who, as directors of, and shareholders in, VL&W, felt that Associated was privileging the Lloyd lease in its development efforts. In a high-profile case, Ralph Lloyd was also sued for the manner in which he had acquired the mineral rights on an adjacent lease.

Getting Sued, Part I

Ironically, by limiting the number of operators in the Ventura Avenue field to maximize ultimate production, Ralph Lloyd and Joseph Dabney increased their exposure to litigation. Where town lot drilling prevailed, as at Santa Fe Springs and other fields in the Los Angeles Basin, landowners could do little legally to prevent myriad operators from draining an oil pool that lay beneath their property from adjacent leases. The only practical response was to lease their land to an operator who was prepared to drill a well immediately and subsequently pump as much oil as possible in the shortest amount of time. And so "flush" conditions prevailed. At Ventura Avenue, the cost of orchestrating the "rational" development of the field, as Lloyd and the professionals employed by Associated, General Petroleum (GP), and Shell understood the term, was measured in legal fees and judgments as well as in labor and materials.

Ralph Lloyd's and Joseph Dabney's subleases with Associated became sources of tension in relations between certain lessors and Lloyd and Dabney, on the one hand, and, simultaneously, between these lessors and Associated, on the other. The geological and topographical conditions under which Associated's crews labored increased in difficulty as they moved their rigs to the leases east and north of the Lloyd lease. These conditions intersected with economic ones associated with nationwide overproduction, persuading Associated to do no more than to meet the express requirements of its agreements with the exception of the prolific Lloyd lease. Under the circumstances, satisfying the expectations of lessors who expected to reap returns comparable to those that Lloyd and Dabney were realizing was all but impossible. Complicating matters, Ralph Lloyd's relatives found cause to challenge the terms of the distribution of the overriding royalty on the Lloyd lease, as prescribed in the contracts executed on 8 August 1922.

Ralph Lloyd continued to act as the de facto representative of all of Ventura Avenue's lessors, as he had done since he and Dabney brought Shell into the field in 1916. But as Associated developed its properties unevenly, Lloyd attracted the ire of lessors. Lloyd's exhortations to Associated's executive managers to treat all leases equitably, however, failed to quell lessors' dissatisfaction. As a result, Lloyd and Dabney became targets of legal action that threatened to derail the "rational" development of the field.

The first lessor to sue Ralph Lloyd, however, no longer had a royalty interest in the field. On 10 February 1927, Ira Gosnell and Lena Bowyer sued Lloyd and the Lloyd Corporation on behalf of the estate of their father, T. B. Gosnell. Two years earlier, Gosnell, then seventy-six years old, had sold the mineral rights to his property to Lloyd for $72,000. Recall that Dabney and Lloyd had leased the Gosnell property from T. B. and Caroline Gosnell and their children in October 1913 and had subleased it to Shell as part of the June 1916 agreement.

Throughout the first half of 1925, Shell's crews struggled with saltwater intrusion into all four wells that they were drilling, even as Associated was setting off Ventura's boom from its wells on the adjacent Lloyd lease. Meanwhile, Gosnell's royalty income from five wells that Shell had completed since 1922 was dwindling, from a monthly high of $3,100 in 1923 to $269 in April 1925.[5]

By June 1925, T. B. Gosnell was eager to sell his mineral rights. For years, he had paid close attention to Shell's activities. A frequent visitor to the oil field, Gosnell also called on Ralph Lloyd in his downtown Los Angeles office to review the daily reports that the latter received from Shell. During these office visits, Gosnell filled notebooks with Shell's drilling and production data and discussed Ventura Avenue field operations with Lloyd or, in the latter's absence, anyone else who happened to be around. He also followed the industry in newspapers and trade publications. Once Associated brought in Lloyd No. 5 in October 1922 and began to develop the Lloyd lease more intensively, as we have seen, Gosnell increasingly complained about Shell's performance on his lease and the size of his royalty checks. Sometime prior to June 1925, Gosnell offered to sell his mineral rights to Lloyd for $150,000. Lloyd demurred.[6]

On 2 June 1925, T. B. Gosnell, accompanied by Ethel, his ex-wife, with whom he had been living since 1921 in Los Angeles pursuant to an arrangement by which she was to receive his property upon his death, met with Lloyd to negotiate the sale of the mineral rights to his property. Gosnell opened the parley at $100,000. Lloyd countered with $72,000, payable in monthly amounts of $1,000. Gosnell balked, insisting that Lloyd pay at least $15,000 up front. Lloyd agreed to put down $15,000 and pay $1,000 per month thereafter. After mulling over the offer for several days with Ethel and his attorney, Gosnell accepted the offer. On 17 June, Gosnell and Lloyd met at the Pacific-Southwest Trust and Savings Bank Building on the corner of Sixth and Spring Streets in downtown Los Angeles and executed a sales agreement.[7]

Subsequently, Shell enjoyed rather more success in developing the Gosnell lease. Just two months after T. B. Gosnell and Ralph Lloyd concluded their agreement, Shell drilled wells that ultimately tapped a deeper, more productive strata of oil sand. In August 1926, Lloyd received about $38,000 in royalties—the highest amount that Shell had paid to date on the lease. It was clear to Gosnell that Lloyd was receiving significantly more royalty income from the lease than he ever had realized. And so, in September, he and Ethel asked Lloyd to advance $18,000 on their agreement. Lloyd obliged.[8]

By February 1927, Ralph Lloyd had paid T. B. Gosnell $51,000 of his $72,000 obligation. Gosnell, however, no longer could bear his seller's remorse. On 9

February 1927, Ethel Gosnell, Ira Gosnell, and Lena Bowyer, acting legally on Gosnell's behalf, but likely with his full consent, given subsequent court proceedings, entered into a pact to recover Gosnell's interest in the lease, which they valued at $3 million. For her part, Ethel would receive one-fifth of any settlement. Asserting that their father was incompetent, Ira and Lena secured an appointment as guardians of his estate the following day and promptly filed suit in Ventura County. Seeking to rescind the 17 June 1925 agreement, they charged Lloyd with deceit, fraud, and misrepresentation. Further, they alleged that T. B. Gosnell lacked the mental capacity to enter into the contract, and that, knowing this and also Shell's impending plans to develop the property, Lloyd breached an alleged fiduciary responsibility by pressuring the septuagenarian to sign away his mineral rights.[9]

Before Superior Court Judge Edward Henderson, Ralph Lloyd and his attorneys countered that T. B. Gosnell was fully competent to attend to his business affairs when he agreed to sell his mineral rights on the advice of counsel. Gosnell's only infirmities, Lloyd contended on the stand, were physical: he walked with a limp and spoke with an impediment. Judge Henderson agreed that Gosnell "was not a person of unsound mind nor likely to be deceived or imposed upon by artful or designing persons" and "acted with full knowledge, voluntarily and with advice and services of counsel." Further, the plaintiffs introduced no evidence that a fiduciary responsibility existed between Lloyd and Gosnell. And so the judge ruled in Ralph Lloyd's favor. To date, Lloyd had earned some $350,000 in royalties from the lease. The children appealed. In May 1932, the Second District Court of Appeal upheld Judge Henderson's ruling.[10]

As Ralph Lloyd noted in the wake of the resolution of the case, it was not unusual for sellers of mineral rights to suffer remorse in light of subsequent development of their property. Indeed, he and Dabney found themselves in T. B. Gosnell's position when they sold an option on thirty-five acres in what became the Inglewood field in the Los Angeles Basin. As of July 1932, the property had produced 12.5 million barrels of crude oil, meaning that he and Dabney each had lost some $1.5 million. He and Dabney had to live with their decision; Gosnell would have to live with his.[11]

Fighting for the Price of Ventura Avenue Oil

Oil production in the Los Angeles Basin—so overwhelming that in 1923 it accounted for 20 percent of global output—had sent the price for a barrel of benchmark 27° API crude oil produced at Signal Hill plummeting, from $1.50 to $0.68 per barrel from June 1921 to December 1923.[12] Over the next three years, prices recovered, as automobile-driven demand surged. At the same time, production in the Los Angeles Basin flagged as relentless town-lot drilling damaged reservoirs. Discoveries of oil pools in California, Oklahoma, and Texas sent prices in California falling by one-third in 1927. In California, the development of deeper zones of production at Santa Fe Springs and Signal Hill together with the discovery of gigantic fields, namely, Elwood in Santa Barbara County and Kettleman Hills in the San Joaquin Valley, renewed the pressure on prices. Industry leaders who earlier had dismissed the idea that government intervention might offer a remedy for a weak market now reversed their stance, initiating a dialogue that would lead to action at the state level. In this context, Ralph Lloyd rallied Ventura Avenue's lessors to the end of obtaining the highest possible price for their royalty oil. In so doing, he demonstrated that symbiosis characterized relations between lessors and major operators in much the same way that it governed relations between majors and independents.[13]

Ralph Lloyd and Joseph Dabney sought a price for Ventura Avenue royalty oil that equaled the highest prices posted in the state—those paid for the high gravity oil pumped from the Signal Hill and Rosecrans-Dominguez fields in the Los Angeles Basin.[14] In the context of softening prices, Lloyd and Dabney chose an approach that required the former to gain, and retain the cooperation of Associated's common lessors who were increasingly competing for the major's attention.

Historically, the crude oil found in Ventura County was heavy and similar in kind to petroleum extracted in the San Joaquin Valley and most often used as fuel oil for locomotive engines. The prices that operators received for it reflected its low specific gravity, which typically ranged from 10° API to 18° API. In contrast, the oil that Associated and Shell produced at Ventura Avenue was lighter, falling in a range from 26° API to 30° API.[15] This compared favorably to the 27° API oil pumped at Signal Hill and the 32° API oil produced at Rosecrans-Dominguez on which refiners based the premium prices that they paid to producers. In the fall of 1925, Lloyd began negotiating a series of oil sales agreements with Associated and Shell to secure these premium prices for Ventura Avenue's crude oil.[16]

Provisions in Ventura Avenue leases regarding the price of royalty oil were vague, allowing both lessor and lessee plenty of room to maneuver. The Lloyd lease, for instance, stated that "the value at the town of Ventura" would determine the purchase price. The Gosnell, Hartman, and McGonigle leases pegged the price at either "market price" or "market value" or both.[17] In the wake of the completion of Associated's wells that marked a breakthrough in production in the field, Associated and Shell continued to set the "market price" of a barrel of Ventura Avenue crude oil based on what Standard Oil Company of California and Union Oil Company were paying for petroleum produced from shallow wells and oil seeps found in largely inaccessible parts of Ventura County, resulting in prices that were 10–15¢ less than the Signal Hill price for similar-gravity oil. For Lloyd and Dabney, this was unacceptable.[18]

Ralph Lloyd first negotiated with Shell, seeking a price equal to the Signal Hill price plus a premium for crude oil that was lighter than 27° API. He also sought an annual royalty that applied to all of Ventura Avenue's lessors as a unit. As he noted to his sister Roberta, if the lessors succeeded in establishing a price for the field acting as a group, it would "probably be without parallel in the history of the oil business in California." Realizing the goal required all the lessors to stand as one. As Lloyd wrote to Ira Gosnell, who had yet to sue him: "We could hardly get a hearing if we were shot to pieces in our control of the different properties as is the case in most of the California fields." In November 1925, Shell agreed to a month-to-month sale of royalty oil at Signal Hill prices for 14–32° API oil and Rosecrans-Dominguez prices for 32° API and higher oil. Lloyd did not get the annual contract that he sought, but the agreement suggested what Ventura Avenue's lessors might achieve with a united front.[19]

When, by the end of the year, Lloyd negotiated a two-year sales agreement with Associated that provided for the sale of royalty oil at the highest Los Angeles Basin prices plus a 7.5 percent bonus for each of the leases covered in the contract, Lloyd and Dabney turned to Shell for even better terms. Ultimately, Shell offered to enter into a two-year contract that provided a 7.5 percent premium over Los Angeles Basin prices, but only for the Taylor lease. With the price of oil rising in early 1926, Lloyd and Dabney blinked, accepting a deal that treated the Taylor and Gosnell leases differently.[20]

The fall in oil prices during 1927 created a yawning gap in expectations between lessors and operators regarding the terms of new sales agreements. Indeed, the two interests were unable to bridge their differences and no sales agreement was executed until the fall of 1928.

In this round, Ralph Lloyd first approached Associated to conclude a deal that he could take to Shell. Both sides agreed that 1927 was a dismal year for the industry. It was just about their only point of agreement. For his part, Lloyd was far more optimistic about the prospects for the industry in the coming year than Associated's executives. Convinced that falling output in the most important fields would stabilize the California market, Lloyd sought to maintain the 7.5 percent premium over the highest Los Angeles Basin prices in a new sales agreement. William F. Humphery, who became Associated's chairman of the board in August 1926 when Tide Water Associated Oil Company was created as a holding company for Associated and Tide Water Pipe Company and the company's president in 1927, contended that the market was oversupplied and would remain so for the foreseeable future. He countered with a 3.5¢ per barrel premium.[21] Further, he insisted that Associated's lessors sign a five-year deal on these terms. Now "is a time for cooperation," he wrote, "not only on the part of the producers, but on the part of all who are interested in the industry." He regretted that there was "such a wide difference of opinion on the value of the royalty oil."[22]

Associated's insistence that prices posted at Ventura, not Signal Hill, serve as the basis for the price it paid to its Ventura Avenue lessors for their royalty oil convinced Ralph Lloyd that the company, in collusion with Shell, was bent on "breaking the price for Ventura oil." With oil from all three of California's producing regions flooding the market, time seemed to be on the side of the operators. When, on 15 August 1928, Standard Oil Company of California increased the prices it offered to purchase gasoline-bearing light crude oil in the state, Associated's prices for Ventura Avenue royalty oil fell below the posted prices at Signal Hill. Lloyd urged Alice Grubb, who owned the mineral rights to the Taylor lease, not to accept anything less than the Signal Hill price for her royalty oil. Conceding that they had "had quite a struggle with the companies in order to hold up the oil value," he nonetheless was adamant that "by standing together I am sure we can protect ourselves against depreciation of the fine oil that is produced at Ventura."[23]

In September 1928, William Humphery arrived in Los Angeles to discuss the price of Ventura Avenue royalty oil with Ralph Lloyd. With statewide production for the year now on course to match 1927's output and with prices stabilized for the moment, the two men met halfway, agreeing on a price of royalty oil equal to the highest Los Angeles Basin price plus three percent. A sales agreement, effective 1 January 1929, was the last one concluded between the parties before crude oil from East Texas swamped the national market.[24]

Controlling Oil and Gas Production at Ventura Avenue

Well before the State of California acted formally in 1929 to limit crude oil production by targeting natural gas waste at the wellhead, the three major operators at Ventura Avenue cooperated to conserve natural gas pressure in the reservoir. Ralph Lloyd supported their efforts because they promised to lengthen the productive life of the field, but nonetheless he was quick to caution Associated, which controlled the leases in which he was directly interested, that natural gas conservation provided no excuse for not meeting the terms of its leases. In pressing his leasehold interests, Lloyd embodied the contradictions in statewide efforts to stabilize an industry plagued by overproduction.

After nearly two years of blowing more natural gas into the air from their Ventura Avenue operations than their counterparts were doing anywhere else in the state, operators acted to prevent the dissipation of a resource that was the lifeblood of their operations. In December 1926, Associated, GP, and Shell formed the Ventura Conservation Committee and charged the engineers whom they appointed to it with coordinating a reduction in the field's ratio of natural gas expended to crude oil produced. Operators used natural gas to bring oil to the surface. To operators, this gas was not "wasted"—a point of controversy between regulators and producers. But without additional infrastructure to deliver natural gas by pipeline to customers in the Los Angeles Basin—where the overwhelming majority of potential users were located—the only practical means of lowering the overall gas-oil ratio was to target wells with high ratios for action. In their first attempt to do so, the three majors reduced the amount of gas that they flared at the wellhead by 7.23 million cubic feet per day without decreasing appreciably the amount of oil that they pumped. Achieving further reductions in the gas-oil ratio would require the companies to bring less oil to the surface, however. In May 1927, Associated, GP, and Shell agreed to reduce their production collectively by one-third. The resulting decrease in the field's daily output by 15,700 barrels promised to reduce overall gas production by 60 million cubic feet per day.[25]

Associated, GP, and Shell had the means to sell or store any petroleum that they pumped. But for the moment, at least, they could neither sell nor store the natural gas that they produced at Ventura Avenue. They had every incentive to produce surplus oil, even if it meant blowing the associated gas into the air.

After only four months, the operators ended their scheme. Daily oil and gas production at Ventura Avenue jumped immediately by 8,500 barrels and 40 million cubic feet, respectively.[26]

Within a month, California's oil and gas supervisor asked Associated, GP, and Shell to renew their conservation efforts. Because they agreed with Ralph Lloyd that maintaining gas pressure in the reservoir held the key to long-run maximization of oil production, the three operators had an economic incentive to coordinate their conservation efforts. And so, under a second proration plan effective 25 October 1927, they agreed to reduce total gas production by 25 million cubic feet per day.[27]

Achieving this goal on technical criteria alone—targeting wells with the high gas-oil ratios, for instance—easily could result in setting inequitable crude oil production quotas. As Shell's chief production engineer explained, a small operator with only high gas-oil-ratio wells would have to cut its crude oil production by a greater percentage than a large, adjacent operator with a mix of high and low gas-oil-ratio wells. Ventura Avenue's major operators enjoyed little success in achieving the goal of the second proration plan, foreshadowing the obstacles that interested parties would face in reaching agreement on statewide conservation measures. On 7 March 1928, Associated, GP, and Shell scrapped it in favor of a third, more direct, approach that prescribed a 20 percent cut in crude oil production and charged an engineering subcommittee with developing a detailed plan that targeted wells with the highest gas-oil ratios for action. On the subcommittee's recommendation, operators shut in twenty-four wells entirely.[28]

At the same time, momentum had gathered behind a statewide policy to restrict the amount of natural gas that could be emitted to the air at the wellhead. By 1929, California operators collectively were blowing more than 620 million cubic feet of natural gas to the skies every day, prompting a public response. In May, Governor C. C. Young signed into law a bill that industry leaders had proposed. The Oil and Gas Conservation Act empowered the Department of Natural Resources to determine whether "unreasonable waste" was occurring or was about to occur, and, once it had made such a determination, to direct the California oil and gas supervisor to order the responsible parties to cease and desist. The oil and gas supervisor could sue uncooperative parties. Supporters of the law, which became effective on 31 August, hoped that by giving State officials the tools that they needed to force operators to cooperate, the industry would solve the overproduction problem. The law, however, did not have an answer for lessors, such as the Hartman family, who insisted that operators meet or exceed the drilling obligations in their leases, glut or no glut.[29]

Getting Sued, Part II

A suit filed by the Hartman family, charging Associated with failing to perform as a prudent operator, bared the tensions between Ventura Avenue's leading operator and its common lessors in the context of statewide conservation efforts. Successful conservation outcomes depended not only on cooperation among operators. Lessors mattered too. The Hartman's legal action also showed how competition between lessors strained the bonds of cooperation that Ralph Lloyd had forged among them in their efforts to secure the highest possible price for their royalty oil.

Protecting a property from drainage was one of several implied covenants of an oil lease under the prudent operator standard. To be sure, *Brewster* and other early cases allowed room for an express provision to displace an implied covenant if it addressed the same matter. All of Associated's leases and subleases obligated the company to drill a certain number of wells within a defined period of time. At the same time, the instruments allowed the operator to drill as many wells as it wished. This provision addressed another implied covenant, namely to develop the property to an extent that was reasonable under prevailing conditions. Express drilling provisions in the leases did not displace this implied duty, much less the implied covenant to prevent drainage. Yet Associated's managers equated their fulfilling the express requirement to maintain a minimum of drilling rigs in operation with their satisfactory performance of the duty to prevent drainage, thereby exposing the company to legal action.[30]

The members of the Hartman family grew increasingly impatient as they watched Associated drill wells on the Lloyd lease without, in their view, drilling a sufficient number of offset wells to protect their property from drainage. At the same time, the terms of the lease that Dabney and Lloyd had executed in October 1913 expressly required Associated to drill only ten wells. And so the company's executives insisted all along that they were doing more than enough to satisfy the Hartmans. By adhering to a policy of abiding by the express provisions of the lease agreement, however, they ensured that Associated would not placate the Hartmans, who would anchor their legal argument in the implied duties of the prudent operator.

As a beneficiary of Associated's intensive development of the Lloyd lease, Ralph Lloyd nevertheless adopted a mediating role in the protracted dispute. For he saw himself as conductor of an orchestra, featuring both operators and lessors whose performances had to be harmonized to the end of maximizing production in the long run across the entire field. For years, he had expressed more confidence in the potential of the leases that lay to the north and east of the Lloyd lease than did the geologists of the major operators, and he continued to do so. More than reputation was at stake. Were Associated to prove the Hartman lease to be as productive as Lloyd thought it might be, then perhaps the company would invest in extending the eastern limits of the field on property that he and Dabney had leased.

When Shell surrendered its sublease on the Hartman property in October 1924, Ralph Lloyd stood alone among the parties with an interest in the Ventura Avenue field in believing that the ranch had substantial potential as a producing property. For, as he cautioned Katherine Hartman, the geologists of the companies operating at Ventura Avenue were convinced that the anticline broke off abruptly south of her property and so concluded that little would be gained by drilling on land that lay to the north of its apex. F. W. Hertel, Associated's resident geologist minced no words: the Hartman lease was "*very* inferior" to the Lloyd lease.[31] Still, Lloyd counseled patience, urging the aging matriarch of one of Ventura's pioneer families not to sign a lease with a company that likely would make hash of a prospect well, thereby ruining the chances of another California major taking another look at the ranch as a potentially producing property. At the same time, the Athens and Rosecrans-Dominguez fields in the Los Angeles Basin were attracting the attention of the industry: "When oil again becomes scarce and capital is justified in taking the risk," he assured her, "then I am sure I can get operations started again on your property." Meanwhile, in the interest of proving his judgment to be correct, he convinced his partners, Joseph Dabney, A. M. Buley, and E. J. Miley, to pay $10,000 for a fifteen-month extension on the requirement in the 1913 lease to drill a well, beginning 1 January 1925, thereby buying Lloyd time to interest Associated in testing the property.[32]

As this amendment to the lease coincided with Associated's breakthrough on the Lloyd lease, Lloyd was able to persuade Associated to sublease the Hartman property, which the major did on 20 January 1925. But L. J. King, Associated's general superintendent in Ventura, hesitated to commit his company's resources to develop the property until his crew completed Lloyd No. 15, which would indicate the potential production on the Hartman lease and help him to establish the location of the first exploratory well. When, in October, Associated completed the well under enormous gas pressure, Lloyd felt vindicated, writing to the Hartmans, "Against the opinion of practically all the geologists of the different companies . . . this well is demonstrating my contention that . . . there is oil on your property." With this showing, Associated moved ahead on the Hartman Ranch, completing its first well in June 1926.[33]

Proving the Hartman property convinced Ralph Lloyd that Ventura Avenue constituted "even a larger project than any one of us realize," putting Associated in better position than any other operator to develop one of state's fields "to get the maximum of production at the minimum of cost." [34] Doing so demanded that Associated maintain high natural gas pressure in the reservoir as it developed the Hartman and Lloyd leases in concert. In hindsight, Lloyd told Associated's chief geologist, he and Dabney held the Hartman lease after Shell surrendered its sublease back to them to prevent a company other than Associated from stepping in and "practically [ripping] open the very bosom of the Ventura Field and [bleeding it] of enormous quantities of gas." He hammered home the point: "A wise use of the gas by proper drilling and locating of the wells" was crucial to production maximization and cost minimization.[35]

Both geological and topographical conditions, on the one hand, and public policy on the other, motivated Associated to do no more than adhere strictly to the express provisions in the lease. At the same time, Associated sought to placate the Hartman family. In the fall of 1928, William Humphery asked for concessions on drilling obligations from its other common lessors in return for drilling wells in excess of lease requirements on the Hartman property. Lloyd and Dabney were willing to allow Associated to postpone the drilling of exploratory wells that tested the productive limits of the field to the east so that Associated might devote its resources to satisfying the Hartmans. They were also willing to postpone the drilling of offset wells along the Hartman-Lloyd lease line—but only to conserve gas pressure in the reservoir, in keeping with the recommendations of California's oil and gas supervisor. At the same time, as the de facto representative of Associated's common lessors, Lloyd insisted that the company consider each of its Ventura Avenue leases separately and develop and protect each one of them in an equitable manner whenever drilling demonstrated commercial production. In January 1929, he served notice: "We are only expecting from you for our leases the customary drilling and protection that are given to other properties . . . by other interests." Conservation of natural gas pressure, of course, was critical to producing the most crude oil as possible from the field in the long run, but Lloyd and Dabney would not let

Associated (or Shell, for that matter) use the front-page issue as an excuse not to adhere to the implied duties and express provisions of its leases.[36]

On the face of it, Ventura Avenue's lessors had little cause for complaint in the performance of Associated, GP, and Shell. Having achieved a reduction in the gas-oil ratio for the field, the three operators ratcheted up the intensity of their development efforts. As of 1 September 1929, when the state began to enforce the conservation law, crews were drilling forty-three wells, up from sixteen at the beginning of the year. Production of crude oil reached 65,000 barrels per day in August, on the way to 21 million barrels for the year—a level of output that operators would not achieve again for another two decades. To be sure, the 135 million cubic feet of natural gas that operators were "wasting" on a daily basis attracted the attention of regulators, but 1929 was *the* banner year of the interwar period for the field. Yet Associated would be unable to placate the Hartmans merely by slightly exceeding the express drilling requirements of the lease.[37]

In the context of operators' reinvigorated drilling programs, the Hartmans increased their pressure on Associated to develop their property more intensively than the lease expressly required because the family had decided to develop a residential subdivision in Ventura. Fearing a slowing of work on their lease—the highest gas producer in the field—in the wake of the passage of the state conservation law, the Hartmans informed Ralph Lloyd for the first time of their willingness to sue for damages related to drainage from their property. Seeking to head off the Hartmans, Lloyd recommended that Associated immediately resume drilling a well that it had shut in and spud a new one on the Hartman lease. He and Dabney would postpone the drilling of an offset well on the Lloyd lease until Associated brought in the new well.[38]

Associated refused to consider the suggestion. The operator was already exceeding its contractual drilling obligations to the Hartmans in the interest of maintaining "friendly relations" among lessors, President William Humphery noted. Passage of the conservation law alone justified not accommodating the latest request of the Hartman family. Argued Humphery: "We should all try to prevent unnecessary drilling." In a statement that perhaps he came to rue, he insisted that Associated was more concerned with maintaining the goodwill of its common lessors than the consequences of any threatened litigation.[39]

When, in 1933, the Hartman Ranch Company sued Associated for breach of the implied covenant in the 1913 lease to protect the property from drainage, the company had completed eleven wells on the Hartman lease. In the same period, that is, from 1925 to 1933, Associated drilled seventy wells on the Lloyd lease. To 1 January 1933, Associated produced 41,721,567 barrels of crude oil from the Lloyd lease and 3,727,537 barrels from the Hartman lease. To be sure, the Lloyd lease was nearly ten times larger: 1,425 acres, compared to the 148 acres that comprised the Hartman lease. And so, for purposes of comparison before the Superior Court in Ventura County, the Hartman family's expert witness selected a portion of the Lloyd lease of comparable size and location. In this area, Associated drilled twenty-three of its seventy wells. From 1 March 1929 to 1 March 1933—the period for which the Hartmans sought damages—Associated produced 9,036,493 barrels from these wells and 2,269,087 barrels from the Hartman lease. In November 1933, the jury returned a verdict in favor of the Hartmans for $593,700—an amount equal to almost 20 percent of Associated's net income for the year. The company deposited $600,000 in an escrow account on appeal.[40]

Ralph Lloyd wanted to maintain the goodwill of the lessors whose property bordered the Lloyd lease—not just the Hartmans but also the McGonigles and his relatives who controlled VL&W, the family business. After all, he was the indirect target of their growing impatience with Associated. For this reason, Lloyd repeatedly pointed to the language in the lease agreements to cajole Humphery to develop these properties more intensively. For the better part of a decade, Associated would fail to satisfy its lessors. Indeed, it would lose its momentum in the field, its "cash cow," never to regain it.

From Conservation to Voluntary Curtailment

To implement California's conservation law, R. D. Bush, the State's oil and gas supervisor, circulated a draft Cooperative Gas Conservation Agreement to operators, lessors, and lessees at Santa Fe Springs, Seal Beach, Signal Hill, and Ventura Avenue—the four fields with the most pressing waste problems—and solicited their comments. In it, the State proposed to pool natural gas production within a field and fill supply contracts on a pro rata basis. Natural gas that could not be distributed to Southern California Edison or other users would be injected into depleted producing zones or stored. The draft agreement did not prescribe how operators might otherwise dispose of their surplus gas, but operators who could not dispose of surpluses by one of these two means faced little option but to curtail operations, even as officials insisted that realizing such an outcome was not their intention. During the week of 9 September 1929, James S. Bennett, legal adviser to Department of Natural Resources Director Fred G. Stevenot on oil and gas issues in Southern California, filed an injunction to restrain operators at Santa Fe Springs—the worst offenders—from unreasonably wasting natural gas, that is, blowing it into the air. In it, he set a limit of 2,500 cubic feet of gas per barrel of crude oil produced as the basis for calculating excessive gas production, although both Bennett and Stevenot explained that the measure was a provisional one and applied only to Santa Fe Springs. In short order, Bennett acted similarly against operators in the other targeted fields.[41]

Ralph Lloyd opposed the conservation law because it aimed to limit production in the field at a time when the Hartmans and other lessors were clamoring to increase it. Indeed, as he complained to C. W. Norton, his man in Portland, it "would practically stop all oil development at Ventura" and make it impossible "to ever develop the field on a large scale." He insisted that operators and lessors would cooperate in the absence of government interference to conserve natural gas in the interest of lengthening the productive life of the field. Specifically, and perhaps with a view to extending the proven limits of the field in an eastward direction, he suggested that operators and lessors could agree to avoid new development along the apex of the anticline that divided the Hartman and Lloyd leases. The reaction of the Hartman family to this recommendation is unrecorded.[42]

Beginning on Saturday, 21 September 1929, Supervisor Bush held hearings in Ventura on the draft agreement. He heard testimony from Ventura Avenue's operators and lessors and public officials, including C. C. Brown of the California Railroad Commission and the deputy district oil and gas supervisor. Bush was especially interested in learning how operators proposed to meet the expectations of the law without severely cutting into output, given the fact that some of the field's older wells were registering gas-oil ratios as high as 15,000 cubic feet per barrel of crude oil produced and recent wells had required as much as 45,000 cubic feet of natural gas to bring their first barrel to the surface. Witnesses for the oil companies prepared Supervisor Bush for their recommendations by explaining in unison that Ventura Avenue was unlike any field in California. And so, as former Shell general superintendent William C. McDuffie, who now supervised production operations for Pacific Western Oil Corporation, put it, "The Avenue field always must have a higher gas-oil ratio than other fields." The operators therefore agreed that applying a single gas-oil ratio across the field, whatever the number, would be an unfair and inefficient means of compliance. Rather, they argued, a dynamic program of identifying and shutting in high-ratio wells would reduce natural gas waste without appreciably curtailing oil production. Joseph Jensen, Associated's chief petroleum engineer, explained that Associated, GP, and Shell had demonstrated the efficacy of this approach. In fact, Jensen averred, by selectively killing high-ratio wells, operators could recover fully 80 percent of the oil in the field without wasting any natural gas. Jensen, McDuffie, Lloyd, and others cautioned, however, that shutting in wells brought in under high gas pressure could damage them irreparably. For doing so required pumping the well bore full of muds. Testified Jensen: "Nobody can forecast what the result will be. The experience in the past has been that many wells so shut in have never returned to their original state of productivity when reopened."[43]

On 3 October, Supervisor Bush issued an order for Ventura Avenue that pleased none of the interested parties. It allowed each well to generate 5,000 cubic feet of waste gas per day. Further, operators could not use more than 20,000 cubic feet of natural gas to produce a barrel of crude oil in identified areas of high pressure; in all other areas of the field, they could use up to 10,000 cubic feet of gas to produce a barrel of crude oil. It was self-evident to Bush that wells producing above the 10,000-cubic-foot limit were wasting natural gas unreasonably, even if operators were not releasing it into the air. Drilling campaigns might continue under the order, but as operators completed wells, they would have to shut in older wells to comply with the order. Operators could store as much gas as they wished. Moreover, they could continue to supply natural gas on a pro rata and seasonal basis to Los Angeles, as they had been doing. At the same time, they could no longer distribute natural gas to Santa Barbara and San Luis Obispo Counties; operators in the newly prolific Elwood field would meet that demand. Complying with the order, according to attorney Bennett, would reduce daily gas production at Ventura Avenue by half. At the same time, crude oil production in the field would decline by no more than 25 percent. The intentions of regulators aside, the order threatened to derail operators' developmental drilling programs, which would please no lessors.[44]

The response of Associated, GP, and Shell to Supervisor Bush's order demonstrated that California's seven majors were no less dismayed by the implementation of the conservation law than the independents who organized the Independent Oil Operators of Southern California when the two groups of operators failed to agree on the distribution of so-called waste gas generated in the four offending fields.[45] Speaking for Shell, Julian P. Beek declared that the order was "unreasonable and arbitrary," and that the oil and gas supervisor had no legal standing to issue it. That is, the law, as Beek interpreted it, targeted as "unreasonable waste" only natural gas blown into the air, not gas "naturally lost in the conduct of business." Therefore, fixing gas-oil ratios for the field was unwarranted. Further, as Shell saw it, the law granted the oil and gas supervisor no authority to address natural gas distribution. Associated's legal representative, Harrison Guio, stated that the company also believed that Bush lacked the authority to issue "a blanket order fixing quantities of gas production." On 11 October, the district deputy oil and gas supervisor posted notices in the field. Before the order went into effect five days later, however, the three majors followed through on promises to appeal the order if the State moved to enforce it.[46]

Ralph Lloyd led the protests of Ventura Avenue lessors against the order. To C. W. Norton, he was succinct: "We think the law is unconstitutional and its application arbitrary and unjust." And so he and his fellow lessors found common ground with independents who contested the constitutionality of the law. Through his attorney, Frederick W. Kincaid, Lloyd asserted that operators who complied with the order would breach their leases and that California had no right to put them in such a position. In any case, he called on Associated, GP, and Shell to disregard the order. Compliance would "be treated as a direct break of agreement with us." Lloyd soon softened his stance, informing the three operators that he and his fellow lessors would abide by any agreements that the companies might make with the State for a two-month period, beginning 1 December, as long as they continued to meet their lease obligations.[47]

Ralph Lloyd's stance provoked legal advisor James Bennett, who charged the oilman with trying to keep a "stranglehold" on the field. He made it clear that the State would not tolerate it: "We have heard time and again that oil companies could not do this or that because of lease terms. But we will take care of Mr. Lloyd."[48]

The fight over the conservation law shaped Ralph Lloyd's perspective on "government interference" in the market. He would never wholly abandon his associationalist views and would continue to be critical of America's antitrust regime for keeping "the elements of a great industry apart" when cooperation was needed. And he would argue that government should step in to enforce agreements reached among private parties by mutual consent. Yet his inclination that private enterprise should handle problems in the market to the greatest extent possible was clarified and reinforced by the interjection of the State of California into the oil market. Lloyd would move ever closer to a conviction that laissez-faire should prevail across all sectors of the economy. With Bennett pressing on, he complained to Norton that the State was "trying to usurp the right to control our oil fields and even the production of every individual well." In this and other statements on the conservation regime that the State of

California sought to implement, Lloyd developed the vocabulary that he would mobilize in an unyielding critique of the New Deal, its tax policy in particular.[49]

In December 1930, the California Supreme Court ruled that the conservation law was constitutional. But for the sponsors of the legislation it was a hollow victory. Operators deployed new technology to reduce the amount of gas they blew into the air—the definition of "unreasonable waste," the court held—and built infrastructure to deliver more of the natural gas that they captured to urban consumers. Thus, they were able to "absolutely swamp the Los Angeles Basin and other fields" with oil without wasting gas, as the California Oil and Gas Association noted.[50] The California Supreme Court denied that the law intended to stabilize the oil market by limiting production, but such an outcome, however indirect the means of achieving it may have been, was certainly what the industry's seven majors initially had in mind when their lawyers crafted the bill. The editors of *California Oil World* summed up the problem: "Drilling can not be stopped—it is difficult even to regulate it."[51] The situation at Ventura Avenue showed that lessors advocating their interests compelled operators to drill, thereby undermining the cooperative efforts of the latter on behalf of conservation.

Voluntary Curtailment and the Development of the Ventura Avenue Field

Early in 1929, California's oil industry formed the Central Committee of California Oil Producers and established a voluntary curtailment program with the approval of policymakers in Sacramento. Operators agreed to allocate actual production on a field-by-field basis as a percentage of "actual and effective potential" production, that is, the amount of crude oil that operators could produce if they pumped their wells continuously at maximum flow. The committee hired so-called umpires to estimate statewide demand and establish production "allowables," or quotas, for individual fields and their operators. The logic of curtailment, as Ralph Lloyd saw it, required operators to maintain their developmental efforts to compensate for normal declines in production to protect their "allowables." That is, curtailment of production would be achieved through the coordinated shutting in of wells upon their completion rather than the mandated cessation of developmental drilling, however detrimental the consequences of shutting in completed wells. The logic of curtailment provided the basis of Lloyd's relations with Associated during the 1930s.

The industry suspended the voluntary curtailment program with the passage of the Oil and Gas Conservation Act—at which time Ventura Avenue's operators had failed to reach agreement on curtailment, as we have seen. In the spring of 1930, California's producers revived it when it was becoming clear that the conservation law would not achieve the cuts in production that it had promised. Umpires Neil Anderson and H. P. Grimm set the state's daily allowable production at 609,000 barrels, effective 1 March, a figure that held steady for the quarter. Their decision allowed operators at Ventura Avenue to produce about 45,000 barrels per day—roughly half of its potential output and a considerable drop from the salad days of the summer of 1929.[52]

Ralph Lloyd was adamant that operators not curtail development along with production. He spoke for all of Ventura Avenue's lessors in insisting that Associated and Shell, in particular, develop the field at least at the pace expressly demanded by their leases. As he did on many issues, as we have seen, Lloyd rallied the Ventura Chamber of Commerce to support his position, leading business owners and a committee charged with preparing a survey of local employment prospects for municipal officials on a tour of the field. He also led efforts among independents in resisting a reduction in the state's allowable to 550,000 barrels per day—a fight that he lost when the umpires fixed allowable production at that level in September 1930. Nevertheless, the logic of curtailment continued to frame his Depression-era relations with Ventura Avenue's largest operators.[53]

Of course, given his leasehold interests, Ralph Lloyd was financially more interested in Associated's field operations than in Shell's. At the same time, he kept one eye firmly fixed on the latter's operations under voluntary curtailment, for Shell set the bar that Lloyd used to evaluate Associated's performance.

Shell's relatively more intensive development of its leases prompted Lloyd to urge Associated to defend its properties with vigorous and persistent drilling from the fall of 1928 onward. At that time, he and Joseph Dabney had agreed to a three-month, renewable postponement of drilling obligations on the VL&W lease and on the west end of the so-called Dabney-Lloyd lease that lay adjacent to the difficult McGonigle lease, but to the east of the field's proven area, so that Associated might defend its western front by drilling more than a dozen wells. To keep pace with Shell and meet its lease requirements to drill offset wells, Lloyd estimated that Associated would have to deploy between twenty and forty rigs in the field during 1929. Similarly interested lessors across the state played no small role in negating the efforts of industry leaders and regulators to solve California's overproduction problem.[54]

With California's conservation law in legal limbo at the end of 1929, Ralph Lloyd anticipated the reinstitution of the suspended voluntary curtailment program. He criticized Associated's failure to keep pace with Shell, noting that it was maintaining only ten rigs in operation to Shell's twenty-three, suggesting that Shell understood better than Associated the logic of curtailment. In reply to Associated's assertion that its present drilling program was protecting the interests of both the company and its lessors, Lloyd noted that the operator was neither developing deeper zones of production on the Lloyd lease nor preventing Shell from draining the property from the Gosnell lease. For the moment, Associated could claim 75 percent of Ventura Avenue's "actual and effective potential"—a share that would surely diminish on present trend. J. G. Jenkins, general manager of Associated's producing division, returned from surgery early in 1930 to clarify and defend his company's position: Associated would base its drilling program on the potential of its properties rather than on the number of Shell's rigs in operation. Lloyd reiterated what he had repeatedly told William Humphery, namely that Associated needed to do more to protect its properties from operations on adjacent leases. He forewarned: "The period of time during which [the company] is privileged to drill for oil [is] limited." Were it to fail to meet its obligations, the operator "might find itself in an unexpected position."[55]

Once crude oil from East Texas flooded the market in the spring of 1931, sending the price of Ventura Avenue oil plummeting to 28–35¢ per barrel, Ralph Lloyd privately threw his support behind the beleaguered voluntary curtailment program. With the industry producing 25,000 barrels of crude oil in excess of the daily statewide allowable of 500,000 barrels, he also helped to organize the Oil Producers Sales Agency of California, of which he was named president. The agency aimed to coordinate production among independents and bargain with refiners to market it. By redressing the imbalance between the sellers and buyers in the crude oil market, Lloyd hoped that the organization's plan would "save the situation."[56] Ironically, even as he complained to Portland banker John C. Ainsworth about "unethical and war-like competition" in the industry to explain his failure to proceed with the construction of his Holladay Park hotel, Lloyd and his colleagues in the agency acted to preempt calls for mandatory proration, toward which Oklahoma and Texas were heading, convinced that such a regime would produce only lawsuits and inaction.[57] For their part, both Associated and Shell slashed their drilling budgets. Associated, however, became the more conservative operator of the two, exacerbating the company's relations with its lessors.

Associated's retrenchment also brought tensions between Ralph Lloyd and his relatives to a boil.[58]

Getting Sued, Part III

Tensions among Lloyd family members simmered since the deaths, in 1922, of Lewis M. Lloyd and Warren E. Lloyd. In the early 1930s, the heirs of Warren Lloyd threatened to sue Ralph Lloyd to recover dividend income that had flowed to him as a result of the agreement drafted by their father. For its part, VL&W sued to improve its overriding royalty position under the 8 August 1922 lease agreement that it concluded with Ralph Lloyd and Joseph Dabney.

In 1931, Caroline Alma Lloyd, her sons, Edward and Paul, and her daughter, Estelle, charged that the agreement under which Ralph Lloyd had acquired his brother's VL&W stock was illegal and invalid. As a result of the agreement, which Warren Lloyd, as a member in good standing of the California Bar had drafted, Caroline and her children together held only the 250 shares of VL&W stock that Warren Lloyd had received in the wake of the death of his father. Litigation was averted when Ralph Lloyd agreed to place eighty-one of his 900 shares of stock in escrow with the company. While Ralph Lloyd lived, Caroline and her children would receive the dividends on the shares; he would retain the voting rights attached to them. Upon his death, the shares would be distributed to Warren Lloyd's heirs. Ralph Lloyd also paid Caroline and his nephews and niece $214,365 as compensation for the dividend income that they would have received on the shares, had they owned them in the decade between Warren Lloyd's death and the settlement of the dispute.[59]

A lawsuit brought by VL&W against the Lloyd Corporation and Joseph Dabney's South Basin Oil Company almost as soon as the original lease on the Lloyd lease expired affected the development of the Ventura Avenue field directly. Because it focused on the terms of the 8 August 1922 agreements, the litigation became inextricably linked to efforts on the part of Associated to negotiate a new master lease agreement that would provide the company with more flexibility in its operations. The case shows how a dispute among lessors might alter the trajectory of oil field development.[60]

The complaint, filed on 17 January 1934, pertained entirely to royalties that would be generated by production on the Lloyd lease from wells drilled between 31 October 1933 and 8 August 1942. Ralph Lloyd's sisters, Roberta Dobbins Lloyd and Eleanor Lloyd Smith, his father's widow, Rosemary Dobbins Lloyd, and his brother Warren's heirs, charged him with violating his fiduciary responsibility by not presenting them with the full panoply of options available to them regarding the extension of the October 1913 lease. In fact, VL&W might have parleyed with Associated directly, obtaining in full the overriding royalty that remained with Lloyd and Dabney (who died in September 1932) under the agreement. Rather than receive the standard one-eighth royalty to which it agreed, VL&W might have obtained the entire one-fifth royalty that Associated paid to its lessors. For this reason, Ralph Lloyd's siblings and in-laws, as directors of VL&W, also removed the oilman as president and general manager of VL&W.[61]

In Ralph Lloyd's mind, there was no justification to revisit the August 1922 agreement. At the time, he noted, VL&W was mired in debt. Moreover, the development of the Ventura Avenue field was by no means assured. As of August 1922, he noted, Associated and State Consolidated Oil Company had invested between them some $2 million in developing the Lloyd lease without realizing a penny of profit. Including GP and Shell, operators had invested some $4 million in the field to that date without appreciable success. By comparison, VL&W had invested a meager $14,000 in the ranch that Lewis Lloyd had sold to Mariano Erburu. There was no material difference in the positions of the directors and shareholders of VL&W, on the one hand, and T. B. Gosnell, who watched in dismay as Shell developed his property after he sold his mineral rights to Ralph Lloyd, on the other.[62]

In a Ventura courtroom presided over by the same Superior Court Judge Henderson who had heard the Gosnells's case against him, Ralph Lloyd contended that he had explained all the negotiating options available to VL&W at the aforementioned impromptu meeting of its directors held during a family picnic shortly before Warren Lloyd died (see Chapter 1). His sisters Roberta and Eleanor and a close friend of Warren Lloyd disputed the story. Proceedings established only that Warren had advised his relatives—erroneously, Henderson ruled—that changing the royalty provisions of the October 1913 lease in a modified lease would be "legally dangerous."[63]

On 30 January 1936, Henderson ruled in favor of the plaintiffs, even as he opined that the members of the Lloyd family were indebted to their brother and uncle because his "vision, ability, faith, and unceasing endeavor" had placed all of them "beyond immediate want." The judge also absolved Ralph Lloyd of any violation of fiduciary responsibility. For there were no damages to restore. The judge decided, however, that the plaintiffs were entitled to the entire one-fifth royalty paid by Associated on production from wells drilled on the Lloyd lease after 31 October 1933. At a meeting of VL&W's board of directors, held on 17 February 1936, Ralph Lloyd served notice that the defendants intended to appeal. He retained former California Supreme Court Chief Justice Louis W. Myers, a partner in the prestigious Los Angeles corporate law firm O'Melveny, Tuller & Myers, to represent him.[64]

Thus the royalty dispute remained unresolved at a time when "the future prosperity of the [Ventura Avenue] field" hung in the balance, "dependent upon the cooperation of all factors necessary in conquering nature." Operators, the Lloyd Corporation now among them, were struggling to develop reserves of crude oil at depths that were testing the limits of drilling technology and within fractured strata that were confounding sophisticated petroleum engineers and crack drilling crews. The contracts at the heart of the dispute contained clauses that, in the view of both Ralph Lloyd and Associated's managers, were needlessly restricting operations and raising the already astronomical cost of drilling wells.[65]

Ralph Lloyd's Decision to Become an Independent Operator

For almost five years, Ralph Lloyd implored Associated to keep pace with Shell's relatively more intensive program of development in the western part of the Ventura Avenue field by drilling more wells than its leases called for and extending the proven area of the field eastward. Under the capital spending policy that it pursued as the Depression deepened, however, Associated adhered strictly to its express lease requirements. In response, Lloyd reentered the field as an independent operator. In this case, a major passed responsibility for defining the limits of a field back to the originator of the project—a variation on the interwar pattern in Texas documented by historians Roger W. Olien and Diana Davids Hinton.[66]

By retrenching its operations, Associated failed to meet the expectations of its lessors. At the end of 1929, Associated could claim at least 75 percent of Ventura Avenue's "actual and effective potential" production. Four years later, Associated's share had dropped 52 percent, produced from ninety-eight wells. Shell was catching up to its San Francisco–based competitor, holding a 39 percent share of "actual and effective potential" production, from 111 wells. Lloyd and the other lessors expected better results from Associated's top management. After all, Ventura Avenue was the crown jewel in the company's portfolio, accounting for almost half of its output in the state.[67]

Associated did not ignore its most valuable producing property. In fact, it continued to devote the "major portion" of its production budget to drilling operations in Ventura Avenue. By 1932, however, management calculated that it could satisfy its lease requirements by keeping a total of four rigs active throughout the year across the Hartman, Lloyd, and VL&W leases. At the same time, almost half of its drilling projects overall deepened existing wells on the Lloyd lease in the belief that substantial pools of petroleum lay in untapped zones.[68]

The Great Depression notwithstanding, Associated sought to increase its reserves—the "foundation of any [oil] company's stability and almost certainly of whatever prosperity it may enjoy," as Southern Pacific Railroad vice chairman and former Associated president Paul Shoup observed.[69] Associated President Humphery concurred with Shoup's recommendation that the company increase its reserves in part by acquiring oil properties "wherever there was a reasonable certainty of getting our money back with good prospect for profit." For Associated was enjoying no success in discovering fields of significance in the state. Indeed, its geological department was coming to the conclusion that the chances of doing so were "being rapidly exhausted." On the recommendation of his directors, Humphery reduced Associated's investment in exploring unproven areas and focused the company's limited resources on development drilling. However sensible it may have been in ensuring the long-term financial viability of the company, Associated's capital spending policy displeased lessors sitting atop undeveloped and unproven areas of the Ventura Avenue field.[70]

Ralph Lloyd initially indicated that he was prepared to reenter the Ventura Avenue field as an independent operator when, in August 1933, Shell's engineers called for parity with Associated in the next voluntary curtailment schedule. To counter Shell's bid, which reflected the company's recent success in extending the western limits of the field on the Taylor lease, Lloyd offered to defray half the cost of a well on Associated's VL&W lease that would test the eastern limits of the field. Indeed, he offered to drill the well in the name of the Lloyd Corporation under the supervision of Associated's engineers. In the interest of protecting his royalty income under the voluntary curtailment regime, Lloyd demanded that Associated take immediate action to defend its 5,000-barrel-per-day allowable advantage over Shell.[71]

Associated's persistent lack of enthusiasm in testing the eastern limits of the field convinced Lloyd to exercise his drilling option on the still-unproven Dabney-Lloyd lease that lay adjacent to the VL&W lease. In April 1929,

Associated had asked for a modification of the forfeiture clause in the lease to set aside eighty acres without a further requirement to drill, owing to "the apparent depth to which it would be necessary to drill . . . in order to encounter producing horizons" and the poor chances of success.[72] Ralph Lloyd and Joseph Dabney (until his death) accommodated Associated's requests for drilling postponements, but refused to modify the clause. Associated relinquished its leasehold interest and rights in the property when Lloyd refused to relax the lease's drilling requirements, in keeping with a policy of restricting capital expenditures to investments that were "absolutely essential to the Company's progress." In response, Lloyd announced plans to drill a test well near the VL&W lease line, at a cost of $250,000, to prove an area of "mutual benefit."[73] He estimated that the Lloyd Corporation would have to drill at least four wells to prove the property, suggesting the risks that he was prepared to assume in drilling wells at depths of 9,000 feet or more—almost twice the depth of Associated's breakthrough wells on the Lloyd lease only a decade earlier.[74]

Notwithstanding the circumstances that motivated him to return to the field as an independent operator, Ralph Lloyd first negotiated with Associated on the terms of a working agreement, covering two exploratory wells, one of which would be the third well that Associated never drilled under the terms of the lease. When the two sides failed to come to terms, Lloyd turned to Shell for financial assistance.[75]

In December 1934, Shell agreed to share in the development of the Dabney-Lloyd lease as well as the Joseph Sexton and William Sexton properties that lay further to the east and which Lloyd and Dabney held in fee, if the Lloyd Corporation demonstrated their commercial potential. In return for a cash consideration and participation in the cost of drilling the initial well, Dabney-Lloyd No. 3, Lloyd agreed to use Shell's facilities to process crude oil produced from the leases—the provision that apparently persuaded Shell to enter into the contract. Fifteen years earlier, Shell had expressed no interest in these three leases as potentially producing properties and there was no reason to think that its engineers had changed their minds.[76]

Associated's president was "tremendously disappointed" to learn of the deal with Shell: "We are not happy in the thought that we could not keep you entirely at the family fireside, as we have regarded you part of the 'Associated family.'" William Humphery thought that Associated enjoyed "a preferential position" with Ralph Lloyd. Hence, news of the deal was "embarrassing." Humphery's comments illustrate the value of Lloyd's leasehold interests in the field, which enabled him to play one major off the other.[77]

At the same time that he set out to compete against Associated, Ralph Lloyd solicited the company's cooperation in defining the limits of the field. He was convinced that reentering the field as an independent operator was the right move. As he put it: "We have never lost faith in ultimately proving the eastern end of the field to be of great value from a productive standpoint."[78] Yet Lloyd also needed the major's cooperation. After all, he still depended on production from the properties on which Associated was operating to generate cash flow. Economic self-interest recommended that the independent and the major jointly plan and execute the development of the eastern half of the field. As Lloyd put it to Associated's drilling superintendent, "I realize [that] your success in large measure is our success, for a great part of our oil property is under your control as the development operator." Lloyd promised: "If we are successful in our attempt to discover oil . . . I assure you that it will be our desire to cooperate with you in every possible way and to endeavor to create a community of interest for a happy and cooperative environment."[79]

Ralph Lloyd facilitated Associated's cooperation by making concessions on the spacing of wells on the VL&W lease and their location along the mutual border between the Dabney-Lloyd and VL&W leases; reducing the pace he expected of Associated's drilling operations; sending his drilling teams to Associated's engineers for advice; and consulting executive managers in San Francisco on developing the field. Indeed, the ink was barely dry on his working agreement with Shell when Lloyd assured Associated that he would coordinate his exploration work with its spacing program on the VL&W lease. By spacing wells every 680 feet in an easterly direction along the common border until it reached the Joseph Sexton lease, the Lloyd Corporation would save Associated the cost of drilling one offset well under difficult conditions. VL&W, he promised, would agree to relax the requirement under the 8 August 1922 agreement that the operator drill an offset well every 600 feet. Even as he reentered the Ventura Avenue field in competition against Associated, Lloyd emphasized that he sought cooperation with the major in extending its proven area.[80]

For its part, the Lloyd Corporation needed all the help that the major operators were willing to provide. Less than two years into its drilling program, Ralph Lloyd confronted the challenges and "staggering" cost of drilling deep wells. In July 1936, operators at Ventura Avenue were drilling sixteen wells to depths of at least 8,000 feet across ten leases. Lloyd estimated that completing all of them would cost up to $6 million. Doing so would not make business sense, however. All but four of the wells, in his view, were "practically failures and may be considered even worse than failures when one takes into consideration the heavy water cut that may be damaging the formation for future production." Half-a-dozen wells showed no production at all. Lloyd wrote them off as "junk," as it would cost more to "redeem" them than it would to drill a new well. As operators had found to be the case two decades earlier, "an irregularity and lack of continuity in the formation [and] a bewildering maze of faulting and subsurface disturbances" were frustrating drilling crews. Lloyd was confident that there was a substantial amount of crude oil to tap at these depths. Yet he concluded that the intersection of human, mechanical, technical, and financial factors made drilling below 8,000 feet unprofitable. Indeed, the field was "at a crisis as far as deep oil production [was] concerned."[81] Little wonder, then, that Shell surrendered its rights under the working agreement, leaving the Lloyd Corporation to explore Ventura Avenue's unproven areas with whatever indirect assistance Associated might provide.[82]

The "crisis" in the field coincided with Associated's push to modify certain clauses in the contracts pertaining to the Lloyd and VL&W leases to give it more flexibility and time to develop the two properties. As Paul Shoup noted, the Ventura Avenue field remained the crown jewel in Associated's portfolio: it could lay claim to "the best prospect for the longest life of any of the California fields." With reserves that were "quite substantial in proportion to the capitalization," the company could benefit from additional investment in the field. It was in position to do so. Notwithstanding the difficult circumstances under which it had operated in the California market, Associated had maintained "fair earnings" and payments of dividends on its common stock to Tide Water Associated Oil Company, its holding company parent, and had retired its bonded debt. Associated sought to negotiate a new master agreement with the Lloyd Corporation, South Basin Oil Company, and VL&W in the interest of extracting "all the petroleum and its by-products underlying [the Lloyd and VL&W leases] that may be economically produced."[83]

As part of a process to merge Associated and Tide Water Pipe Company, which had continued to operate independently since 1926, into Tide Water Associated Oil Company to alleviate the tax burden imposed on holding companies by the Revenue Act of 1934 and the Public Utility Holding Company Act of 1935, Associated's managers reviewed the company's Ventura Avenue contracts and recommended amendments to the 1913 and 1922 leases.[84] Eliminating the 250-foot spacing requirement on wells drilled after 31 October 1933 was critical. For if Lloyd's assessment of the condition of many of the wells being drilled was correct, Associated would have to drill new wells near the locations of abandoned wells to produce petroleum at deeper depths.[85] Most importantly, Associated needed additional time to develop its properties. With his production department designing a development program on the assumption that the leases would expire in 1942, William Humphery called on his lessors to free his company from the spatial and temporal restrictions now in place. Otherwise, Associated would intensify its development of its most lucrative property, drilling a well on every available site outside of the 250-foot radius surrounding existing wells on the Lloyd lease and exerting its right to "deepen, straighten, re-drill, and recomplete existing wells" on it, as the company's field superintendent advised. Orderly development would become disorderly development in short order as Associated sought to maximize its return on investment in the few years that remained on the lease in which it was most interested.[86]

Ralph Lloyd agreed with Associated's president on the need to revise the existing contracts. Indeed, he had driven home the necessity of doing so at the same meeting of VL&W's board of directors, held on 17 February 1936, during which he advised his siblings and in-laws of his intention to fight the Superior Court decision against him.[87] A modification of the leases, however, required the Lloyd Corporation, South Basin Oil Company, and VL&W to resolve their outstanding legal dispute pertaining to their 8 August 1922 agreement.

Ralph Lloyd was adamant that he and Joseph Dabney's heirs preserve an interest in the 7.5 percent overriding royalty on the Lloyd lease. As an initial gambit, he proposed directing three percent to VL&W, thereby increasing its total royalty to 15.5 percent, and leaving 4.5 percent to split between the Lloyd Corporation and South Basin Oil Company. He tied his proposal to a modification of the Lloyd and VL&W leases that would drop the spacing requirement on wells and extend Associated's right to drill until it was no longer in its interest to develop the field. In return, Lloyd asked Associated to contribute 1.5 percent, or half, of the overriding royalty that would flow to VL&W on wells drilled under the original 1913 lease, that is, on wells drilled prior to 31 October 1933.[88] In the negotiations that would ensue, Lloyd would agree to adjust the share of the overriding royalty that he expected to flow to his company and South Basin Oil Company. Ultimately, in the interest of obtaining the lease modifications that both he and Associated sought, Lloyd would concede a greater share of the overriding royalty than he initially proposed.

For its part, VL&W retained a petroleum engineer, C. R. "Rolf" McCollum, to review Associated's production records and advise it on what would constitute an equitable settlement. McCollum concluded that VL&W should receive a 17.56 percent royalty, leaving less than 2.5 percent for the Lloyd Corporation and South Basin Oil Company to split between them. With McCollum's report in hand, VL&W's directors voted to negotiate a new lease with Associated that provided their company with at least 85 percent of the one-fifth royalty that the major would continue to pay under any modified agreement.[89]

Associated resisted Ralph Lloyd's suggestion that it contribute cash to a final settlement. As William Humphery put it, Associated's "progressive development" of the field under a modified agreement alone would produce "proportionate benefit to all parties." Moreover, Humphery made clear Associated's resolve to protect its drilling rights, should a new agreement not be forthcoming. In the interest of securing the modifications that it sought before the leases expired, however, Associated complied with Lloyd's recommendation.[90]

In the spring of 1937, VL&W's directors came under pressure to settle the royalty dispute. In his discussions with Ralph Lloyd and representatives of VL&W on the terms of an agreement that addressed both the Lloyd and VL&W leases, William Humphery reiterated his company's willingness to intensify its drilling on the Lloyd lease under its existing agreements with the Lloyd Corporation, South Basin Oil Company, and VL&W and go to court to protect its right to do so. At this juncture, Ralph Lloyd threw his support behind the Associated president. He promised that the Lloyd Corporation and South Basin Oil Company would join in any litigation as plaintiffs, as "our interests run parallel." Further, he continued to seek at least 4.5 percent as his and Dabney's heirs' share of the overriding royalty. In this context, VL&W offered to settle the dispute on the basis of splitting the one-fifth royalty on the Lloyd lease so that it would receive 16 percent and the Lloyd Corporation and South Basin Oil Company each would receive two percent. Given his 30 percent interest in VL&W, Ralph Lloyd agreed with Homer D. Crotty, a partner in the law firm of Gibson, Dunn & Crutcher and his son-in-law (by way of marriage on May 12, 1934, to Ida Hull Lloyd, his youngest daughter), that resolving the matter outweighed any benefit that might accrue to him and Dabney's heirs by pressing further for tenths of one percent on the overriding royalty. On this basis, the three parties negotiated the terms of a master agreement. It would take another year of protracted, if intermittent, negotiations to produce an agreement that satisfied Associated's executive managers, but resolving the royalty dispute removed a material impediment to achieving that end.[91]

In June 1938, the Lloyd Corporation, South Basin Oil Company, and VL&W settled their outstanding litigation with an agreement that provided Associated the right to operate on the Lloyd and VL&W leases for ninety-nine years from

1 January 1938. To resolve the royalty dispute, Associated agreed to pay over two years: $188,500 to VL&W, $116,000 to the Lloyd Corporation, and $20,500 to South Basin Oil Company. In return, Associated obtained the concessions that William Humphery had demanded in the spring of 1936 as well as the relaxation of the requirement to drill offset wells every 600 feet along the VL&W lease. Associated secured the flexibility it desired to develop the Lloyd and VL&W leases as its engineers recommended, relieving the company of the need to litigate its drilling rights. Together with a three-year agreement on the sale of royalty oil from the leases that was effective 1 January 1936, the modified agreement softened the blow to Associated of the California Supreme Court ruling, on 26 November 1937, in favor of the Hartman Ranch Company on the money damages portion of its lawsuit against the company. For Ralph Lloyd, who for more than two years had delicately balanced his roles as sublessor and lessee of valuable drilling rights, the agreement catalyzed his efforts to extend the eastern limits of the field as an independent operator.[92]

Ironically, New Deal tax policy provided additional financial incentive for Ralph Lloyd to reenter the field as an independent operator. For targeting personal holding companies as a vehicle of tax avoidance substantially lowered the "staggering" cost of drilling a deep test well. President Roosevelt, with the encouragement and technical support of Treasury Secretary Henry Morgenthau Jr. and his staff, decided to reform the tax system to check the growth of corporate economic power and the accumulation of individual wealth. The impetus driving the redistribution impulse was Senator Huey Long's "Share the Wealth" program and other challenges from the political left. The Revenue Act of 1935 established a graduated corporate income tax. At the same time, it raised surcharges on high incomes that boosted marginal rates for individuals far above marginal rates for corporations. Further, on the belief that corporations were retaining earnings to avoid the taxation of dividends at individual income tax rates, the Revenue Act of 1936 revised the corporate income tax and imposed a graduated tax on undistributed profits. Secretary Morgenthau blamed a subsequent shortfall in revenue on widespread tax avoidance. The following year, the administration redoubled its tax reform efforts, taking aim at entities that high-income individuals allegedly were using to avoid taxes. In response to President Roosevelt's exhortation, on 1 June 1937, to "stop these evil practices," Congress passed a revenue act that closed "certain serious loopholes" identified by a special Joint Committee on Tax Evasion and Avoidance and steeply raised the surtax on the undistributed net income of personal holding companies.[93]

As a corporation that was closely held—that is, one in which five or fewer people held a majority of outstanding shares—and received 80 percent or more of its income from annuities, interest, dividends, royalties, rents, or sales of stock and securities, the Lloyd Corporation met the definition of a personal holding company. The Revenue Act of 1934 imposed a tax on personal holding company net income for the first time, but excluded rents. The law left no room, however, for a business owner to argue that retained earnings were the mainspring of an operating business. The Revenue Act of 1936 taxed the first $2,000 of undistributed personal holding company profits at eight percent and retained earnings above $2,000 at 48 percent. The Revenue Act of 1937 eliminated a number of deductions and raised the marginal surtax rate to 75 percent. Perhaps most importantly for Ralph Lloyd's investment decision-making, the law also expanded the definition of personal holding company income to include rents, on the thesis that personal holding companies were using nominal rental income to shield substantial passive income.[94]

Assuming, then, a net income of $300,000, and given a federal corporate tax rate of 19 percent and a 75 percent marginal surtax rate on the undistributed profits of his personal holding company, Lloyd calculated that spending $200,000 on an oil well cost him only $46,062.50 on an after-tax basis. Allegedly punitive taxation of Lloyd Corporation income helps to explain Ralph Lloyd's investment in oil exploration in the second half of the 1930s.[95]

And so, in the face of long odds, Lloyd persevered. On 26 March 1938, the Lloyd Corporation began drilling Lloyd Corporation No. 1 on the former Dabney-Lloyd lease under the supervision of Mark Justin Dees, Lloyd's son-in-law.[96] As of May, the contractor retained by the Lloyd Corporation had drilled the well to a depth of 6,000 feet. Five months later, Lloyd was hopeful, but still hedged his bets: "At present we are in the experimental period in our attempt to become a producing company of moment." In another three to four months, he would be able "to determine the degree of success that we may expect." The completion of the well in 1939 confirmed Lloyd's long-held belief that the productive area of the field extended beneath this property, if not beneath the two Sexton leases that lay beyond it.[97]

In July 1939, Ralph Lloyd reported to Humphery that "we are going to have a lot of oil on our own," which, when combined with the production from Associated's leases, "will make for a very attractive and permanent supply . . . one of the most reliable supplies that could be procured anywhere in the State of California."[98] Lloyd's decision to invest his capital to explore an area that both Associated and Shell wrote off was validated. Moreover, Associated, as a division of a merged Tide Water Associated Oil Company, was following Lloyd's lead and developing its eastern leases more intensively under the new lease agreement. Alfred S. McGonigle, for one, knew whom to thank: "We realize the developing of our east end has been due to your pioneering and discovering production where the Tide Water Associated claimed no commercial production would be obtained, and we appreciate that fact and hope you will continue to have fine success with your wells."[99]

The Lloyd Corporation continued drilling wells along its common border with the VL&W lease. By the end of 1940, it had seven wells in production, which were contributing to a revival in the field. A decade later, the company ranked among California's top twenty-five producers, with an average daily output of 4,643 barrels of crude oil from an average of thirty-one producing wells. It was receiving $5 million per year in royalty income and sales of the oil and gas that it was producing. At the end of 1950, the company had assets of $19,960,278 and total capital and surplus earnings of $17,242,074, up from $7,054,496 and $5,435,154, respectively, as of 31 March 1936. Thus it had the financial means to participate in a Shell-led "deep zone" drilling campaign in the field in the early 1950s, involving the drilling of wells in excess of 12,000 feet in depth at an average cost of $325,000, which propelled Ventura Avenue's production to its all-time peak of 29.9 million barrels in 1953. Thereafter, as production declined, the Lloyd Corporation determined the eastern productive limits of the field by testing its Sexton Ranch properties and developed the nearby Oxnard field in competition and cooperation with Standard Oil Company of California. Success from petroleum extraction as an independent operator provided Ralph Lloyd with the confidence and capital to revive his dream of developing his real estate portfolio on Portland's East Side.[100]

Chapter 6

Depression-Era Commercial Real Estate Development and Management

It seems quite evident that 'westward the course of empire takes its way,' as far as development is concerned in the city of Los Angeles.

—Ralph B. Lloyd, 1938

With the dénouement of his Holladay Park hotel project at hand, Ralph Lloyd turned his attention to Los Angeles, making a series of speculative real estate acquisitions along Wilshire Boulevard, hailed by local boosters as the "Fifth Avenue of the West." In May 1931 and January 1932, he acquired six lots at the southeast corner of Wilshire and San Vicente Boulevards on the Los Angeles side of the city's border with Beverly Hills and four lots at the northwest corner of Wilshire Boulevard and Beverly Drive in the so-called Business Triangle in Beverly Hills. The latter property—"destined to be the outstanding corner on Wilshire Boulevard west of Western Avenue in the years to come," according to real estate advisor R. V. Morrison—lay within a block also bounded by Dayton Way and Rodeo Drive. Over the next three years, Lloyd would acquire most of the remainder of the nearly vacant block. With these acquisitions, the small businessman established a beachhead in a market that offered him the opportunity, even under turbulent economic conditions, to develop commercial real estate as a profitable and sustainable investment vehicle for the capital that he was accumulating from petroleum extraction. Convinced that "westward the course of empire takes its way," as far as commercial development in Los Angeles was concerned, Lloyd intended to develop these valuable holdings. His commitment to do so, however, would waver once he committed a substantial portion of his surplus capital to oil field development, as the last chapter described.[1]

Ralph Lloyd was buying into a market that had been soft for some time. The construction sector lagged the overall economy throughout the second half of 1920s, even if it suffered less than agriculture and other "sick" industries that had not recovered from World War I–propelled booms.[2] While the national economy grew 9.8 percent from 1925 to 1929, the sector as a whole shrunk by more than four percent—though, in terms of the value of contracts let, the commercial construction segment fared better, advancing 6.1 percent.[3] Market watchers were hopeful that 1929 would mark a return to "normalcy." Nevertheless, the industry posted disappointing results at the national and local levels. The $93 million in building permits issued by the City of Los Angeles for the year was 8.5 percent below the mark set in 1928 and 37.8 percent below the tally logged in 1925—the last year that the City had registered an uptick in permit valuations. The types of buildings that Lloyd was interested in constructing on his newly acquired parcels—offices, stores, and theaters—all experienced double-digit increases in permit valuations for the year, however.[4]

The poor overall results of 1929 notwithstanding, national building industry leaders were convinced that "a period of normal and stable building had commenced" and that 1930 would be a good one.[5] Economist George Eberle, a long-time analyst of real estate trends in Los Angeles, harbored no such illusions for the local market, however. He expected building activity for the year to be "considerably below normal."[6] Eberle's prediction was spot on. The value of building permits issued by the City of Los Angeles fell 20.3 percent, to $74.1 million. In number, however, permits declined by only 3.5 percent to a level that was only 7.5 percent below 1929.[7] At the same time, national output, as measured by GDP, declined 12 percent. In 1931, the year during which Lloyd negotiated the purchase of his Wilshire Boulevard properties, the bottom began to fall out of the local market. Los Angeles suffered a 44.4 percent decline in the value of building permits issued from the year before to its lowest level since 1919. The number of permits issued fell only 17.3 percent, but, as was the case with valuations, now stood at a level not seen since 1919. Were it any consolation to local boosters, Los Angeles retained its third-place ranking, behind New York and Baltimore, in building permit valuation per capita.[8] If the depths to which the national and local economies would eventually plummet were not fully evident at the time of Lloyd's Wilshire Boulevard transactions, the trend was clear.

In any case, Ralph Lloyd was always willing to accumulate properties even if their potential development lay in the distant future, his ability to do so a result of his control of valuable mineral rights in the Ventura Avenue field. Even the "severe and expensive struggle in our Portland operations" was not enough to alter his conviction that, whatever the circumstances, "eventually there is a turn in the road that often brings the most surprising results," as far as real estate development was concerned.[9] And property accumulation during hard economic times, especially in Los Angeles, was a shrewd investment, according to bankers, real estate brokers, and other interested members of the local growth machine. Less than three weeks after Lloyd closed escrow on his northwest corner at Wilshire Boulevard and Beverly Drive, the *Los Angeles Times* reported that "men who have been through other depression periods" agreed that "right now fortunes based upon Los Angeles real estate are awaiting the investor and that opportunities today are greater and have more guarantees than ever before."[10] J. F. Sartori, founder, in 1888, of the Security Trust and Savings Bank and now president of the Security-First National Bank of Los Angeles, averred that "the present recession in prices is no more severe than others which have occurred in the past." Moreover, he counseled, each of several "dormant periods" in the regional real estate market, spanning his business career, was followed "by periods of renewed activity during which values have recovered and increased."[11] Orra E. Monnette, vice-chairman of Bank of America, was unequivocal: "Recovery is certain as that the sun shines and real estate will further be crowned as king."[12]

Were he simply a speculator seeking to buy at the bottom of the market, Ralph Lloyd was not far off in the timing of his Wilshire Boulevard acquisitions. In 1932, building in Los Angeles collapsed, with the valuation of permits issued plummeting to $17.5 million.[13] Since 1929, the local building industry had contracted 81.2 percent—almost twice as much as the decline in national output, which had fallen 43.3 percent. Nevertheless, Lloyd insisted that he accumulated his Wilshire Boulevard properties in anticipation of a revival "in the business of building buildings."[14]

When Ralph Lloyd accumulated these properties, there was seemingly nothing that would slow the transformation of the Wilshire Boulevard, which was destined to become "the most famous lane in modern civilization," as the editor of *Wilshire Topics* (probably James Lee) enthused only four months before Lloyd made the first acquisition. It was a conclusion, Lee blustered, "founded on the predictions of statisticians, engineers, and city planners who have with mathematical accuracy conducted exhaustive analyses of certain growth."[15] And of all the lots that comprised the 400-odd blocks on both sides of the almost sixteen-mile-long thoroughfare that stretched from downtown to the Pacific Ocean, Lloyd's northwest corner at Beverly Drive was "generally considered" to be "the finest piece of property on the boulevard."[16] Given the purchasing power of Beverly Hills and its strategic location, the city's Business Triangle remained woefully underdeveloped relative to Westwood Village to the west and the so-called Miracle Mile to the east, real estate brokers were

Figure 15. Wilshire Boulevard at Beverly Drive, Beverly Hills, looking east, circa 1935. The California Bank Building occupies the northeast corner of the intersection. To the left is the Beverly Theatre (demolished 2005). To the right, the Warner Theatre (demolished 1988) occupies the southeast corner of Wilshire Boulevard and Canon Drive. Ralph Lloyd's block intrudes into the photograph at the lower left. Photo undated. (Security Pacific National Bank Collection, Los Angeles Public Library.)

quick to note.[17] Nevertheless, the eight-story Beverly Wilshire Hotel (1927–1928) on the south side of Wilshire Boulevard and the California Bank Building (1928–1929) across Beverly Drive provided early, if isolated, evidence that the site had the potential to fulfill the aspirations of boosters like Lee and pay off spectacularly for Lloyd (figure 15).[18] Warner Brothers provided additional assurances when it opened a theater on Canon Drive on the south side of the boulevard—a site that it paid $200,000 to acquire in February 1930 (figure 15). Real estate advisor Morrison predicted that, in time, Lloyd's lots would "probably be surrounded by greater purchasing power than any corner of Wilshire Boulevard outside of downtown." It was the "outstanding corner in Beverly Hills . . . where all of the large businesses and chain stores . . . and department stores will have to come."[19] Depression or no depression, it seemed like only a matter of time before Lloyd would develop his Business Triangle property in keeping with Lee's declaration that there was "something of awe in the thought of the Boulevard, its splendor, and its potentialities.[20]

When Ralph Lloyd moved his family into an elegant 5,590-square-foot, two-story residence on the northeast corner of Hobart Boulevard and Seventh Street in 1923, Wilshire Boulevard was zoned exclusively for residential use (figure 16). The pre–World War I idea that the thoroughfare would materialize as a grand avenue of lavishly landscaped homes was fading, however, and the city's financial elite was beginning to decamp to more secluded and exclusive areas, such as Bel Air, Beverly Hills, Hancock Park, and San Marino. (The Lloyds would follow suit in 1930, as we shall see.) Residential development along Wilshire Boulevard in the 1920s may have been mostly middle class in character, as architectural historian Richard Longstreth observes, but the seven-bedroom, four-bathroom house that John L. De Lario designed for his nouveau riche cousin was more in keeping with the earlier vision for the boulevard. Before the Lloyds had time to settle into their home, however, efforts to transform Wilshire Boulevard into the "Fifth Avenue of the West" gained momentum.[21]

Organized as the Wilshire Boulevard Association, property owners successfully lobbied the City of Los Angeles to widen the streetcar-less boulevard and rezone the blocks between Westlake (now MacArthur) Park and Western Avenue for commercial use. A "Fifth Avenue of the West" also required tall buildings. And so the association urged its members to line not just its eastern portion, but the entire boulevard with height-limit (thirteen-story) office and retail towers. Just imagine. With the construction of Bullock's Wilshire, a major department store, and the Pellissier and Wilshire Professional Buildings,

Figure 16. The Lloyd residence, 694 S. Hobart Boulevard, Los Angeles, constructed 1923; John L. De Lario, architect; S. M. Cooper, builder; no longer standing. Photo undated. (Photo by E. W. Barker. The Huntington Library, San Marino, California, LCR drawer 2, box 2.)

the section of the boulevard between Westlake Park and Western Avenue became recognized as an important new business district. Had Ralph Lloyd walked the long block from his house to the intersection of Hobart and Wilshire Boulevards at the end of the decade and cast his gaze both eastward and westward, however, he would have seen scant evidence of a burgeoning "Fifth Avenue of the West." Commercial development remained "strikingly unfocused," as Longstreth describes it, and scattered among apartments, churches, gas stations, single-family homes, and vacant lots. While the construction of Bullock's Wilshire may have dealt a blow to downtown's department store monopoly, developments along the seventeen blocks between La Brea and Fairfax Avenues from 1924 to 1931 provided a better blueprint for Lloyd's ambitions. Indeed, before he acquired his Wilshire Boulevard properties, the oilman would relocate the headquarters of the Lloyd Corporation to one of the most prominent buildings to rise along the boulevard during this time. Between 1932 and 1941, he would compete with the owners of parcels located at major intersections along this stretch, promoted as the "Miracle Mile" by real estate agent A. W. Ross.[22]

With dreams of establishing a business district in keeping with Wilshire Boulevard's destiny, as editor Lee and other boosters envisioned, Ross began accumulating vacant land between Fairfax and La Brea Avenues in the same year as the Lloyd family moved into its Hobart Boulevard residence. The dream was slow to materialize. In 1925, the City of Los Angeles zoned the entire stretch of the boulevard for residential use—a decision that the U.S. Supreme Court upheld in May 1927. As a result, Ross was required to apply for zoning variances on a case-by-case basis. Thus his development plans gained traction only in 1928, when Desmond's, a clothing store, chose the Miracle Mile over a location in Hollywood for its first branch store. Transformation was seemingly assured with the erection of the Wilshire Tower Building. Modeled on the just-completed City Hall—the tallest building in Southern California at the time—the project was the first of its type to materialize along the boulevard. Taking its cue from Desmond's, a competitor, Silverwood's established a branch store in the building, again choosing the Miracle Mile over Hollywood. In short order, the E. Clem Wilson Building (1929–1930) and the Dominguez-Wilshire Building (1930) were completed at La Brea Avenue and in the block two streets west of it, respectively. With one- and two-story buildings between each of them, the four freestanding structures commanded a presence individually that attracted prominent retailers to their ground floors and collectively to this stretch of the boulevard, and drew shoppers who allegedly possessed "the highest class purchasing power" in Los Angeles.[23]

The surge in development along the Miracle Mile persuaded Ralph Lloyd to vacate his downtown office in the Bank of Italy Building. Seeking to escape "the gas fumes that are becoming so intense in the downtown district," Lloyd signed a five-year lease for the top (tenth) floor of the newly completed Dominguez-Wilshire Building. Designed by Morgan, Walls & Clements, architects for the Pellissier Building and his stillborn Portland hotel, the top floor enabled him and his staff to escape the worst effects of the emissions generated by the thousands of vehicles that passed by the building each day (figure 17). *Wilshire Topics* featured the edifice—which attracted customers to the flagship Myer Siegel department store on its lower levels with air-cooled fitting rooms, terrazzo floors, and stylish aluminum furniture—on its cover. Lloyd sent a copy of the magazine to C. W. Norton, his man in Portland. In his cover letter, Lloyd noted, perhaps wistfully, that the building reminded him in "a small way" of his hotel, which existed only in model form (figure 18). Indeed, the building's similarity to the hotel in massing and appearance may have caught Lloyd's eye, enticing him to seek out the property's management.[24] In any case, as Richard Longstreth notes, the building offered potential tenants a more usable office floor plan than the Wilshire Tower Building, its primary competitor in the area.[25]

Rather than establish Wilshire Boulevard as a "Fifth Avenue of the West," that is, a congested street at the bottom of a canyon walled by skyscrapers, the Dominguez-Wilshire Building and its sister structures on the Miracle Mile created a place, Longstreth explains, "that was easy to reach, that always seemed busy but never crowded, a place permeated with natural light and air."[26] As a thoroughfare that offered both the consumption amenities and array of choices of the actual Fifth Avenue and the feel of the Main Street of a much smaller town, Wilshire Boulevard was fulfilling the expectations of its most enthusiastic boosters.[27] Ralph Lloyd had ample opportunity to develop a sense of the place from his offices and during his commute from his new palatial home in the hills above the Business Triangle.

In 1930, the Lloyd family joined the elites who fled in increasing numbers into exclusive, all-white neighborhoods during the interwar period. The Lloyds moved into a three-story, thirty-five-room Italianate villa at 962 North Alpine Drive on a site carved out of the sprawling estate of Thomas Thorkildsen, the "Borax King," who had ceded the property to his wife in a pre–World War I divorce. Designed by John De Lario in keeping with his designs for luxury

Figure 17. Ralph Lloyd on a terrace on the tenth floor of the Dominguez-Wilshire Building, Los Angeles, on his seventieth birthday, 1945. (The Huntington Library, San Marino, California, Photos drawer 1, box 4.)

Figure 18. Dominguez-Wilshire Building, 5410 Wilshire Boulevard, Los Angeles, constructed 1930; Morgan, Walls & Clement, architect. Photo undated. (Security Pacific National Bank Collection, Los Angeles Public Library.)

homes in nearby Hollywoodland, the mansion changed levels—seven in all—with the contours of the sloping site, which was lushly landscaped by Russell L. McKown and Alfred C. Kuehl (figure 19). On a clear day, a colonnaded open patio and terraced garden would have provided Lloyd with an oblique view of the Business Triangle and beyond as he enjoyed breakfast before setting out for the office. De Lario's masterpiece, which interior designer Genevieve Butler lavishly appointed, offered Lloyd a place to retreat and reflect when his Portland hotel project faltered and the oil market collapsed.[28]

No doubt, the ongoing transformation of Wilshire Boulevard into the early 1930s was persuasive in Ralph Lloyd's decision to accumulate property west of the Miracle Mile. At the same time, local perceptions aside, the Los Angeles economy collapsed in line with other cities on the Pacific Coast and not materially out of line with the major metropolises that lay to the east of the Rocky Mountains. Shannon Crandall, president of the California Hardware Company, a leading regional distributor of building materials, thought that the city "did not drop as low as other sections of the country."[29] Basing his observation on the value of building permits, Crandall was technically correct. Los Angeles's building industry fared better than the four metropolises in the East with more people than the City of Angels. The latter's 83.9 percent drop in the value of building permits issued between 1929 and 1933 nonetheless was severe and hardly could have been a source of comfort for a wholesale supplier of building products (table 1). In terms of retail sales, a broader measure of general economic performance and one of particular interest to the speculator in commercial real estate, Los Angeles declined less than Chicago, Detroit, and Philadelphia, but slightly more than Boston and New York (table 2).[30] Perhaps based more on his personal experiences than on a consideration of the data, Lloyd believed that Los Angeles's building sector performed better than Portland's throughout the Depression.[31] In terms of the decline in both retail sales and the value of building permits, however, the performances of the two cities were nearly identical (table 1 and table 2). In terms of retail sales, Los Angeles fared little better or worse than its regional counterparts (table 2).[32] True, local boosters could take comfort in the fact that, in absolute terms, Los Angeles ranked second only to New York in terms of the value of building permits issued at the nadir of the Depression. Nevertheless, the environment in which Lloyd set out to develop his Wilshire Boulevard corners was a difficult one; it would deteriorate further before it would recover.

Figure 19. The Lloyd residence, 962 North Alpine Drive, Beverly Hills, constructed 1929–1930; John L. De Lario, architect; no longer standing. Drawing, south elevation. (Homer D. Crotty Papers and Addenda, The Huntington Library, San Marino, California, box 63 [Addenda], folder 5.)

Regardless of economic conditions, the Wilshire Boulevard Association of Beverly Hills, of which Ralph Lloyd was named a director and one of four vice-presidents in 1932, remained "vitally interested" in protecting real estate values along the boulevard. To ensure that developers attracted "the right type of merchants," the association's officers combed through the *Beverly Hills Citizen* and other publications for building permits filed with the City Council. The association stood ready to act, should anyone propose a project that threated property values. It paid particular attention to Lloyd's corner at Wilshire Boulevard and Beverly Drive because its officers considered it to be "the most valuable and important one on the boulevard."[33]

Ralph Lloyd attracted scrutiny when he negotiated with San Francisco–based Owl Drug Company and S. H. Kress, each one a chain operation with dozens of outlets across the country, for a store on the first floor of a proposed building on a portion of the property. The site was large enough to accommodate a department store as well, and so the Lloyd Corporation queried the interest of Bullock's and Sears, Roebuck & Company in opening Beverly Hills branch stores. Neither company was contemplating another store in Los Angeles, however. At the same time, negotiations with the chains and perhaps other potential tenants—the record is unclear—progressed far enough for the Lloyd Corporation to file a permit for the construction of a building to accommodate them.[34]

Ralph Lloyd's growth machine colleagues in the Wilshire Boulevard Association responded swiftly in terms reminiscent of the protests lodged a decade earlier by business owners against Lloyd and the Bank of Italy over their plans for the Hollywood Gateway Tract, as we have seen. Norman Stirling, secretary of the organization, minced no words in explaining that there were "certain stores which are known to be the means of developing a business center," and drug and five-and-dime chain stores were not among them. Were such businesses to occupy the "most valuable and important [corner] on the boulevard," the "main business section of this city will undoubtedly locate elsewhere, and our opportunity for increasing land values on the boulevard will be lost forever." What Frederick M. Kincaid, the Lloyd Corporation's lawyer, may have replied when he responded to the missive by phone is not known, but ultimately no chain would open a store on the corner while Lloyd held the property. Ralph Lloyd may well have acceded to the wishes of local boosters and ended his pursuit of the chains. Whatever the oilman's response may have been, Owl Drug Company subsequently opened a store nearby on the northeast corner of Western Avenue and Wilshire Boulevard.[35]

Table 1
Building Permits for Selected Cities, 1929–1937 (thousands of dollars)

	1929	*1931*	*1933*	*1934*	*1937*	*% Down*	*% Up*
Los Angeles	93,020	41,422	15,396	14,968	64,014	83.9	327.7
San Francisco	33,426	21,442	58,198	7,112	20,359	n/c	n/c
Portland	15,504	7,156	2,385	2,358	8,688	84.7	268.5
Seattle	29,101	12,483	1,953	2,318	7,003	93.3	258.6
Chicago	210,798	66,694	5,729	10,176	35,957	97.3	527.6
Detroit	48,369	23,435	4,053	8,889	53,412	91.6	1,217.8
Boston	58,834	33,968	7,562	9,382	22,187	87.1	193.4
New York	942,298	362,863	86,570	96,662	315,544	90.8	264.5
Philadelphia	104,406	35,265	12,099	8,284	35,794	92.1	332.1

Note: Percentage changes for San Francisco not calculated; 1933 included permits for the Golden Gate and San Francisco Bay Bridges.

Source: U.S. Department of Commerce, Statistical Abstract of the United States (Washington, DC: GPO, various years).

Table 2
Net Retail Sales for Selected Cities, 1929–1939 (thousands of dollars)

	1929	*1931*	*1933*	*1935*	*1939*	*% Down*	*% Up*
Los Angeles	875,775		453,340	593,902	782,842	48.2	72.7
San Francisco	474,683		254,075	298,371	383,554	46.5	51.0
Portland	208,601		105,865	147,413	183,561	49.2	73.4
Seattle	252,169		129,096	163,185	208,537	48.8	61.2
Chicago	2,127,520		990,084	1,216,706	1,514,829	53.5	53.0
Detroit	890,189		369,936	543,690	665,565	58.4	79.9
Boston	672,760		374,805	439,121	490,396	44.3	30.8
New York	4,272,633		2,245,801	2,847,332	3,192,594	47.4	42.1
Philadelphia	1,122,168		514,456	656,744	766,622	54.1	32.9

Source: U.S. Department of Commerce, Statistical Abstract of the United States (Washington, DC: GPO, various years).

At the same time that he held talks with the chains, Ralph Lloyd approached Van de Kamp's Holland Dutch Bakeries about leasing his corner at San Vicente and Wilshire Boulevards. The company, founded in 1915 in Los Angeles by Lawrence L. Frank and Theodore J. Van de Kamp, was regionally famous for the blue windmills that adorned the roofs of its coffee shops and bakeries. Negotiations on the basis of a ten-year lease of a $15,000 building did not come to fruition. The Lloyd Corporation, however, successfully petitioned to have the property rezoned from unrestricted residential to restricted commercial status.[36]

With potential lessees for buildings on his Wilshire Boulevard properties evaporating as the economy spiraled downward, Ralph Lloyd accumulated additional lots to improve his chances of attracting first-class tenants to these sites once economic recovery was assured. Between December 1933 and September 1935, he acquired all but two of the remaining lots in the block bounded by Wilshire Boulevard, Beverly Drive, Dayton Way, and Rodeo Drive as well as the northeast corner of Beverly Drive and Dayton Way. Lloyd's conservatism in developing properties did not extend to acquiring unimproved property. In the meantime, Lloyd strove to keep his struggling tenants in his existing buildings.[37]

Problem Leases

More than anything else as a developer, Ralph Lloyd abhorred vacant buildings. The depreciation was bad enough, but add in wear and tear and property taxes and "you have a picture that is not attractive."[38] Lloyd did not construct a building unless he could be assured, at least on paper, of realizing a reasonable return on it. To realize a reasonable return, it was imperative that he kept his lessees in the buildings that he had erected or modified to their specifications.

Owing to the timing of Associated Oil Company's breakthrough in the Ventura Avenue field, most of Lloyd's lessees occupied their premises just before the Depression exerted pressure on their businesses. For instance, in July 1928, International Harvester entered into a ten-year lease at $600 per month for a 41,767-square-foot warehouse constructed by L. H. Hoffman at the corner of East First and Oregon Streets in Portland.[39] Later that year, MacDonald & Driver completed a three-story, reinforced concrete building at 3500 South Hope Street, Los Angeles, for the Fisk Rubber Company, a Massachusetts-based manufacturer of automobile, bicycle, and motorcycle tires, which signed a ten-year lease at $1,755 per month.[40] In 1929, the Lloyd

Corporation constructed a domed restaurant for Willard's Far-Famed Chicken Steak Dinners at the southwest corner of Los Feliz Boulevard and Hillhurst Avenue in the upscale Los Feliz district of Los Angeles. On 18 March 1929, Willard's signed a ten-year lease, payable at $375 per month for five years and $475 per month thereafter.[41] Two months later, Earl B. Staley, owner of a truck equipment business, signed a ten-year lease at $360 per month for a building at the corner of East Second and East Irving Streets in Portland. Designed by Charles W. Ertz and built for $17,445, the building featured large plate glass display windows that extended along both streets from a corner entrance that faced the intersection. A mezzanine level above the lobby provided office space.[42] A year later, Lloyd put Lyon Motor Company, a Chrysler dealership, and an Associated service station into another Ertz-designed building at the corner of East 16th Street and Sandy Boulevard (figure 20).[43] While none of these leases approached the Sears store on Portland's East Side in scale, together they demonstrated Lloyd's determination to invest his royalty income from a depleting asset in commercial real estate. By 1931, all of these leases were in trouble.[44]

With deflation tightening its grip on the national economy, many of Lloyd's lessees pleaded for relief. In response, Lloyd typically offered temporary reductions of 10–20 percent on the understanding that tenants would make up the difference before their leases expired.[45] In no case, however, would a lessee repay its rent in arrears. In most cases, it would not even sustain its monthly payment at the reduced rate. Of the aforementioned cases, only International Harvester stayed in its building until America entered World War II—and only after the Lloyd Corporation lowered the rent by one-third.[46]

The Great Depression simply overwhelmed the businesses of Lloyd's lessees. The experience of Willard's was typical. In July 1932, its owner asked Lloyd to reduce the rent by $125 per month, to $250, for one year. Perhaps hoping against hope that economic conditions would improve, Lloyd acceded to the request, but only for six months. In January 1933, as the nation awaited the inauguration of President-Elect Roosevelt and as the economy continued to collapse, Willard's asked Lloyd to reduce the monthly rent still further, to $150, a level at which the restaurant could hope to cover its fixed costs. Lloyd agreed to reduce the rent to $200 per month for six months; Willard's agreed to repay the $1,620 in arrears over the last four years of the ten-year lease. Willard's asked for, and Lloyd granted, two more extensions at the reduced rental. Beginning in July 1934, at its five-year mark, the lease called for payments of $475 per month. Willard's sent $400 to the Lloyd Corporation. Lloyd accepted

Figure 20. Associated Oil Company service station and Lyon Motor Company Chrysler dealership, East 16th Street and Sandy Boulevard, Portland, constructed 1930; Charles W. Ertz, architect. Photo undated. (Photo by News Pictures, Ltd. The Huntington Library, San Marino, California, LCB drawer 1, box 1.)

the payment and allowed the restaurant to continue to occupy the building at that rate until its business recovered. Meanwhile, back rent continued to accumulate. When, in April 1935, the restaurant's proprietor informed the Lloyd Corporation that he would never be able to repay the accumulated rent, Ralph Lloyd chose to sell the property.[47]

In many cases, Ralph Lloyd was left abruptly with a vacant building or portion of a building on his hands. In early 1931, for instance, Harry W. Lyon, the owner of the aforementioned Chrysler dealership, informed the Lloyd Corporation that he could no longer pay the rent and vacated his portion of the building. Associated continued to operate its service station until May 1933.[48] On 28 July 1932, the Adcox Auto and Aviation School declared bankruptcy, leaving the building at 1300 NE Union Avenue (now Martin Luther King Jr. Boulevard) vacant for the first time in two decades.[49] In November 1932, Fisk Rubber Company, which had entered into receivership, paid $26,334 to cancel its lease (though it paid $750 on a month-to-month basis to continue occupying the building).[50] In July 1933, Studebaker, which also had entered receivership, opted out of its $1,371-per-month lease on a sales showroom at 150 West Jefferson Boulevard, Los Angeles, that the Austin Company had erected a decade earlier.[51] With his leases in varying degrees of distress, Lloyd conceded that his commercial real estate portfolio had become an "enormous burden."[52]

Ralph Lloyd scrambled to identify viable businesses to lease his vacated buildings. Unsurprisingly, finding a new tenant whose business might thrive in a deteriorating economic landscape proved to be a challenge. For example, the Fisk tire people signed a one-year lease effective 1 April 1934 on the building formerly occupied by Studebaker at $600 per month. Just four months later, the company vacated the agreement. Under a lease arrangement reached on 9 November 1934, Sheperd Tractor and Equipment paid $500 per month and occupied the premises at least through November 1941, but no later than September 1944, when General Electric opened an appliance outlet in the building.[53] Unable to collect any back rent from the bankrupt Adcox Auto and Aviation School, the Lloyd Corporation put the Latture Company, sellers of builders' equipment and machinery, in the vacant building at 1300 NE Union Avenue in October 1933 under a one-year lease based on a percentage of gross sales. Before the year was up, the company abandoned the premises. In October 1934, the Lloyd Corporation leased the building to the newly organized L. L. Adcox Trade School for a paltry $75 per month. Even this sum proved to be beyond the financial means of the tenant, and so, in April 1936, the landlord terminated the lease to make room for the Portland Mercantile Company, with which it was negotiating the terms of a five-year lease. The deal soon fell through, however, prompting Franz Drinker, who replaced C. W. Norton as Ralph Lloyd's man in Portland in June 1932, to exclaim, "The Adcox building seems to be a jinx!"[54] Ultimately, as we shall see, the Lloyd Corporation would find a more stable tenant for the Adcox building in the Bonneville Power Administration.

Between 1933 and 1937, the economy recovered spectacularly. Real GNP increased by an average of eight percent per year—one-third in all. The principal source of this performance was the U.S. government's abandonment of the gold standard and, along with it, the devaluation of the dollar. With monetary policy unshackled from its "golden fetters," the Roosevelt Administration also pursued an expansionary fiscal policy. Robust growth ensued. By mid-1936, Southern Pacific Railroad vice chairman Paul Shoup could stand before a Portland audience and declare: "The renewed business activity in the Northwest is impressive. . . . I have been in Portland Union Station two evenings around nine o'clock and the crowds remind me of old times. . . . More people from the East have passed through Portland this summer than during any previous summer in the past five years."[55] Recovery provided the real estate speculator, even a relatively conservative one such as Ralph Lloyd, with new opportunities for profitable development. The recovery was halted abruptly in 1937, however, when the Federal Reserve Bank and the Roosevelt Administration reversed course and adopted contractionary monetary and fiscal policies, respectively. In 1938, real GNP fell five percent. Recovery and recession had a whipsaw effect on a commercial real estate developer such as Ralph Lloyd, who found himself once again with problem leases on his hands after putting tenants in buildings on terms based on the assumption that recovery would be sustained.[56]

As the example of Frank Chevrolet shows, Lloyd also potentially exposed himself financially by insisting that his architect design a distinctive structure that would help the dealership compete in the marketplace. From the mid-1920s, architecture was increasingly viewed as a critical factor in marketing a business. It was imperative in outlying areas, such as the Walnut Park district on Portland's East Side, where the Chevrolet dealership would be located, that owners invest in improving the appearance of even modest commercial buildings. This held true for automobile dealerships as much as it did for women's apparel stores. In this instance, capitalizing the enhanced value of the building in the lease added to Lloyd's exposure to a faltering economy.[57]

Late in 1935, Franz Drinker approached owner Herbert L. Frank about the dealer's possible interest in a new used car showroom at the southwest corner of NE Union Avenue and NE Sumner Street. With a net worth of about $50,000, assets of $100,000, and little short-term debt, Frank Chevrolet, established in 1927, offered the Lloyd Corporation the financial stability that it sought in its tenants, but had found to be in short supply in recent years. John C. Ainsworth's United States National Bank vouched for Frank's ability and integrity. Ralph Lloyd was willing to build a showroom for the car dealer, if he could obtain the terms he sought.[58]

Lloyd, however, based his expectations on the more favorable Los Angeles market, and so established a wide gap between the parties. As was the policy that he only recently had developed for that market, Lloyd sought a return on investment of 14 percent on a building that he could amortize within ten years. To protect the lease from the inflation that was accompanying general economic recovery, he also tied Frank's rental payments to a percentage of gross sales. Based on a review of four trade publications, he determined that, above some level of volume, Frank should pay a rental equal to two percent of gross sales.[59]

Initial negotiations used Charles Ertz's preliminary plans and specifications and estimate of $33,500 to construct the building. On that basis, Herbert Frank offered to pay $350 per month. Before he took an offer to his boss, Franz Drinker persuaded the dealer to accept $400 per month with a percentage of gross sales provision. The offer did not come close to meeting Ralph Lloyd's expectations.[60]

After reviewing Ertz's preliminary plans and specifications and factoring in the inevitable overruns, Lloyd estimated that the dealership would cost some $35,000 to construct. On this basis, at $400 per month, he informed Drinker that he could obtain a better return on investment from parking cars on a vacant lot. He instructed his man in Portland to press Frank to accept a minimum monthly rental of $500.[61]

Franz Drinker took pains to point out to his boss that major dealers in downtown Portland were paying at most 1.5 percent in their percentage clauses. Herbert Frank expected his sales for 1935 to amount to just over $500,000. Acceding to a monthly rental of two percent of gross sales would result in Frank paying an outlandish $10,000 to rent the building for the year. Drinker also noted that Frank had to offer more inducements to potential customers because of his East Side location. The automobile dealer offered to pay $450 per month—based on a 12 percent return on investment on a $35,000 building, plus three-fourths of one percent on gross sales above $750,000.[62]

Once again, the oilman compared Portland unfavorably with Los Angeles. Complained Lloyd: "One of the great troubles with Portland is that the property owners in the city are seemingly willing to rent their properties at prices not in line with similar structures in other cities." In making a counteroffer that included the concession in the percentage clause that Drinker sought, Lloyd added that Portlanders "have no proper conception of the value of property and the necessity of a certain revenue in order to justify improvements."[63]

Ultimately, Ralph Lloyd and Herbert Frank agreed to terms on a ten-year lease for a building and three adjacent lots that Frank utilized to sell used cars: one percent on gross sales, with a minimum payment of $450 per month until sales reached $650,000, at which point Frank would pay a straight one percent on gross sales. While negotiations continued through December 1935, Lloyd sent his architect back to the drawing board.[64]

For Ralph Lloyd found Charles Ertz's design to be wholly unsatisfactory. In particular, the front of the building was not attractive enough to catch the eye of passing motorists. He wanted Frank Chevrolet to be the "outstanding agency building" in Portland. To this end, he assigned Claud Beelman to work with Ertz, who had moved to Los Angeles a few months earlier because his son had matriculated at the University of Southern California. Together with partner Alexander Curlett, Beelman had contributed to the Los Angeles skyline with more than a dozen height-limit buildings, including the stunning Eastern Columbia (1929–1930) and Sun Realty (1930–1931) Buildings, whose dramatic and vibrant façades, featuring colorful terra-cotta tiles, zig-zag and chevron patterns, and dominant vertical lines, evoked the modernity embedded in the Art Deco movement.[65] Two years earlier, Lloyd had retained Beelman, whose partnership with Curlett dissolved in 1932, to design and consult on buildings that he planned to build in Beverly Hills. To keep within his budget, Lloyd instructed the architects to use existing buildings as models. The Ford Motor Company, he noted, had recently built "some beautiful" agencies in Los Angeles, photographs of which he shared with his Portland office. The collaboration produced a sleek Streamline Moderne exterior, executed in white stucco with chrome trim, that captured the ethos of technological progress that was animating a decade of remarkable innovation amid national economic distress (figure 21). Ertz estimated the cost of the redesigned building, based on one of the Los Angeles dealerships that had captured Lloyd's attention, to be $36,000, or only $2,500 more than the estimated cost of his initial design. The design impressed Herbert Frank, who moved some way toward meeting Lloyd's requirements upon reviewing it.[66]

Figure 21. Frank Chevrolet, 5131 NE Union Avenue (now NE Martin Luther King Jr. Boulevard), Portland, constructed 1936; Claud Beelman and Charles W. Ertz, architects. No longer standing. Photo undated. (The Huntington Library, San Marino, California, LCB drawer 1, box 1.)

With lease in hand, the Chevrolet dealership occupied one of only a handful of buildings that the Lloyd Corporation constructed for private clients in Portland in the 1930s. All of them were located outside the company's core holdings around Holladay Park, which Lloyd continued to reserve for a revived hotel project.

Recession hit Herbert Frank's business little more than a year after he opened his showroom at 5131 NE Union Avenue. Citing a 50 percent drop in sales, Frank asked Lloyd for relief in the summer of 1938. Confident that the economy soon would rebound, Lloyd refused to offer a rental holiday. Rather, he counseled patience: the Lloyd Corporation was contemplating "some material improvements" in the Walnut Park area that "should result in an increased drawing power" for the Chevrolet dealership. Despite his numerous setbacks in the real estate market, Lloyd was more able than his lessee to await an improvement in business conditions.[67]

Herbert Frank was "very much surprised and disappointed" by his landlord's unsympathetic response. All across Portland, he noted, lessors had granted relief to their tenants who dealt in automobiles: Frank was now paying twice as much for floor space as his competitor who was operating on the most desirable location on the more attractive West Side. As someone who may have been familiar with Ralph Lloyd's frequent promises to Portlanders dating from the oilman's Holladay Park acquisitions in 1926, Frank was less than impressed with Lloyd's promise to "improve" Walnut Park. Should any such improvements materialize and benefit his business, Frank promised to acknowledge Lloyd's contribution to increasing the pace of economic activity in the district. In the meantime, he was entitled to "consideration." Negotiations stalled when Ralph Lloyd stood pat. Nevertheless, Frank stayed in the building.[68]

The relatively short duration of the sharp recession and strong, even spectacular, economic growth nationally from 1938 to 1942, spurred by a return to expansionary fiscal and monetary policies, helped to convert a problem lease into a profitable one for both parties. Indeed, in 1947, the Lloyd Corporation completed a second building for Frank at a cost of $102,930 at 5225 NE Union Avenue in the block immediately to the north of the dealership. As of 1949, Frank Chevrolet was doing nearly $3 million in business annually, selling both automobiles and trucks, and paying $1,717 per month in rent. At the same time, Lloyd's long-promised improvements for Walnut Park were materializing, including a two-story store and office building on NW Union Avenue directly across from the site of the theater, store, and office project that he had abandoned in the late 1920s. Postwar economic conditions increased Lloyd's confidence that he would obtain his expected return on investment from developing commercial real estate for private clients.[69]

During the unstable 1930s, Los Angeles offered Ralph Lloyd the most attractive market for new commercial real estate investment on the Pacific Coast. Nevertheless, his willingness to develop his holdings in the metropolitan area with his own money even as the market rebounded diminished once he committed millions of dollars to oil field development. The following cases illustrate how the risk-reward calculus that persuaded Lloyd to renter the Ventura Avenue field as an operator mid-decade changed his approach to developing commercial real estate for private clients, on his Beverly Hills block in particular.

Remodeling for Profit

Remarkably, Los Angeles's central business district was becoming outmoded even before some of its most prominent retail structures, such as the Eastern Columbia Building, were completed. Having established branches in Hollywood, along Wilshire Boulevard, and other outlying areas, retailers determined that their flagship stores required refurbishment if they hoped to continue attracting customers who, importantly, were finding it easier to park their vehicles at branch locations and were becoming increasingly hostile to a private street railway system that, in theory, provided the most convenient means of travel to downtown. In the context of unprecedented economic contraction, remodeling both exteriors and interiors became the primary means by which businesses attempted to enhance or protect their brands and increase their trade. Landlords and real estate speculators, too, stood to benefit from upgrading properties for existing or new tenants. Lloyd was no exception. Rather uncharacteristically, given his generally long-term horizon for commercial real estate investment, his contribution to imparting the appearance of newness to downtown Los Angeles involved "flipping" a property for a "reasonable profit" within three years of acquiring it.[70]

The southeast corner of Sixth and Olive Streets missed out on Los Angeles's interwar building boom. A two-story Class A concrete and steel building that faced Pershing Square had occupied the entire 132-by-150-foot site since 1904. In May 1929, while Lloyd still worked out of his office in the Bank of Italy Building a block away to the south, King C. Gillette, founder and president of the Gillette Safety Razor Company and the site's owner, leased the property to a syndicate headed by Wade Hampton, president of the Western National

Bank. Hampton's investment group planned to erect a height-limit office building in the summer of 1931, when the leases in the building expired. Across Olive Street, the James Oviatt Building, an Art Deco masterpiece designed by Walker & Eisen, had recently opened. The Pacific Mutual Building, one of the most imposing of the sixty-four office buildings erected in Los Angeles during the 1920s, dominated the northwest corner of Sixth and Olive Streets. Together with the Biltmore Hotel, the Beaux Arts edifice defined the eastern border of Pershing Square. Hampton's syndicate retained John and Donald B. Parkinson, designers of recently completed City Hall and soon-to-open Bullock's Wilshire, as architects of a similarly impressive structure. The Depression put paid to the project, however, leaving the two-story building an orphan among distinctive tall buildings. In April 1931, citing ill health and age, Gillette stepped down as president of his company. On 9 July 1932, he died at his Hollywood home. Late in 1934, when Lloyd was mulling over the property's commercial potential, the building was dilapidated, with a tired tile and artificial stone exterior that cried out for rehabilitation.[71]

When Ralph Lloyd looked to acquire the property, ten stores on the ground floor and the second floor were under lease. The tenants included, most notably, the Los Angeles Fish Grotto, a restaurant, and the Cinderella Roof, a locally renowned dance hall that had occupied the second floor at least since 1909. In all, the leases were generating annual rental income of some $58,000. Some of the leases provided for annual increases in rent, but, as was the case with Lloyd's problem leases, the recent trend was a downward revision in monthly payment obligations. In July 1934, for instance, Maurice Bernstein, proprietor of the Fish Grotto, obtained a six-month decrease in his rent, from $1,500 per month to $1,000. Remodeling the building would be justified only if Lloyd could secure increases in rents from existing tenants and capitalize the space on the second floor.[72]

Through his Property Service Corporation, Lloyd acquired the property for $500,000. The Penn Mutual Life Insurance Company held a $400,000 mortgage that matured on 1 May 1936, but the company offered to refinance it. On 1 November 1934, Property Service Corporation signed a ten-year, five percent promissory note to Havelock C. Boyle, an intermediary, who promptly assigned the loan to the insurance company. The note was repayable in quarterly payments of $5,000, beginning on 1 February 1935, with a balloon payment of $205,000 due on 1 November 1944.[73]

Ralph Lloyd bought the building with the intention of remodeling the exterior and converting the unheated second floor to offices, which would involve constructing a new entrance, lobby, and stairway, and installing an elevator. The project expanded with the addition of a granite entrance for Bernstein's Fish Grotto and remodeled spaces for both the restaurant and the Union Pacific Railroad, a new tenant that operated a ticket office. Lloyd retained Claud Beelman for the project. The architect estimated the project to cost between $42,000 and $45,000, not including a heating system for the second floor and radiators for several ground floor stores that also remained unheated; the entrance for the Fish Grotto; and partitions to divide the second floor into 17,800 square feet of rentable office space. With these additional features, Beelman's estimate ranged from $48,600 to $50,800 (figure 22).[74]

The remodel was executed between September 1935 and July 1936. After cladding the exterior in Texas Cordova limestone, Lloyd let a $60,000 contract to Pozzo Construction for the conversion of the second floor into some of "the most desirable" office space in the city. The contractor was also responsible for the granite entrance and tenant improvements for the Fish Grotto. Pozzo completed this work in February 1936. Remodeling the ticket office, which was undertaken separately, was completed in the summer of 1936. According to Lloyd, local Union Pacific Railroad executives were pleased with the result, deeming the new ticket office their "finest" in America.[75]

Remodeling for profit, however, required capitalizing the investment with increased rental income. As he had negotiated with Frank Chevrolet, Lloyd sought to lease the remodeled Sixth and Olive Building on the basis of a percentage of gross sales and a minimum rental, with the larger of the two determining the rental. Indeed, in the context of inflation, Lloyd adopted this approach for long-term leases on all of his properties. Bernstein's Fish Grotto, for instance, paid either $1,250 or seven percent of gross sales per month. In the summer of 1936, its payments ranged from $1,600 to $2,000. Among the ground floor tenants, only the Union Pacific Railroad successfully resisted this approach. Lloyd had to be content with an increase in rent, to $1,395 per month. In all, Lloyd negotiated a minimum of $63,660 in annual rent from his ground floor tenants.[76]

Before he leased the second-floor offices, however, Ralph Lloyd "flipped" the property. Drilling for oil at Ventura Avenue was depleting his reserves, as his royalty and rental income no longer provided cash flow sufficient to fund his investments on a pay-as-you-go basis. And so, having committed capital to testing the eastern limits of the Ventura Avenue field, Lloyd decided to sell the building. He held preliminary discussions with the Santa Fe Building Company, an arm of the Santa Fe Railway, which was looking for a site to

Figure 22. Alterations to building, southeast corner of Sixth and Olive Streets, Los Angeles; Claud Beelman, architect. Drawing 1935. (The Huntington Library, San Marino, California, LCR drawer 5, box 3.)

construct a $1.6 million height-limit office building (and would therefore demolish the just-remodeled structure). Lloyd had no ambivalence about the prospective destruction of the building. He felt that downtown Los Angeles needed improvements on the scale of the proposed building, especially now, "when every constructive project helps us along the road to ultimate recovery." When Santa Fe executives opted for a nearby site at Sixth and Main Streets, Lloyd approached Union Pacific Railroad executives, who agreed to purchase the remodeled building (for use, not demolition) for $1.025 million, providing Lloyd with the "reasonable profit" that he sought.[77]

At the time of the sale of the Sixth and Olive Building, Lloyd continued to search for a department store to anchor the development of his Beverly Hills block. Given the demand of oil field development on his capital, he sought terms that would minimize his risk in any project.

Developing a Beverly Hills Block

Robust business recovery encouraged Ralph Lloyd to consider the development of the crown jewel in his Los Angeles real estate portfolio as the remodeling of the building at the corner of Sixth and Olive neared completion.[78] Between 1935 and 1938, he negotiated with executives of three of the city's most prominent department stores for a branch store on his Beverly Hills block. Had he succeeded, Lloyd would have moved closer to realizing his goal of finding in real estate development a means to convert surplus capital generated by depleting petroleum assets into an investment that provided stable and sustained returns. Given his commitment of capital to oil field development and the treatment of rental income under New Deal tax policy, however, Lloyd apparently drove a harder bargain in each instance than his potential tenant was willing to accept.

In recovery, Los Angeles outpaced the Pacific Coast, in building permits, if not in retail sales (table 1 and table 2). Monthly building permits issued by the City regularly exceeded in value those issued by the other leading cities of the region combined, as California Hardware Company President Shannon Crandall often remarked.[79] The *Los Angeles Times*—with little doubt *the* organ of the Southern California growth machine—went further, observing that, in 1936 alone, the City of Los Angeles permitted $10,322,231 more in construction than Berkeley, Oakland, Phoenix, Portland, Sacramento, and San Francisco combined.[80] In that year, the local construction industry rebounded after five miserable years. The $62,355,541 in permits issued by the City of Los

Angeles nearly doubled the volume of 1935 and quadrupled the nadir year of 1934.[81] And while Los Angeles's recovery may have lagged its East Coast counterparts in percentage terms, the city remained second only to New York in the value of permitted building volume throughout President Roosevelt's first two terms in office. California Hardware's Crandall could "hardly understand our doing twice the amount of building Chicago is doing," but was delighted that building remained "quite brisk in this locality," notwithstanding a "slackening up somewhat" in 1937.[82] Indeed, with the national economy growing at a robust 10 percent rate annually between 1938 and 1941, competition from Sears and chain stores posed a greater threat to Crandall's business than a possible relapse in national economic performance.[83] On the face of it, Los Angeles offered ample opportunity in the second half of the 1930s for Ralph Lloyd to develop his Beverly Hills block.

Yet "regime uncertainty," associated with New Deal tax policy in particular, diminished the small businessman's appetite for commercial real estate investment. To be sure, Lloyd's complaints regarding his "enormous tax burden" predated Roosevelt's presidency.[84] And so he hoped that the incoming Democratic administration would provide relief: "If governmental authorities do not make some radical changes and savings in the cost of government, I am fearful that we have a hard road ahead of us."[85] Moreover, his concerns were not limited to federal tax policy. In 1937, for instance, he complained bitterly to the Los Angeles County Board of Supervisors that "the tax rate in Los Angeles County has mounted so high that its burden is acting as a hindrance in industrial development and is reducing the standard of living of the individual taxpayers."[86] Two years later, he urged Supervisor John Anson Ford to lower taxes by adopting the recommendations of the Los Angeles County Citizens Committee on Local Governmental Budgets to eliminate wage and salary increases, reduce a planned increase in maintenance costs, eliminate land acquisitions, and cut Works Progress Administration funding by two-thirds.[87] Lloyd also complained about Portland's property taxes, instructing Franz Drinker to urge both the Chamber of Commerce and the East Side Commercial Club to convince municipal officials that increases in rates were hindering local growth and development.[88] Of course, Lloyd was hardly the only businessman to complain about New Deal tax policy. For instance, in noting that taxes equaled 80 percent of California Hardware Company's net income in 1939, Shannon Crandall called attention to an "excessive tax burden" from which there was seemingly no prospect for relief.[89] Indeed, as president of the National Wholesale Hardware Association, he identified "the steadily mounting tax burden" as "the Nation's Number One problem."[90] For his part, Paul Shoup, whose general views on taxation were well publicized, identified "increased taxation related directly to income" that did "not leave all that could be desired in the way of cash or credit to promote expansion of the industry" as a chief problem of the Southern Pacific Railroad.[91] Yet the U.S. government's treatment of the personal holding company, as we have seen, had a particular impact on Lloyd's business decision-making, causing him to rethink commercial real estate as an investment vehicle.

As the Roosevelt Administration turned the screws on closely held business enterprises, Ralph Lloyd refrained from embarking on major private projects in both Los Angeles and Portland. In the spring of 1936, for instance, he stood on the sidelines, awaiting the outcome of a congressional debate on the treatment of undistributed profits. At the behest of Treasury Secretary Henry Morgenthau Jr., the president had called for a tax on retained earnings to compensate for the U.S. Supreme Court's invalidation of the processing tax prescribed by the Agricultural Adjustment Act of 1933 and Congress's override of his veto of a bonus bill for World War I veterans. Before President Roosevelt acted, Lloyd insisted that he stood ready to launch several projects, even as he committed capital to oil field development.[92] Subsequently, he pointed to the provisions of the revenue bill that Roosevelt signed into law as the reason why he declined to buy the Alexandria Hotel in downtown Los Angeles and revive the construction of his Portland hotel.[93] In the wake of the passage of the Revenue Act of 1937, Lloyd declared that he was shelving his plans for his Beverly Hills block, where they would remain until Congress passed a less "destructive and confiscatory" tax bill.[94] Indeed, in declining an offer to acquire another property in downtown Los Angeles, he declared: "We are out of the market, owing to the effects of the New Deal. With the present tax laws, we cannot even improve the properties we already own."[95] In 1939, Lloyd rejected an offer to build and lease a building in Ventura for Montgomery Ward's, indicating that he had no funds for real estate investment and adding, "nor can I accumulate any." He was adamant: "If the present rate of taxation keeps up, we with others will eventually be eliminated from further activity in the business world."[96] In the fall of 1940, Lloyd declined an opportunity to acquire the Willamette Iron and Steel Corporation, which had been operating in a limited capacity since 1933, at which time a bondholders' committee had organized the firm in the wake of foreclosing on its bankrupt predecessor. Pleading "the inability to keep your money even if you make it," he lamented that the U.S. government "apparently is going to see to it that out of earnings you cannot grow."[97] New Deal tax policy was one of the reasons why Lloyd's vision for his Beverly Hill block did not materialize.

Before he committed capital to testing the eastern limits of the Ventura Avenue field, Lloyd demonstrated his interest in developing the block in keeping with its enviable location. In November 1933, that is, six months after the economy began to show signs of recovery under the New Deal, he retained Claud Beelman to design a new home for the Victor Hugo, an elegant French bistro located at 628 South Hill Street, that would be "one of the finest restaurant buildings on the American continent." Charged by Lloyd with developing plans and specifications for a building that would cost between $75,000 and $100,000 to erect, Beelman evoked the Renaissance with an offering that featured an arcuated entrance, windows framed by decorative columns and semicircular pediments in the manner of the piano nobile of the Palazzo Farnese, and a balustrade reminiscent of the Petit Trianon. The interior featured an oval dining room that overlooked an Italian garden and included a reception hall, private dining room, and catering facilities. Completed by Pozzo Construction in December 1934, Lloyd let the $90,000 building to Hugo and Violetta Aleidis, the restaurant's proprietors, on a ten-year lease at a monthly rental equal to six percent of gross sales, with a minimum rental of $1,200. Located midway between Wilshire Boulevard and Dayton Way, the glamorous venue soon became a favorite of the Hollywood set (figure 23).[98]

Encouraged by the opening of the Victor Hugo and ongoing economic recovery, Ralph Lloyd pursued department store executives who expressed interest in expanding their retail empires beyond their downtown confines. None of the negotiations for a branch store located on one of his Wilshire Boulevard corners concluded with the successful execution of a lease.

During 1935, Lloyd negotiated with the Broadway, which had recently expanded into three previously unused floors of its nine-story emporium on Hollywood Boulevard that it had acquired in March 1931. Now company executives expressed an interest in capturing the trade of the posh Beverly Hills district, in keeping with a strategy of using branch stores to upgrade its customer base.[99] In the context of Lloyd's decision to reenter the Ventura Avenue field as an operator, negotiations for a six-story emporium on the corner of Wilshire Boulevard and Beverly Drive stalled by the fall of 1935. Of primary importance to Lloyd was structuring the lease to ensure that the Lloyd Corporation would recover its investment of $826,000 to construct the proposed building and capitalize the $400,000 estimated value of the land beneath it.[100] Ultimately Broadway executives decided that the cost of constructing the

Figure 23. Victor Hugo Restaurant, 233 North Beverly Drive, Beverly Hills, construction completed 1934; Claud Beelman, architect; Pozzo Construction, builder; no longer standing. View from Wilshire Boulevard. Note the use of the vacant lots between the restaurant and Wilshire Boulevard for parking. Photo dated 1937. (Works Progress Administration Collection, Los Angeles Public Library.)

store that Lloyd proposed was prohibitive. They opted instead to expand their Hollywood store, which they executed in 1938 with a six-story addition that added more than 52,000 square feet of retail floor space.[101]

Ralph Lloyd's pursuit of a department store tenant for one of his strategic corners revived in the spring of 1937 when the May Company launched a search for a suitable location along Wilshire Boulevard for a store that would replace its flagship store at Broadway and Eighth Streets—the first time nationally that a major department store sought to abandon its downtown location. When he got wind of the company's intentions to erect a store that would attract affluent customers from Beverly Hills and Hollywood, Lloyd called attention to the accessibility of his properties not only to Beverly Hills, but to Hollywood, Culver City, and, in the future, the San Fernando Valley, via Laurel Canyon and the Cahuenga and Sepulveda Passes. Lloyd was willing to discuss a long-term lease of the land (that is, a ground lease), a long-term lease of the land with a building constructed by the Lloyd Corporation, or even the sale of the property (notwithstanding the tax implications). Only the second approach would require the Lloyd Corporation to inject a significant amount of capital into the project. The May Company would pay monthly rent on a percentage-of-sales basis, with a minimum amount guaranteed, as Lloyd was now demanding of all of his tenants.[102]

It is unclear the extent to which the May Company considered any of three corners that Lloyd offered (two of which were located on his Beverly Hills block) before company executives selected the northeast corner of Wilshire Boulevard and Fairfax Avenue as the site for a store with more than 270,000 square feet of floor space. Commanding the intersection of the two major thoroughfares with a rounded, mosaic corner element set against a striking, concave, polished black granite frame, the store made its greatest appeal to customers with the largest surface parking lot associated with a single retail outlet constructed in America to date.[103] It could do so because the area around the site remained undeveloped. As Lloyd conceded, the May Company chose the site because it could get "practically an unlimited amount of parking for automobiles at that location."[104] Perhaps in this case the matter of parking alone was decisive. At the same time, the May Company's decision coincided with both Lloyd's announcement that he was shelving his plans to develop his Beverly Hills block because of confiscatory New Deal tax policy and the spudding of Lloyd Corporation No. 1 at Ventura Avenue. Owing perhaps to these demands on his capital, Lloyd offered May Company executives lease terms that were less attractive than his competitors, prompting them to look elsewhere for a site.

Lloyd pitched J. W. Robinson Company at the same time he approached the May Company, with similar results. Unlike the May Company, Robinson's was not seeking to abandon its downtown quarters. Indeed, in 1934, it modernized the interior of the store and resurfaced its exterior walls with Art Deco features.[105] To meet the company's requirements for 100,000 square feet of merchandising and storage space, Lloyd initially offered to build a store costing not more than $500,000 on his Rodeo Drive corner, with parking provided in the area between the building and Dayton Way. As was the case with the May Company, Lloyd offered lease terms that would minimize the landlord's risk in the project. Robinson's would pay a monthly rental equal to 3.5 percent of gross sales on a twenty-five-year lease and advance a payment of $150,000 at three percent interest. Moreover, the lease would contain a "recapture provision" that would reimburse the landlord, should 3.5 percent of gross sales not provide a "reasonable return" on the value of the building and the land. Moreover, Robinson's would maintain the building throughout its leasehold.[106]

The corner prospectively offered Robinson's a significant competitive advantage over the branch locations of both Coulter's and the May Company along the Miracle Mile. Argued Lloyd: "It seems quite evident that 'westward the course of empire takes its way,' as far as development is concerned in the city of Los Angeles. Building progress and business activity are constantly moving westward . . . note how development has traveled like wildfire to the west but has been static in its development to the east."[107] Robinson's would reap the benefit of the westward migration of retail markets. Lloyd asked the company's president to imagine "the possibilities of the most beautiful store, with the finest location for ample parking, that one could desire."[108]

Notwithstanding the attractiveness of the site, Robinson's ultimately stood pat on its remodeled downtown store. As late as September 1938, J. W. Robinson expressed an interest in a smaller, 50,000-square-foot store on the corner of Wilshire Boulevard and Beverly Drive. In response, Lloyd proposed a rental payment that would provide a return of 10 percent on the building and eight percent on the land, the total cost of which equaled an estimated $400,000.[109] At this point, the parley apparently stalled. Because Robinson's chose not to build elsewhere, it is unclear the extent to which the terms of Lloyd's proposal factored into the department store's decision-making. After failing to secure a department store tenant, Lloyd leased the lots that fronted Rodeo Drive to the Brown Derby Cafe across the street and held out hope for a store and office building at the corner of Beverly Drive.[110] All of Lloyd's Wilshire Boulevard corners remained undeveloped when Japan attacked Pearl Harbor. Turning his attention to the commercial development of Portland's East Side, he sold all of his lots in his Beverly Hills block in October 1944 for $800,000 and two residential lots in Hancock Park.[111]

In the spring of 1935, W. G. McCarthy, the owner of the Beverly-Wilshire Hotel, sold his holdings in the block on the south side of Wilshire Boulevard, between Camden Drive and Rodeo Drive, to a New York syndicate headed by Beverly Hills resident E. L. Cord, of motor car fame, for $1.3 million. In short order, construction began on a five-story building for W & J Sloane, high-end furniture dealers, which would relocate to Beverly Hills from downtown Los Angeles upon its completion later in the year. Cord, who two years earlier had completed a 32,000-square-foot mansion, Cordhaven, on his sprawling eighteen-acre estate on North Hillcrest Road, also lined up tenants for specialty shops on the ground floor of an adjacent four-story building, the upper floors of which provided additional space for Sloane when it was completed in 1936. Together, the two buildings, representing an investment of $525,000, helped to fulfill Lloyd's prophecy that "westward the course of empire takes its way," as far commercial development in Los Angeles was concerned. Saks Fifth Avenue soon followed, opening its first store in the West in a neoclassical, four-story building two blocks to the west at Peck Drive. The Business Triangle was finally beginning to materialize as the upmarket retail center that Beverly Hills boosters had always envisioned. But Lloyd would contribute little to its development, other than to convert the Victor Hugo into a fashionable salon for Adrian, a designer of high-end women's apparel, when the restaurant went bankrupt in January 1941.[112]

A handwritten, anonymous post card that landed in Ralph Lloyd's in-box in the wake of the Wilshire-Rodeo-Camden deal noted that Cord had already lined up seven "Class-A" tenants for his planned four-story building: "Further proof that ACTION (!), and not mere talk nor 'plans,' get RESULTS. L.A. people love DO-ERS!"[113] The implication, of course, was that Lloyd was failing to seize the opportunity before him to develop his holdings as others were clearly able to do. As the Southern California economy rebounded strongly, other owners found it profitable to build for tenants along Wilshire Boulevard—for high-end retailers in Beverly Hills as well as for numerous chains along the Miracle Mile.[114] Why not Lloyd, who held perhaps the most enviable position in the local market?

In short, Ralph Lloyd was a relatively risk-averse developer in a highly competitive market. Given federal tax policy and the other factors that motivated him to commit substantial capital to Ventura Avenue field development, Lloyd's appetite for real estate investment diminished even as he maintained his interest in real estate development. Even so, as the examples above illustrate, he was willing to build for potential tenants. In return, he required them to minimize his investment risk. In the absence of data on comparable transactions, it is unclear the extent to which the lease terms that Lloyd offered to satisfy his risk-return criteria were uncompetitive. That he failed to secure a department store tenant for any of his Wilshire Boulevard properties, however, is suggestive, if not dispositive. In depression—and war—federal and state agencies authorized to invest public capital in office construction would fund several of Lloyd's most important building projects and lay the foundation for the construction of one of America's largest regional shopping centers.

Chapter 7

Public Capital and the Development of Portland's East Side

We at BPA have dealt with Mr. Lloyd since our beginning in 1938 and we have come to look upon him as a godfather to our administration building needs.

—Dean Wright, 1951

In May 1939, Ralph Lloyd surely would have stood with the 64.8 percent of corporate executives who responded to a poll conducted by *Fortune* magazine, agreeing with a statement that policies of the Roosevelt Administration had held back economic recovery significantly.[1] By then, federal tax policy in particular had converted the one-time supporter of the cartelization of the oil industry under the National Industrial Recovery Act into a strident opponent of government intrusion into the affairs of business.[2] It is ironic, then, that federal and state agencies and the public capital that supported their activities in large part sustained Lloyd's commercial real estate business until the private sector recovered after World War II.

Between 1934 and 1954, the Lloyd Corporation leased existing properties to, or erected new buildings for, the Bonneville Power Administration, the State of Oregon, and the U.S. Department of the Interior on Portland's East Side. The company used Federal Housing Administration mortgage insurance to construct Park View Apartments directly south of the golf course that the Lloyd Corporation completed in October 1932. Ralph Lloyd also found a solution to one of his problem tenants in Los Angeles by leasing space to California's motor vehicle department. And before these agencies became his tenants or financial underwriters, Lloyd looked to the Reconstruction Finance Corporation to fund "public" projects that he proposed for Portland, including a revived Holladay Park hotel.

With an aim to undertake "sound and worthwhile business projects . . . for the relief of existing conditions," Ralph Lloyd submitted a multi-project proposal to the joint City of Portland and Multnomah County committee responsible for vetting local RFC projects soon after President Roosevelt took office. The list included a home for Portland's minor league baseball team, which was currently playing in an aging stadium on the city's northwest side; a National Guard armory; a fire station; a modified, more modest version of the Holladay Park hotel; and the beautification of Sullivan's Gulch. All but the latter effort would require RFC funding. In each of the other instances, the Lloyd Corporation would donate land and provide junior financing. The projects promised to inject between $2 million and $3 million into the local economy.[3] "I have not lost faith in Portland," Lloyd told the *Oregonian*. "If Portland does not succeed, then there is no hope for the rest of the country."[4]

A direct descendent of the War Finance Corporation, which boosted capital markets and loaned money to war industries during World War I, the RFC, created on 22 January 1932, initially made low-interest loans to financial institutions and railroads to the ends of restoring confidence in the banking community and increasing commercial credit. In six months, the RFC made more than $1 billion in loans. The injection of capital, however, failed to achieve the policy goals of the Hoover Administration because bankers parked the majority of the funds that they received in U.S. government securities. And so, reluctantly, President Hoover changed course, pushing a relief bill through Congress that he signed into law on 21 July 1932. The Emergency Relief and Construction Act authorized the RFC to make $1.5 billion in loans to "self-liquidating" public works that, the administration hoped, would employ thousands of workers in short order. To handle the expected flood of requests from local and state bodies, the RFC created a Self-Liquidating Division. It was to this division that the joint City of Portland and Multnomah County committee would forward projects submitted by the Lloyd Corporation that met its approval.[5]

From the outset, the Self-Liquidating Division failed to meet expectations. In part, this owed to the long lead times required to plan any large-scale public works project. More importantly, applicants had to show that their "self-liquidating" projects would generate revenues (for instance, by charging tolls) sufficient to repay the U.S. government in full at above-market interest rates (so that the agency would not compete with private financial institutions) in no more than ten years. As a result, the RFC denied almost all loan applications. By the time President Hoover left office, the RFC had approved only ninety-two applications totaling $197 million and had disbursed only $20 million. Empowered by the Emergency Banking Act of 1933, however, the RFC under Jesse H. Jones became an engine of state capitalism. The early signs for Portland were encouraging. Already, the RFC had approved a municipal market on the downtown waterfront.[6]

Approval for Lloyd's proposal was widespread among the building, business, financial, and real estate communities. Paul S. Dick, president of the United States National Bank, lauded Lloyd's interest in Portland's development as "timely and most constructive." He singled out the hotel for praise: because of its impact on local employment, the project "would be exceedingly helpful to the community."[7] The bank's John C. Ainsworth noted that the reconfigured hotel "would be a valuable addition" to the city and wished Lloyd success in realizing it.[8] Commonwealth Security Corporation's W. H. Hemphill reported that he and his colleagues were "very optimistic about your proposed development on the East Side."[9] F. H. Murphy, president of the Oregon Building Congress, wrote to express his appreciation for Lloyd's efforts "to set the wheels of the building industry in motion." Both he and his trade association recognized and appreciated the Californian's "inherent and abiding faith in Portland, now and for the future."[10] The Chamber of Commerce applauded the proposed program and offered Lloyd its support in seeing it to fruition.[11]

Among public office holders, Multnomah County Commissioner C. A. Bigelow, who was seeking to revive his political career after resigning his Portland City Council seat in August 1930, called Lloyd's proposals "the brightest ray that has come to Portland in the last four years."[12] City Commissioner (and future mayor) Robert Earl Riley, who had survived a recall petition filed in April 1932 against him for malfeasance and negligence, called the proposed program the "most constructive" before the joint RFC committee.[13]

Yet support for the proposals apparently did not run very deep among local officials. On Bigelow's motion, the joint RFC committee unanimously approved Lloyd's program and pledged the assistance of local governments in obtaining the federal funds that the Lloyd Corporation sought. Portland City Council members, Multnomah County commissioners, and the local RFC loan chairman also publicly backed the projects. At the same time, as the Lloyd Corporation's Portland office reported, competition for funds was fierce and officials would privilege their projects, most of which lay to the west of the river, over those of the Lloyd Corporation.[14]

Anti-Lloyd sentiment, undoubtedly lingering in the wake of the suspension of the hotel project, posed an impediment. As a circuit court judge who served as the executor of a Holladay Avenue estate argued privately, Ralph Lloyd made fulsome promises, but never delivered on them. Time and again, the California oilman gave interviews and held press conferences with reporters and secured the support of the Chamber of Commerce, the East Side Commercial Club, and municipal officials. In turn, he received special benefits. It was high time, the judge concluded, for Lloyd to bear the full cost of his projects. In the context of recent recall efforts, these were strong words. It is unclear the extent to which sitting officials harbored similar views and resentments—none of the commissioners who supported the Lloyd Corporation's street petitions of the late 1920s remained in office. Yet such undercurrents could only have worked against Lloyd in his efforts to secure public monies for his projects.[15]

Lobbying Oregon's congressional delegation to throw its weight behind the program failed to achieve the desired result. The Lloyd Corporation cabled Senators Charles McNary and Frederick Steiwer and Representative Charles Martin, seeking their assistance in gaining approval from the RFC. While he was in the nation's capital, Alfred B. Swinerton, chairman of the eponymous San Francisco building contractor, paid visits to both Steiwer and Martin on behalf of Lloyd, who had recently returned from Washington as California's delegate at an oil conservation conference chaired by Interior Secretary Harold L. Ickes. The ineffectiveness of these entreaties owed in part to prioritization. Oregon's delegation was spending its political capital on securing funding for the Bonneville Dam, whose construction, as we shall see, would benefit the Lloyd Corporation handsomely, if indirectly, in time.[16]

The RFC made no decision on Ralph Lloyd's proposed projects before President Roosevelt directed the newly created Public Works Administration to assume responsibility for the functions of the Self-Liquidating Division on 16 June 1933. In short order, the PWA rejected Lloyd's bid to wrap his hotel in the garb of a public project. The agency was also slow to act on Lloyd's other proposals. In late July, the erstwhile developer of East Side property traveled to the nation's capital. The Portland office recommended that Lloyd lobby Oregon's congressional delegation and Secretary Ickes, calling attention to his languishing proposals. It is unclear whether Lloyd acted on this recommendation. Ultimately, Lloyd received no RFC support for his ambitious program.[17]

Although he failed to secure direct public financing for his projects, Ralph Lloyd showed his willingness, albeit in desperate economic times, to turn to public sources of capital in support of private commercial real estate

Figure 24 (*opposite*). Lloyd Corporation Properties, East Side, Portland, 1 February 1948. (The Huntington Library, San Marino, California, LCR drawer 2, box 5.)

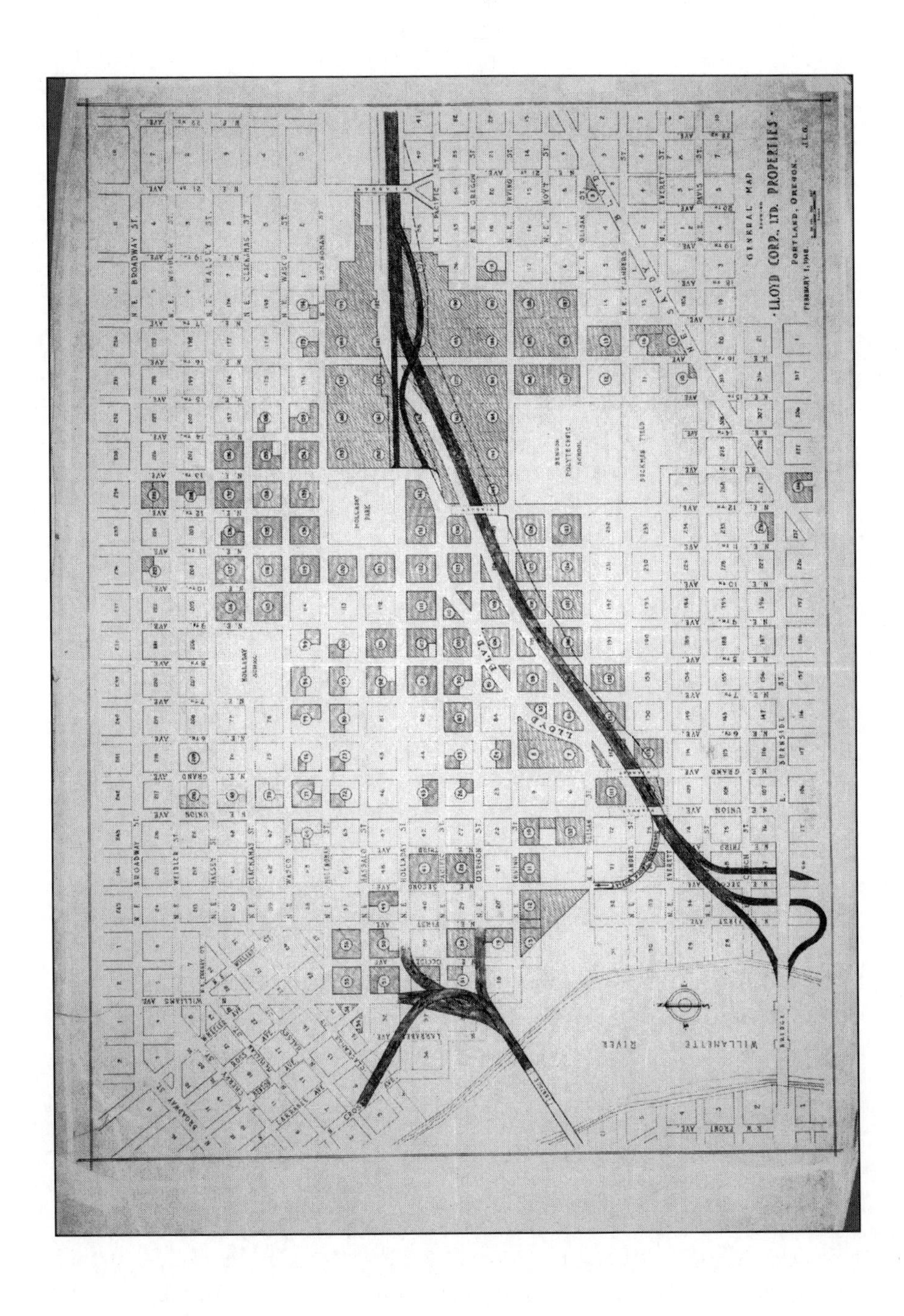
GENERAL MAP
LLOYD CORP., LTD. PROPERTIES
PORTLAND, OREGON
HOLLADAY PARK
LLOYD BLVD.
WILLAMETTE RIVER

development. As the decade wore on, he would enjoy success in constructing projects for public agencies because they proved to be more able and willing than private enterprises to meet the lease terms that the oilman sought. His earliest success consisted of replacing a delinquent private tenant with a state agency that continued to lease the building until the Lloyd Corporation sold it.

Meeting the Needs of State Motor Vehicle Divisions

After the Fisk Rubber Company vacated its distributorship at 3500 South Hope Street, Los Angeles, in April 1934, Ralph Lloyd secured a stable tenant in the California Division of Motor Vehicles. Without an interruption or reduction in rental payments, the DMV occupied the premises for more than a dozen years. Effective 1 May 1934, the DMV leased the building for five years for $1,250 per month, that is, $500 more per month than the Fisk company had been paying on a month-to-month basis since it had cancelled its lease in October 1932, as we have seen. In 1939, the DMV renewed the lease on the same terms. Five years later, it signed a five-year lease that provided for monthly payments of $2,130 through June 1945 and $2,280 thereafter. In October 1946, the Lloyd Corporation sold the property for $145,000. Three years later, it received $126,252 of the $313,000 that the DMV paid to the new owners when the State of California condemned the property in an eminent domain proceeding.[18]

The Lloyd Corporation found a stable tenant in Oregon's DMV counterpart too, though it took more than a decade before it housed the State Auto License Division in an Ertz-and-Burns-designed building at 625 NE Oregon Street (block 83 on figure 24).

In response to a request for proposal published by the State Board of Control in the spring of 1937 for a 65,000-square-foot building that could be expanded to 100,000 square feet, with a total cost of land and structure not to exceed $700,000, the Lloyd Corporation initially offered to construct a three-story, 100,000-square-foot building on two blocks that lay on the west side of the Benson Polytechnic School, located south of Sullivan's Gulch (figure 24). To create a 200-by-460-foot site with ample room for parking and the maintenance of state vehicles, as the State of Oregon desired, the Lloyd Corporation agreed to arrange the vacation of the portion of NE 11th Avenue that separated the blocks. Claud Beelman, whom Lloyd retained for the project, advised that the building could be constructed to his plans and specifications for $500,000. A hopeful Ralph Lloyd wrote to Governor Charles H. Martin and the Board of Control that he could not "think of any objectionable feature applicable to [his proposal] that would bar it from being recommended without qualification."[19]

The Lloyd Corporation's proposal was apparently well received when Franz Drinker and his colleagues in the Portland office presented it to the Board of Control on the last day of June and then met with Governor Martin and other state officials to discuss it. In the wake of the violent so-called Little Steel strike that had swept across the Midwest in May—following on from the bloody General Motors sit-down strike in Flint, Michigan, that had resulted in the automaker's reluctant recognition of the United Auto Workers local as a bargaining unit under the National Labor Relations Act—the State of Oregon was reluctant to assume responsibility for the construction of a building. And so the Board of Control asked the Lloyd Corporation—were it to land the contract—to construct the building and turn it over to the State. The State of Oregon would finance the project through the sale of bonds secured by either rental payments or a portion of its industrial accident fund. In response, Lloyd agreed to accept State-issued rental certificates as payment.[20]

The Board of Control failed to award a bid by mid-1938, however, by which time Ralph Lloyd had lost patience with—as he saw it—foot-dragging on its part. The oilman felt that, by now, his offer to deed the two blocks to the State had expired. Exasperated and frankly more concerned with drilling deep wells in the Ventura Avenue oil field, as we have seen, Lloyd dropped the matter.[21]

Once post–World War II economic conditions stabilized, the Lloyd Corporation constructed a home for the State of Oregon's Auto License Division. As 1945 drew to a close, Lloyd reported that he was keen on "getting matters under way" on a $5 million development program that included a building for the Auto License Division and additional office space for the Bonneville Power Administration, which, as we shall see, had become his most important commercial real estate client. On the advice of Commonwealth, Inc., acting as broker, the Board of Control expressed an interest in block 71, which lay immediately to the south of the former Adcox Auto and Aviation School, which now housed some of BPA's engineers (figure 24). In short order, however, the Lloyd Corporation acquired the block immediately to the west of the Adcox building from the Roman Catholic Archdiocese and offered it as an alternative site. Since NE Union Avenue provided the centrality that it was seeking in a location, the Board of Control considered either block to be acceptable. As a result of further consideration and negotiation, however, the Lloyd Corporation ultimately erected a one-story, 9,720-square foot building designed by Charles Ertz and Thomas Burns on block 83 (figure 24). The Auto

Figure 25. State of Oregon, Auto License Division Building, 625 NE Oregon Street, Portland, constructed 1946–1947; Ertz & Burns, architect; no longer standing. (Photo Ackroyd Photography, 28 February 1949. The Huntington Library, San Marino, California, LCB drawer 1, box 1. Reproduced by permission of Thomas Robinson.)

License Division thus joined BPA and others in an office cluster that materialized west of Holladay Park in the decade before the Lloyd Center broke ground, as we shall see. The Board of Control leased the premises, which employees found to be "well suited for their purposes," for ten years at $750 per month (figure 25). While together they did not lease as much space as BPA, the California and Oregon motor vehicle divisions constituted important public-sector clients who helped Ralph Lloyd to bridge the temporal gap between the collapse of his hotel project and its revival as a hotel-office-retail complex anchored by a regional shopping center.[22]

Building FHA Multi-Family Housing

In 1937, Ralph Lloyd contemplated reviving his Holladay Park hotel project. This time around, he was far less enthusiastic about proceeding with it, reporting to the local business community, "I may build a hotel in Portland, not because I want to get into the hotel business, but because I have property that calls for a development of this type. Otherwise, nothing could induce me to touch hotel investments." He added, in passing, "I feel a great deal the same way about apartments." The hotel would remain on the drawing board for more than a decade, at which time, as we shall see, it would be constructed in conjunction with the Lloyd Center. With a $335,000 mortgage backed by the Federal Housing Administration, however, he would build Park View Apartments, a $500,000, ninety-two-unit garden apartment complex comprising nine two-story structures on four blocks located due east of the Benson Polytechnic School (figure 24). The process by which the project was realized would be contentious and, for Ralph Lloyd in particular, exasperating. Nevertheless, its completion and profitable operation demonstrated both Lloyd's tenacity and determination to develop his East Side holdings and his willingness to use public capital to secure a foothold in the local commercial real estate market.[23]

The garden apartment was embraced by FHA administrators as a solution to slums, tenements, and urban blight. Indeed, the idea of the garden apartment was rooted in social reform. The deterioration of urban life in late Victorian Britain gave rise to a movement, led by Ebenezer Howard and others, to create "garden cities" that combined elements of town and country living. As distilled in America by urban planners Clarence S. Stein and Henry Wright, under whose auspices the Regional Planning Association of America was formed in 1923, in collaboration with Alexander Bing, Catherine Bauer, Benton MacKaye,

Lewis Mumford, and others, the garden city concept featured so-called superblocks—large parcels with limited or no through streets; garage courts that centralized parking; large, open, and consolidated green spaces; and units that turned toward courtyards, gardens, and parks. These characteristics found expression in Sunnyside Gardens, a low-rise complex developed in Queens, New York, from 1924 to 1929, and Radburn, New Jersey, a much larger version of the type whose realization was halted by the Great Depression. The principles developed by Stein and Wright, which nevertheless became known as the Radburn Plan, resonated with New Dealers who sought to address the plight of both the poor who endured deteriorating living conditions and the unemployed who had lost their homes to foreclosure. They resonated in particular with housing officials who also sought to eliminate speculation from the rental housing market, which, they believed, was responsible for the poor planning and shoddy construction that plagued multi-family structures built during the 1920s.[24]

Resurrecting a moribund construction industry and employing its idle workers were foremost on the minds of President Roosevelt, his advisers, and the members of Congress who sought and secured passage of the legislation that created the FHA in June 1934. Under Section 207 of the National Housing Act, the FHA insured first mortgages on multiple-unit housing projects up to 80 percent of their valuation and up to twenty years, amortized by monthly payments. The act reduced the amount of capital that a developer needed to put down, reduced his monthly payments, and extended the repayment period. Before passage of the act, commercial banks and life insurance companies had limited first mortgages on residential projects to half of a property's valuation and demanded repayment in five years. Moreover, the federal government's guarantee of repayment in case of default exerted downward pressure on interest rates, which had hovered above six percent between 1920 and 1934. As amended in 1938, Section 207 permitted for-profit projects, providing Ralph Lloyd with an opportunity to build Park View Apartments. The FHA imposed a maximum interest rate of 4.5 percent, plus a one-half percent mortgage insurance premium that the lender paid to the agency on large-scale housing projects. As amended in 1939, however, Section 207 incorporated a prevailing wage provision, which resulted in the number of multiple housing units built with FHA insurance plummeting in 1940, the year in which Park View Apartments was financed, by more than 73 percent.[25]

To ensure, as the FHA's deputy administrator put it, that developers constructed "honest merchandise designed to meet the needs of broad classes of the people in a manner to hold their occupancy and to resist obsolescence," the agency's Large Scale Housing Division set property standards for multi-family projects. In doing so, it embraced the principles embedded in the Radburn Plan. Large-scale projects had to be "cohesive and efficiently organized"; provide a "high degree" of privacy, with a low density of families per acre and ample open space; feature buildings that were "low in height, modest in scale, and domestic in character," yet "inviting in their outward aspect"; and, finally, offer "convenient and comfortable" dwellings.[26] In sum, the ideal apartment project was one "set in what amounts to a privately owned and privately controlled park area."[27] Unsurprisingly, walk-up garden apartments dominated the multi-family housing market insured by the FHA before America entered World War II.[28]

In the interest of attracting homogenous groups of financially stable, middle-class tenants to "distinctly residential areas," the FHA developed detailed standards and recommendations on all aspects of the large-scale housing project.[29] A conservative agency from its inception, the FHA took steps to ensure that its projects were insulated from intrusion by commercial and industrial development. In cases where a developer, such as the Lloyd Corporation, owned additional property in the area, the FHA might require it to adopt covenants on the uses to which such property could be devoted. Developers were keen to meet these requirements to qualify for FHA insurance, especially in a time of economic instability.[30]

If the case of Park View Apartments is any indication, however, developers did not accept any and all of the agency's requirements without question. Moreover, local FHA officials influenced project outcomes by engaging in preliminary discussions regarding prospective projects and negotiating the terms of loan applications once their counterparts in Washington had vetted them. Finally, the FHA was charged with reviving the building industry. The agency had an interest in approving applications, not in rejecting them. To be sure, the FHA exerted substantial influence on the type and character of projects that it backed and thus on the economic, racial, and social composition of the tenants who subsequently occupied them. From a developer's perspective, however, the devil was in the details, as far as making a project "pencil" was concerned. Realizing Park View Apartments required compromise on the part of both the FHA and the Lloyd Corporation.

In May 1938, when Ralph Lloyd was considering building an FHA-insured multi-family residential project east of the Benson School, the rental market on Portland's East Side was not favorable for private development. Metzger-Parker

Company reported a vacancy rate of 10 percent for the 700-or-so units under its management—a figure that was in line with the 8–10 percent rate for Portland as a whole, as published by the Apartment House Association. Further, rents were 25 percent lower than they had been in 1929, yet building costs were roughly equal. Unfurnished, two-room apartments were renting for $30–32.50 per month, three-room flats for $35–40, four-room apartments for $45–50, and five-room flats for $55–60. At the same time, the FHA, which was interested in adding 300 units or more to Portland's apartment housing stock, was using $15 per room per month as a metric for local developers to use for its projects.[31] Eventually, the FHA would adopt a policy of insuring projects "whose rental range is low enough to be acceptable to a large segment of the renting public."[32] As of 1938, however, the agency set a rental floor that overpriced the local market.

An influx of BPA employees boosted the East Side rental market and helped to persuade Ralph Lloyd to pursue the FHA project. Even as he retained Charles Ertz to design a garden apartment complex to meet the federal housing agency's standards, he negotiated a lease to put BPA engineers in the vacant Adcox Auto and Aviation School building and construct an administration building for the federal agency at the corner of NE 8th Avenue and NE Oregon Street, as we shall see. Frank Drinker ensured Lloyd that he would have little problem letting his FHA-insured apartment units. For BPA surely would stimulate additional demand for housing for its staff by the time that the project was completed. Prospective BPA-driven demand for apartment housing on the East Side was a key factor in Lloyd's decision to move ahead with the Park View Apartments project. Thus, BPA assisted Lloyd as a real estate developer indirectly even as its demand for office space for its engineers and managers benefited the oilman directly.[33]

Ralph Lloyd sparred with FHA officials on three fronts, namely the composition of the project, financing, and covenants governing properties surrounding the proposed site. Before concluding an agreement with the federal housing agency, Lloyd would step aside out of frustration with the process or for reasons of health, or both, allowing Richard R. Von Hagen, his son-in-law (by way of marriage, on 8 October 1936, to Lulu May, the youngest daughter) to conclude negotiations with the FHA as president of Park View Apartments, Inc., the Oregon corporation that the Lloyd Corporation organized to operate the housing project.

The Park View project originated in Los Angeles, where Ralph Lloyd and David W. Yule, the Lloyd Corporation's internal auditor, discussed it with local

FHA officials. Based on these initial meetings, Lloyd charged architect Charles Ertz with developing a proposal. After one or more internal reviews of Ertz's design and budget, Franz Drinker and Thomas Burns, Ertz's design partner, presented the project to the FHA office in Portland. Local officials approved the layout of a $500,000 project, featuring nine two-story brick buildings in the "modern French idiom" arranged around a central square, with each building facing an interior courtyard. Agency officials apparently did not object to the inclusion of a rear, street-facing entrance to each building. They balked, however, at Ertz's inclusion of nineteen five-room and twelve four-and-a-half-room units that raised the average monthly rent per unit well above the $40 maximum set by the FHA. Officials expected the architect to reduce the size of these units. In addition, Burns concurred with FHA officials that a number of Ertz's three-and-one-half-room units should be reduced to three rooms to align the project with local conditions and reduce its average monthly rent.[34]

In the wake of this meeting, the Lloyd Corporation applied for a $350,000, FHA-backed mortgage loan for a proposed ninety-two-unit project that retained the five-room apartments. It would contribute the site and place $17,000 in an escrow account. While he awaited word from the agency, Franz Drinker learned from his contacts that local officials had recommended a number of changes to the plans and rental schedule before they forwarded the application to the FHA office in Washington. Curious to learn first-hand how a comparable project had met FHA expectations, Ralph Lloyd's man in Portland traveled to Seattle.[35]

In setting standards for design, construction, and materials, the FHA sought both to maximize local acceptance of a project and to minimize maintenance and repair costs over the life of its buildings.[36] These standards, however, did not prevent the construction of unattractively designed and cheaply built housing, particularly in the context of the FHA's belated drive to make the housing that it financed more affordable. In January 1939, Drinker visited Edgewater Park, a 304-unit project sited on Lake Washington that *Architectural Forum* used to show architects how "to shave construction and operating costs in low rent garden apartments."[37] He walked away from his tour unimpressed.

Designed by John Graham Sr., father of the architect who would design the Lloyd Center, and William Painter, Edgewater Park was roundly criticized within the local real estate community, reported Drinker, because it resembled an army barracks and obviously was constructed of low-end materials. Composed of eighteen two-story buildings arranged around and within a U-shaped access street, the project grouped four families to a floor in three- and four-room apartments around a public stairwell. Four three-room flats comprised the typical configuration. At the FHA's insistence, units included no breakfast nooks or dining rooms. The kitchens were needlessly small, in Drinker's opinion, and the only way to reach it in many units was through the front hall. The buildings were located far from laundry facilities, raising the ire of housewives. In sum, according to Drinker, the apartments, which rented for an average of $14.50 per room, offered their occupants "no features whatsoever [to] make them at all attractive." At the same time, he concluded, with proper planning, the Lloyd Corporation could learn from Edgewater Park in the design and construction of the "better apartment house" project in which it was interested and still satisfy FHA requirements.[38]

Franz Drinker returned to Portland to learn that FHA officials in Washington, like their Oregon counterparts, objected to the inclusion of five-room units in the plans, and recommended that the architect reduce them to three-and-a-half rooms. Ultimately, Ertz would remove one room from each of the five-room units, reducing to 342 the total number of rooms in the project. Complying with FHA standards on the design of the project, however aggravating it may have been to Ralph Lloyd, was less "burdensome" than other changes and additions that, in his opinion, "render[ed] the project hazardous from an economic and practical standpoint."[39]

Once it agreed to comply with the FHA's conditions for insuring a loan, the Lloyd Corporation anticipated that the agency would approve its application. It therefore entered the market, seeking a four percent, $350,000 loan when the going rate was close to 4.5 percent. The FHA offered some assistance in stabilizing the market, setting a ceiling of 4.25 percent on its mortgage interest rate by the fall of 1938. By the end of the year, life insurance companies were offering four percent loans. Only in August 1939, however, did the Lloyd Corporation interest New York Life Insurance Company in financing the Park View project at four percent. At the same time, the FHA committed only to a $335,000 loan at 4.25 percent. In addition, the FHA required the Lloyd Corporation to place $24,500 in escrow (rather than the $17,000 that it had offered), set aside cash to pay for change orders, and increase its reserve for debt service by $11,306 (to compensate for the higher interest rate in the loan). The FHA also reduced the number of rooms, as mentioned, imposed rent ceilings, and required the Lloyd Corporation to encumber its adjoining properties with restrictive covenants. Lloyd considered the FHA's commitment to be nothing less than a rejection of the loan application.[40]

Ralph Lloyd asked Portland attorney Robert Treat Platt, whom he had retained to assist him in organizing Park View Apartments, Inc., to advise him in the matter. The California developer complained that the regulations and controls imposed by the FHA rendered the Park View project financially unsound. Still, he was willing to proceed with construction under the terms of the initial application. As a fig leaf, he offered to meet the FHA part way in limiting the use of some of his adjacent properties to single- or multi-family residential construction.[41]

Lloyd also appealed to Paul S. Dick, president of the United States National Bank, and Ross McIntyre, president of the Portland Chamber of Commerce and a director of the bank, both of whom had offered to contact FHA officials on behalf of the project. Together with Platt, the two men met with Jamison Parker, the FHA's director for Oregon. As McIntyre related to Lloyd, the three men "impressed [Parker] with the necessity of doing everything in his power" to get the project approved "because of the standing, not only of yourself and your corporation, but the unusual high class location of the property." Parker wired John E. McGovern, the FHA Rental Housing Division's zone (regional) rental manager in San Francisco, whom, he felt, would able to resolve the differences between the Lloyd Corporation and the housing agency. McGovern agreed to promptly meet with Lloyd's representatives and local FHA officials.[42]

In Paul Dick's office, McGovern reviewed Lloyd's complaints with architect Burns, Dick, Drinker, McIntyre, and Platt, with Parker in attendance. In doing so, he resolved to the satisfaction of Lloyd's attorney that the Park View project could proceed under the FHA's commitment. McGovern explained that the FHA had reduced the amount of the commitment in line with the removal of nineteen rooms from the project. Moreover, the FHA had no objection to a four percent loan, if the Lloyd Corporation could identify a private lender willing to offer a mortgage at that rate. The increase in the amount that the FHA required the Lloyd Corporation to place in escrow reflected a general decision on the part of the housing agency, based on its experience, to increase the amount of working capital supplied by project sponsors as a percentage of the mortgage. The reserve for debt service that Ralph Lloyd found troubling would stand. At the same time, the FHA would allow the Lloyd Corporation to earn interest on the reserve at the same rate as the mortgage and use it to reduce the principal on the mortgage. McGovern also assured his audience that FHA supervision of the disbursement of the mortgage would not impose a burden.

Moving down the list before him, the zone rental manager explained that no material difference existed between the average rental of $14.68 in the FHA commitment and the average rental of $12.98 proposed in the Lloyd Corporation's application: the agency had merely incorporated the cost of gas and electric utilities into the rental figure. Further, the agency would not regulate rents, as it had no legal authority over the operating company. The FHA would interpose itself only if the project spun out of control economically, so to speak. The language in the commitment that so bothered Ralph Lloyd had been inserted because sponsors of projects that the agency had insured to date not always had had adequate financial backing. Finally, McGovern assured his audience that the agency gathered monthly, semi-annual, and annual reports strictly for purposes of policymaking. That is, it did not do so for purposes of supervision and control. Based on McGovern's presentation, Robert Platt concluded that a "thorough misunderstanding" of the FHA's commitment lay at the heart of Ralph Lloyd's distress. He asked Thomas Burns to brief his design partner on the structural changes required of the FHA that accounted for the principal differences between the loan application and the agency's commitment. Charles Ertz, in turn, would convey them to Lloyd. For his part, McGovern joined Ertz in that discussion "to further elucidate any features of the administration's commitment" that were "not entirely clear."[43]

Establishing restrictive covenants on adjacent properties remained contested. With the incorporation of Park View Apartments, Inc., in place, Ralph Lloyd delegated responsibility of reaching agreement with the FHA to Von Hagen, an attorney by trade. State Director Parker had indicated that the FHA would be satisfied with an agreement from the Lloyd Corporation not to develop for business or industrial use blocks that bordered the site on its north and west sides, blocks located immediately south and southwest of the site, and certain property now occupied by the Lloyd Golf Course. Lloyd balked at restricting all of these properties to residential use. As his son-in-law explained, the Lloyd Corporation owned none of the property adjoining the two blocks that lay to the south of the Park View site, making it "inadvisable" to place restrictions on these blocks. Therefore it was willing to encumber only the southern halves of the blocks that lay to the north of the site and the eastern halves of the blocks that lay to the west of the site. Further, the restrictions would remain in effect only so long as the FHA's insurance commitment on the mortgage loan remained in force. Ultimately, Lloyd agreed to restrict the entire blocks that lay immediately to the west and north of the site to residential use. In holding the line against restricting two blocks on the south side of the apartment site in like manner, Von Hagen argued that "the success of this project in no way depends upon the placement of restrictions upon [them], for the hazard of future development to the south inimical to the success of the project would in no way be lessened by such restrictions." He added that the FHA was "fortunate" to have a project protected on two sides, for "we know that there are many such projects located in areas in other cities which have been placed in much less desirable conditions and under circumstances which prevented the [FHA] from obtaining any restrictive covenants on the adjacent property." Indeed, the East Side neighborhood in which Park View was located was, and would remain, desirable in the absence of restrictive covenants. Von Hagen's argument apparently carried the day. For the FHA agreed to confine its restrictions to the properties that lay to the north and west of the site. In doing so, the federal housing agency ceded ground to an insistent developer to move a project forward.[44]

Still, the start of construction was delayed until September 1940 while FHA officials, architects Burns and Ertz, and representatives of the Lloyd Corporation engaged in so-called value engineering, a process by which the parties sought reduce the project's costs without compromising its quality. In doing so, the Lloyd Corporation demonstrated that developers did not necessarily capitulate to FHA officials in the interest of securing the federal housing subsidy for their projects.

At the request of Zone Rental Manager McGovern, the FHA's Zone Architectural Unit in San Francisco vetted the architect's plans and specifications in the spring of 1940 with a view to reducing the project's costs so that, as McGovern explained to Von Hagen, "your cash investment may be reduced and at the same time retain the amenities and general appeal of the proposed rental units." The FHA's commitment of $335,000 equaled 78.5 percent of the agency's Estimate of the Fair Value on Completion for the project of $427,000. As the FHA limited its commitment to 80 percent of its fair value estimate, the agency had little room to increase the amount of the insurable mortgage to $350,000—the amount in the loan application that Ralph Lloyd continued to seek. The Zone Architectural Unit nominated some two dozen features for either elimination or modification.[45]

There ensued a series of meetings and communications between agency officials in Portland and San Francisco and the architects, acting on behalf of the Lloyd Corporation. Burns and Ertz argued vigorously on behalf of retaining features, such as mouldings and oak flooring, which enhanced the aesthetics of the units. They also noted several instances in which incorporating a recommendation would produce little or no savings, such as eliminating ceiling fans. (In this case, the building code required kitchen ventilation: running fan ducts from each kitchen to compensate for the removal of the fans would have increased costs.) Ultimately, the architects, on behalf of the developer, accepted a number of changes offered by the Zone Architectural Unit, such as omitting restrooms and plumbing from the common laundry rooms and using cheaper wallpaper. At the same time, they proposed, and the FHA accepted, a number of cost-saving measures, such as substituting enameled iron toilets for vitreous china units and applying two, rather than three, coats of paint on the interior woodwork of closets and inside kitchen cabinets. After some two months of negotiation, the FHA, the architects, and the Lloyd Corporation found common ground on the final plans and specifications.[46]

The Lloyd Corporation received approval for the project on 28 August 1940 when the New York Life Insurance Company committed to a four percent, $335,000 FHA-insured loan. The final configuration of the Park View project included seventy-two one-bedroom (three-and-a-half-room) units and twenty two-bedroom (four-and-a-half room) units, with eight to sixteen units in each of the nine two-story buildings. Architect Ertz intended each unit to be "as near a private home as possible." The Lloyd Corporation promised that tenants would find them to be "as near ideal from the standpoints of comfort, sanitation, light, ventilation, and recreation as it is possible for housing authorities to devise." The structures in plan occupied approximately 30 percent of the 200,000-square-foot site, the balance of which was dedicated to landscaping, lawn, playing fields, service yards, and eighty-six garages for automobiles, egress for which was provided from NE 16th and NE 18th Avenues. The Lloyd Corporation let the construction contract to local builder Ross B. Hammond, who completed work in June 1941 (figure 26). The units in "Portland's largest and most carefully planned apartment house project of all time," as Ralph Lloyd billed it, rented for an average of $54.59 per month, which was slightly higher than the average of $52.59 per month for the forty-eight projects insured by the FHA in that year. Among its earliest occupants were BPA engineers, reflecting the growing presence of the agency on the city's East Side.[47]

The case of Park View Apartments suggests that both developers and federal housing officials shaped FHA multi-unit residential project outcomes. To be sure, the FHA wielded substantial influence over the project. The Lloyd Corporation as developer, however, did not accept any and all of the agency's requirements without question. It successfully pushed back in the interest of making the project worthwhile financially and marketable upon its

Figure 26. Park View Apartments under construction, 1940–1941; Charles W. Ertz, architect; Ross B. Hammond, builder. Photo undated. (The Huntington Library, San Marino, California, LCB drawer 1, box 1.)

completion. The Lloyd Corporation mobilized the support of local and regional housing officials, who assisted it in securing approval for design features that it sought to implement. The developer also marshaled the support of local business leaders, who hammered home the project's economic benefits to FHA officials. Realizing Park View Apartments required compromise on the part of both the FHA and the Lloyd Corporation.

Building for BPA through Recession, War, and Recovery

As a developer who had committed a substantial portion of his surplus capital to risky oil field development, Ralph Lloyd benefited from the Bonneville Power Administration's relentless and successful drive from its creation in August 1937 to establish and expand a public power domain across the Pacific Northwest. In addition to housing BPA's engineering and drafting departments in the "jinxed" Adcox Auto and Aviation School building, the Lloyd Corporation erected six buildings for BPA staff, five of which constituted the core of an office cluster that materialized immediately to the west of Holladay Park well before the Lloyd Center broke ground. The accretive consolidation of the public agency's operations in Lloyd Corporation structures, sustained through depression and war and accelerated in peacetime, constituted Ralph Lloyd's greatest success as a commercial real estate developer.[48]

Invested by Congress with the responsibility to distribute cheap and abundant hydroelectric power generated by the federal Bonneville and Grand Coulee Dams to agricultural and industrial users throughout the region, BPA Administrator James D. Ross embodied the New Deal ethos of social reform. With missionary zeal, the former superintendent of Seattle City Light, a municipal power company, set out to transform the Columbia Basin. With the enthusiastic support of Interior Secretary Ickes, Ross initiated the construction of a regional power grid, set a uniform, "postage stamp" rate for electricity that undercut the prices of private suppliers, and encouraged the formation of public utility districts. When sudden death, on 14 March 1939, from a blood clot cut short Ross's ambitions, Paul J. Raver, his permanent replacement, fulfilled them. A professor of engineering at Northwestern University who had used his chair of the Illinois Commerce Commission to attack Samuel Insull's public utility holding company empire, Raver worked tirelessly to ensure that public power transformed the region's natural resource–dependent economy, if not its society, in the ensuing years.[49]

Moreover, BPA made a seamless transition to a wartime footing. In the wake of the attack on Pearl Harbor, the U.S. Office of Defense Mobilization federalized the Pacific Northwest's electrical grid by compelling all municipal and private utilities to link their transmission systems to a BPA-controlled distribution network, the Northwest Power Pool. The move entrenched the agency as the primary supplier of electricity to the region. By expanding its transmission network in wartime, BPA accelerated regional urbanization and industrialization. In less than a decade, BPA tripled the region's electric generating capacity. The agency, with the encouragement of New Dealers in the Roosevelt Administration and the Congress, became the Columbia Basin's "indispensable guardian," validating the New Deal argument for public power generation and distribution. At war's end, BPA was poised to extend its geographic reach.[50]

BPA's empire building in depression and war redounded to the benefit of the Lloyd Corporation, as Administrator Ross chose to locate his agency's offices in Portland over Seattle and Tacoma, Washington. And so, as soon as Congress appropriated $3.5 million for the agency on 26 May 1938, BPA officials opened discussions with Franz Drinker on both the terms of a lease for the vacant Adcox building and possible East Side locations for a new administration building. They made it clear that they had no interest in leasing space at the rates on offer within the central business district. At the same time, Ertz & Burns worked on a bid for a building that would meet BPA's requirements for as much as 15,000 square feet of office space and up to 6,000 square feet of garage and storage area. Drinker advised Ralph Lloyd that the project "would certainly be a fine starter for Holladay properties" and "well worth taking [a] chance." The oilman concurred. Constructing an office building for a public-sector client constituted a clear departure from his plans for a hotel, shops, and a theatre around Holladay Park. Those plans, however, remained on the drawing board, at best. After considering at least three locations, Administrator Ross opted for a building on the western half of block 103, which was bounded by NE Oregon and NE Pacific Streets and NE 8th and NE 9th Avenues (figure 24). It offered both a quiet environment—the other two sites under consideration lay along major thoroughfares—and room for expansion, which would be required if, as Administrator Ross expected, Congress expanded BPA's responsibilities to include the distribution of power generated by the Grand Coulee project. Confident "in the permanency of the power development program," as he later put it, yet still hedging his bets, Lloyd authorized Drinker to offer BPA officials a one-story, steel-truss structure that facilitated the reconfiguration of office space if the agency did not renew its lease. Ertz & Burns supplied a bid for a 16,701-square-foot, Class A office building that would rent for $17,836 per year. The prospective building was included in a five-year lease under which BPA also leased the Adcox building as an annex for engineering staff for $8,400 annually. In short order, the Lloyd Corporation secured the support of the Portland Planning Commission for a zoning change for the block, which the City Council promptly approved. In September, BPA officials in Washington approved the construction of the new building.[51]

Soon after he was named Ross's permanent replacement as BPA administrator, Paul Raver acted to consolidate his engineering and drafting departments, now operating in two buildings in Portland. In late November 1939, he met with Franz Drinker to discuss the capacity of the Lloyd Corporation to accommodate his requirements for an additional 20,000 square feet of office space. The 17,000-square-foot Adcox building supplied about 11,300 square feet of usable work space. Converting the remaining space to office use would fall short of satisfying the needs of the BPA administrator. The main building in which BPA's engineers were working included a garage that had been converted into a work area. It was unheated, however, and could not be used during the winter months. Raver laid out his options, as he saw them: he could expand the main engineering building or abandon the Adcox building for larger quarters.[52]

At the same time, Paul Raver elaborated his long-term vision for consolidating his entire staff in what later would be called a corporate office park centered on the new administration building. He foresaw a two-story office building on block 90, across NE 8th Avenue from the administration building, that would provide space for BPA's real estate and purchasing departments, now housed on the West Side. A one-story building would share block 103 with the administration building. On the triangular block bounded by Lloyd Boulevard, NE Oregon Street, and NE 7th Avenue, Raver envisioned a "monumental" headquarters building, replete with an exhibit hall to showcase the work of his men, in keeping with his imperial ambitions for the agency (figure 24).[53]

With "every real estate agent in the city" competing to put BPA staff "in larger quarters," the Lloyd Corporation's Portland office, in cooperation with Ertz & Burns, developed a proposal for a $73,000 engineering building on block 70 (north of the Adcox building) with a ground floor and mezzanine that together would offer the agency 25,200 square feet of leasable space. In discussing the proposal with Ralph Lloyd, Franz Drinker whetted his boss's appetite with an overview of the BPA administrator's vision for Lloyd's holdings just west of Holladay Park. Lloyd traveled to Portland the week of 19 February 1940 to meet with Raver and tour the site of the administration building now under construction.[54]

In the wake of this trip, BPA officials agreed to lease Bonneville Engineering Building No. 1 (as it would be known within the Lloyd Corporation), which architect Thomas Burns estimated would cost $80,604 to construct, for five years at $27,173 per year. The initial proposal was revised to raise the height of the ground floor by four feet to accommodate a balcony for blueprinting equipment. At the same time, Paul Raver asked the Lloyd Corporation to upgrade the Adcox building to the standards of the new building, as detailed in Burns's specifications. Ralph Lloyd agreed to spend as much as $15,000 on the project, which included the conversion of 4,500 square feet of garage space to offices. In September 1940, Drinker secured a supplemental rent increase for the renovated Adcox building, rechristened Bonneville Engineering Building No. 2, to $11,564 per year.[55]

For the moment, Paul Raver set aside his plans for additional buildings on Lloyd Corporation properties. Nevertheless, during the war, he moved staff from West Side locations into the Lloyd Corporation buildings to accomplish part of his consolidation objective.[56] The three buildings in which BPA operated between 1940 and 1 July 1945, when the agency and the Lloyd Corporation reached agreement on a new lease for all three structures, provided more than $4,700 per month in rental income.[57] Given his frustrations in interesting department store owners in his Beverly Hills block and in dealing with FHA officials on the Park View Apartments project, Ralph Lloyd must have seen Raver's BPA as a godsend. Investment of public capital by a federal agency charged with responsibility for regional economic development through hydroelectric power provision afforded the oilman an uncommon opportunity in time of depression and war to capitalize his real estate assets. Moreover, Raver's long-term ambitions for an office cluster near Holladay Park gave Lloyd cause to believe that the development of what he was now calling the Lloyd Center would be possible once normal economic conditions resumed.

With the Axis powers vanquished, Paul Raver moved to extend BPA's mission of economic and social transformation beyond the Columbia Basin, and identified a series of multi-purpose dams on the Snake River as indispensable to achieving the goal of selling federal electricity at cost throughout the region. Indeed, the BPA administrator envisioned a public power domain that encompassed the West. Driven by these ambitions, Raver resumed discussions with the Lloyd Corporation regarding the consolidation of his agency's operations on the East Side. Owing to the exigencies of defense mobilization, BPA employees were now scattered among numerous sites across the city, among them the Port of Portland's Swan Island facility, which had served as the location of one of Henry J. Kaiser's shipyards.[58]

With victory in Europe in sight, Paul Raver also sought funds for a headquarters building, requesting authorization to transfer $4,438,000 in appropriated funds not spent by his agency to acquire a site and erect a building.[59] At the same time, Ralph Lloyd offered to contribute land for a building "of distinction" that would "represent properly the enormous water power that would be under the Administrator's control."[60] Specifically, Lloyd offered to the U.S. Government: two blocks (111 and 122) bounded by NE Hassalo and NE Holladay Streets and NE 9th and NE 11th Avenues that would serve as the site of the building; two blocks (131 and 142) immediately to the east of the building site (and south of Holladay Park) and several lots on either side of Lloyd Boulevard that, together, would provide space for a public park to "enhance the setting for this building"; and parts of the two blocks immediately to the west of the building site that would be devoted to parking (figure 24). Lloyd hammered home to Administrator Raver and Interior Secretary Ickes the proposed site's strategic location "in the geographic and population center" of Portland. He called attention to the proximity of the site to the central business district even as he enumerated the suburban-like amenities of his properties that enhanced its attractiveness, including the golf course, parks, and "ample room" for expansion.[61]

The Bureau of the Budget rejected Raver's request for, as Ralph Lloyd put it, a "fitting monument to [the] majesty" of "the vast power of the Columbia River Basin" under BPA's control, arguing that wartime restrictions banned the construction of public works not justified by the national emergency. The signature building that Raver desired would not materialize on the site for another decade. In the meantime, the Lloyd Corporation met the needs of the BPA administrator with a suite of two-story reinforced concrete buildings that created an office cluster around the administration building at 811 NE Oregon Street.[62]

Paul Raver brought his agency's pressing need for additional and more adequate office space to Ralph Lloyd's attention, initiating discussions that culminated a year later in BPA's acceptance of bids to construct two engineering buildings, known as Bonneville East and Bonneville West, on either side of the administration building. Raver called initially for a single building to be located near the two engineering buildings along NE Union Avenue. The Lloyd Corporation was unable to acquire the additional properties that it would need to accommodate this request, however. BPA officials then considered lots near the administration building and decided that block 90 would serve as a suitable location for a four-story office building (figure 24). The balance of the block in which the administration building was located would serve as a parking lot.[63]

Ralph Lloyd approved the location of the new building. Notwithstanding the apparent permanency of BPA as a client, however, he was adamant that he would not approve a building that was designed specifically for the specialized needs of the agency. He instructed Franz Drinker to offer BPA officials either a building with as few "partitions and divisions" as possible or a typical office building to avoid the "enormous loss [that] would take place when the building is vacated and the whole inside of the building revamped." Agency officials opted for the latter choice. The Lloyd Corporation responded with a bid, which BPA awarded on 2 November 1945, for a 70,776-square-foot, $571,000 building that it would rent for $11,560 per month. A single structure on block 90 became two two-story structures on the eastern halves of blocks 90 and 103 when the Lloyd Corporation encountered delays in acquiring the lots that comprised the southwest quarter of block 90 (figure 24).[64]

BPA officials and the Lloyd Corporation ultimately agreed on preliminary plans developed by Thomas Burns for a 49,073-square-foot building ("Bonneville West") at 729 NE Oregon Street and a smaller, 26,388-square-foot building at 827 NE Oregon Street ("Bonneville East"). Heavily invested in oil field development, the Lloyd Corporation financed the projects with loans from the Mutual Life Insurance Company. On 1 July 1946, it awarded the construction contracts to Ertz & Burns.[65]

As he had done with Frank Chevrolet, Lloyd retained Claud Beelman as consulting architect. Beelman's brief included reviewing the plans and specifications for Bonneville Engineering Building No. 1. He offered a critique that identified structural weaknesses in the support provided for the heavy equipment installed in the building and made a number of recommendations to strengthen the floors in the new buildings, which would serve similar engineering functions. His influence on the design of the buildings was also expressed aesthetically in the treatment of their façades.[66]

The Art Moderne façades of the buildings were characterized by strong vertical lines reminiscent of the Art Deco Farmers Insurance Building, designed by Charles Ertz, that was constructed in 1930 at 1785 NE Sandy Boulevard.[67] Shorn of the decorative details that, for instance, adorned Beelman's Eastern Columbia and Sun Realty buildings, the exterior featured columns that extended from the base of the building to the roof, separating narrow strips of

Figure 27. Bonneville West Building, 729 NE Oregon Street, Portland, constructed 1946–1947; Ertz & Burns, lead architect; Claud Beelman, consulting architect; Ertz & Burns, builder. Photo undated. (The Huntington Library, San Marino, California, LCB drawer 1, box 1.)

windows accented by metal frames and vitrolite. In contrast to the Farmers Insurance Building, the windows did not extend to the roofline. Rather, concrete panels above and below the windows, set slightly inset from the columns, unified the façades. The design muted the verticality of the massing of the buildings relative to the Farmers Insurance Building. But the effect was no less modern (figure 27).

Construction of the Bonneville East and Bonneville West Buildings proceeded in the context of accelerating postwar inflation nationally, from 2.25 percent in early 1946 to a peak of more than 19 percent in the spring of 1947. Owing to the deployment of the trades on Saturdays and on an overtime basis, the Bonneville West Building "progress[ed]fairly well" to its completion in June. But the costs associated with delivering the building on this date were "terrific." Construction of the Bonneville East Building lagged its counterpart by three months. In the meantime, prices did "settle down somewhat," as Franz Drinker hoped, with inflation falling below 11 percent by the time that BPA employees occupied the building. Confident that inflation would continue to subside and having recuperated from a bout with gastroenteritis that had sidelined him for much of the spring, Ralph Lloyd was ready to plan new commercial building projects. And so he was prepared to meet BPA's immediate need for additional office space.[68]

Almost immediately after he moved his employees into the Bonneville East Building, Paul Raver contacted Franz Drinker. He thanked the Lloyd Corporation for the two new buildings, which his engineers found to be wholly satisfactory, and inquired whether it could erect a building of similar size, and adjacent to, the Bonneville West Building as soon as possible.[69] BPA moved its engineers and draftsmen from Bonneville Engineering Building Nos. 1 and 2 into the new buildings. In so doing, it achieved a net gain of less than 30,000 square feet in office space with the completion of the projects.[70] Regional industrial growth fueled by public power generation was producing acute power supply problems that called for "continued speedy installation of new generation" in existing plants and the construction of additional dams and high capacity transmission lines.[71] Raver aimed to address the cost of accelerated program expansion in his agency's budget for fiscal year 1949. Congressional approval for the request, which Raver anticipated, would mean that he would need to hire 250 employees by 1 July 1948. As a result of staff consolidation to date, BPA would have no place for these new hires unless it obtained additional space.[72]

Figure 28. General Petroleum Building, 710 NE Holladay Street, Portland, constructed 1949; Ertz & Burns, architect and builder. (Photo by Photo-Art Commercial Studios, 21 November 1950. Photo-Art Collection, Oregon Historical Society, Negative 142671-3.)

The Lloyd Corporation now also benefited from the return of private capital to the commercial real estate market. Before the completion of the Bonneville West Building, General Petroleum (GP) approached the Lloyd Corporation's Portland office regarding a building to serve as its Oregon headquarters. The California oil major was an early mover among private employers seeking to abandon their downtown Portland offices, which were renting at a premium, for more relaxed and spacious surroundings. With 230 employees housed in the former Oregon Mutual Life Building, GP was "desirous of getting out of the congested area and onto a full block where they can have ample parking space and a model super-service station in conjunction with their offices," Franz Drinker reported.[73] Willing to sign a twenty-five-year lease and build out the interior of a two-story office building at its own expense, GP offered Ralph Lloyd terms about which he only could have dreamed for the better part of the past two decades. Well acquainted with GP's executives, who were planning to erect a headquarters of their own in downtown Los Angeles, Lloyd instructed his man in Portland to inform local GP managers that "we are interested in working out a nice location for them at a place and on terms that will be mutually satisfactory."[74]

GP's desire to abandon its quarters provided evidence to business owners, municipal officials, and planners that their growing concern regarding the obsolescence of Portland's downtown was not misplaced. Those concerns often focused on the challenges faced by retail merchants in enticing potential customers to their stores.[75] Citing increasingly frequent complaints about the difficulty of getting into downtown from the East Side, for instance, the *Oregon Journal* opined that Portland had to adjust to the geography of its postwar demographic and economic growth, now that it had become "a metropolitan area beyond the wildest dreams of its founders."[76] The newspaper's editors called for the extension of the city's retail district beyond its downtown confines—undoubtedly to the delight of Ralph Lloyd, who would soon begin planning seriously for the Lloyd Center, as we shall see. GP's initiative to relocate its Oregon headquarters, however, suggests that office workers, no less than shoppers, were becoming disenchanted with Portland's central business district.

When BPA Administrator Raver contacted Franz Drinker regarding a new office building, negotiations between GP and the Lloyd Corporation were reaching fruition. The parties had agreed on a building, sketched by Thomas Burns, that essentially replicated the Bonneville West Building in plan, but with a glass-dominated façade that the architect executed by doubling the width of the window strips between columns that lay to either side of the main entrance and framing the entrance itself within a rectangular bank of metal-framed windows that was four times the width of the strips on the Bonneville West Building (figure 28). Drinker provided GP with a tour of the Bonneville West Building to demonstrate the "quality of workmanship we propose in the new building." The two sides were in the process of selecting the block that the GP building and associated parking lot would occupy.[77] Paul Raver was well aware that GP had approached the Lloyd Corporation for office space in the vicinity of his new buildings, so he beseeched Drinker to give his request "your most serious attention."[78]

Ralph Lloyd traveled to Portland the week of 17 November 1947 to discuss BPA's requirements with Administrator Raver.[79] Ultimately, the agency accepted the bid of the Lloyd Corporation for a two-story, 52,000-square-foot reinforced concrete building, with basement, on the block (102) that lay immediately north of the Bonneville East Building. Under a five-year lease that provided for annual payments of $129,018, BPA moved staff into the so-called Bonneville North Building in the fall of 1949. With the simultaneous construction of the GP building immediately to the west of it on block 91, the Lloyd Corporation completed a cluster of four buildings of similar design and appearance that constituted a core from which the Lloyd Center would materialize (figure 29).[80] For Paul Raver, all that remained to realize his vision of consolidating his staff on the East Side was a "monumental" headquarters building that would serve as centerpiece of the agency's public power empire.[81]

The Port of Portland's decision to construct a dry dock and ship repair center at Swan Island initiated planning for a building that ultimately would house not only BPA employees, but those of other Interior Department agencies. Together with BPA, these agencies were leasing a total of some 125,000 square feet of temporary office space for about 700 employees. To realize its plans, the Port needed to reclaim the area occupied by these wartime structures. In 1949, the U.S. General Services Administration accommodated the Port by agreeing to vacate its Swan Island premises by 8 March 1952. In the summer of 1950, GSA officials opened negotiations with the Lloyd Corporation for a building that would be located on the two blocks that Ralph Lloyd had offered Paul Raver in 1944. No less than the BPA administrator, Interior Department officials were interested in the consolidation of staff. They estimated that they would need 300,000 square feet of office space to concentrate the employees of their agencies, including BPA, in a single location. Moreover, with the outbreak of war on the Korean Peninsula, they deemed interagency coordination of programs and activities to be vital to the defense of the nation.

Figure 29. Lloyd Corporation Properties, East Side, Portland, July 1954. Together, the newly completed U.S. Department of the Interior Building and six office buildings constructed between 1940 and 1949 constituted Ralph Lloyd's greatest success as a developer. Note the excavated hotel site east of Holladay Park. A portion of Park View Apartments is visible to the right of the golf course. (Photo by Delano Aerial Surveys. The Huntington Library, San Marino, California, LCR drawer 12, box 5.)

And so the construction of a "monumental" federal building on Lloyd Corporation property, once considered to be an extravagance, would proceed as an essential wartime project. As GSA Regional Director O. C. Bradeen put it, by creating space for soon-to-be-displaced federal employees, "the orderly development of the defense program in the Portland area would thus be greatly accelerated."[82]

The need to find a home for displaced Swan Island workers may have created the impetus for the project. Nevertheless, as Bradeen explained, "the ultimate objective was to provide adequate and efficient office space" for BPA, whose employees continued to work at fourteen locations in Portland. To meet the needs of the agency, the Lloyd Corporation proposed to build an eight-story building, comprising approximately 130,000 square feet of office space and some 90,000 square feet of basement, garage, and storage area, and a 20,000-square-foot "special purpose" annex. With the four buildings that BPA occupied in the burgeoning office cluster west of Holladay Park offering 143,419 square feet of usable office space, the new building would offer nearly all the additional space that Interior Department officials predicted that they would need to accomplish their consolidation objective.[83]

The "ultimate objective" of the project provided BPA Administrator Raver with the opportunity to project both the public power and technical capability of the agency in the new building. For all intents and purposes, the building would materialize as the BPA headquarters that he had envisioned more than a decade earlier even if, owing to a reduction in force, BPA would not occupy the entire building. Under an agreement reached between GSA officials and the Lloyd Corporation on 12 April 1951, John Graham Jr., in consultation with Dean Wright, the government's architect, designed a structure that was both monumental in scale and innovative in structure, with a gleaming white, marble façade masking what the American Institute of Steel Construction would call "a remarkable construction job." Upon its completion, all of BPA's employees in Portland would move into the new building. The employees of other federal departments would occupy the four buildings that it vacated.[84]

To assemble the site for the 360-foot-long, fifty-foot-wide building, with a two-story special-purpose wing, 199-by-72-feet in plan, that formed an "L" with the larger structure, and parking, the Lloyd Corporation petitioned the City Council for the vacation of NE Tenth Avenue between Lloyd Boulevard and NE Holladay Street and a change in zone for blocks 111 and 122. Based on the City engineer's recommendation that NE 9th Avenue be widened to eighty feet between NE Oregon and NE Hassalo Streets to handle traffic from

Figure 30. U.S. Department of the Interior Building, Portland, constructed 1952–1954; John Graham & Co., architect and engineer; Dean Wright, federal architect; A. L. Miller, seismic design consultant; Gil Schaller, welding procedure consultant; Ralph & Horwitz, builder. This photograph was the last of a series of photographs commissioned by the Lloyd Corporation to document daily progress in the construction of the building. (Photo by Steffens-Colmer Studio, 30 April 1954. The Huntington Library, San Marino, California, LCR drawer 12, box 5.)

connections to the proposed Banfield Expressway that the State of Oregon would build through Sullivan's Gulch in the coming years, and the City of Portland's desire to widen NE Multnomah Street, the Lloyd Corporation agreed to exchange portions of several lots along the streets that these projects would affect to secure the vacation. The City Council unanimously approved the petition.[85]

Under a construction contract awarded on 28 August 1952, Ralph & Horwitz erected a building that was as distinguished structurally as it was distinctive aesthetically.[86] The architects and engineers reduced the steel content of the frame by more than 10 percent by substituting welds for rivets. In fact, the building was the first major structure in Portland to deploy a welded steel frame. The methods employed to construct the frame, which also helped the building meet the requirements of resisting the seismic forces unleashed by an earthquake in Seismic Zone 2 received widespread praise in the trade press.[87] At a time of increasing congressional attention to alleged extravagance and waste on the part of federal agencies, BPA secured approval from the City Council for a change in the building code to allow the installation of movable partitions with a sixty-minute fire rating in place of the masonry partitions required under the former code. Flexibility in office space provided Interior Department and BPA officials with the capability to respond to changes in their labor forces produced by federal budgets. As it turned out, the new building supplied for the moment all the additional space that the Interior Department required to consolidate its Portland work force. For the number of federal employees in the city dropped by 16 percent between June 1951 and December 1953.[88] The reduction owed in large part to retrenchment in BPA's construction program as a result of the Eisenhower Administration's successful effort to replace federal control of regional hydroelectric power with a "partnership policy" that reinserted private interests into river basin development.[89] BPA employees would occupy only the first six of the eight floors as well as the basement. Staff from other Interior Department agencies would fill the remaining space. Thus the building became known as the U.S. Department of the Interior Building. Federal employees from other departments, including Agriculture, Commerce, Defense, and Transportation, occupied the four Holladay Park office buildings formerly leased by BPA.[90] In April 1954, Ralph & Horwitz completed BPA's new $4.4 million home (figure 30).[91] The following month, Richard R. Von Hagen announced plans for the construction of the Lloyd Center.

Upon the conclusion of the agreement between GSA officials and the Lloyd Corporation to build a BPA headquarters building, Dean Wright, the U.S. government's consulting architect for the project, congratulated Von Hagen and noted, "We at BPA have dealt with Mr. Lloyd since our beginning in 1938 and we have come to look upon him as a godfather to our administration building needs."[92] In fact, the relationship between the agency and Ralph Lloyd, however beneficial mutually, was a godsend for the beleaguered developer who, as we have seen, struggled to lease space to private sector tenants even as the economy rebounded from the worst depression in American history. In April 1954, GSA Regional Director O. C. Bradeen estimated the value of the Lloyd Corporation's real estate holdings in Portland to be $15 million.[93] In terms of fair market value, the seven buildings that BPA was occupying or had recently occupied accounted for more than a third of this amount. Including Park View Apartments and the State Auto License Division Building, public-sector clients or projects funded with public capital accounted for roughly 40 percent of the value of company's Portland portfolio, as estimated by Bradeen. Even that measure falls short of representing the importance of public capital and public agencies to Ralph Lloyd's commercial real estate activities between the collapse of the Holladay Park hotel project and the postwar revival of private markets and private sources of capital. As the project to build GP's Oregon headquarters suggests, by the end of the 1940s, corporate executives once again were prepared to invest in new office space for their employees. At the same time that it was completing GP's building, for instance, the Lloyd Corporation was constructing a Claud Beelman–designed building for Mack Truck International and a second structure for International Harvester on Portland's East Side.[94] From 1938 to 1949, however, BPA was the Lloyd Corporation's most important commercial real estate client.[95] The irony may have been lost on Ralph Lloyd. Yet, as we have seen, he was happy to seize the opportunity to fulfill Paul Raver's desire to consolidate his staff in a cluster of buildings that represented aesthetically the technical expertise of the agency and projected the BPA administrator's regional ambitions. In doing so, Ralph Lloyd materialized his holdings around Holladay Park in a manner that he hardly could have imagined in 1926 and provided a solid foundation on which to proceed with the Lloyd Center.

Chapter 8

The Suburbanization of Urban Space: The Lloyd Center

The realization of a dream of almost a half century duration of a man who dared to believe that Portland's progress would eventually make it the great metropolis of the great northwest, and who gambled millions on his dream.

—W. Joseph McFarland and Richard G. Horn, 1954

Ralph Lloyd "harbored dreams of creating an extension of [Portland's] urban core that would become in effect a 'new city'" with the development of the large block of property in the vicinity of Holladay Park, writes architectural historian Richard Longstreth. Lloyd's efforts to create this "new city" would fail to materialize as he envisioned prior to his death in September 1953, but Richard R. Von Hagen and his managers would "never [lose] sight of the founder's vision of a new city" in realizing the Lloyd Center, a regional shopping center and hotel that opened in August 1960. In this reading, the Lloyd Center represents a case example of real estate developers and department store executives, who were frustrated by aborted and stillborn efforts to revitalize retail cores, "creat[ing] the advantages of downtown *de novo* by combining a regional mall with dense office building and apartment house construction in a centralized setting." Thus the postwar "new city" developed in the context of competition between core and peripheral areas for the custom of shoppers, who were fleeing city centers for suburbs in droves.[1]

The idea of the "new city," of course, was Victor Gruen's. At the same time that the celebrated Viennese-born architect was designing regional shopping centers that drained the commercial life out of downtown shopping districts, he was confidently prescribing the new building type as the antidote to their deterioration. Later, he applied the term to the development of greenfield sites as communities composed of clusters of neighborhood units. Separated by greenbelts and linked by ring roads, clusters of communities anchored by regional shopping centers would constitute "a new type of metropolitan organization," which Gruen called "the 'cellular form' of urban planning."[2]

In a quarter century of developing his East Side holdings, however, Ralph Lloyd never envisioned a "new city" in either sense of the term. In his mind, the development of these properties would constitute a natural extension of the central business district, which was confined by hills to the west and the Willamette River to the east. Rather the idea of creating a downtown *de novo* evolved in the minds of those responsible for developing the Lloyd Center after his death.

The arc of Victor Gruen's career as a prominent urban planner saving America's downtowns by suburbanizing them fell outside of Ralph Lloyd's life. Civic and business leaders began to take note of Gruen's recommendations on remaking their retail cores with the architect's publication, in late 1954, of "Dynamic Planning for Retail Areas," in *Harvard Business Review.* In this essay, Gruen offered a way to redesign urban centers, using what he was learning about the design of (the still largely untested) regional shopping center.[3] His grandiose plan to revitalize downtown Fort Worth, Texas, followed in the wake of its publication. By the time that Gruen was promulgating his idea of the "cellular form" of the metropolis six years later, the cities of Kalamazoo, Michigan, and Rochester, New York, had retained Gruen's services and had implemented large parts of the blueprint that his office had developed for them. Dozens of other cities would consult Gruen on revitalizing and renewing their downtowns during the 1960s. Victor Gruen Associates would also surpass John Graham and Company, architect of the Lloyd Center, and Welton Becket as the leading designer of regional shopping centers.[4]

Richard Von Hagen and his managers paid close attention to what Gruen said about urban planning and downtown renewal. They also absorbed how outside observers characterized the Lloyd Center as it materialized in its East Side setting. Heeding the advice of their consultants, who detailed what the project would have to offer its potential customers if it hoped to compete successfully with downtown retailers, Von Hagen and those who worked for him eventually thought of the Lloyd Center as a "city within a city."[5] Richard G. Horn, whom the Lloyd Corporation hired in May 1954 as project manager, would tell an Urban Land Institute symposium in September 1962 that the Lloyd Center was a "complete city."[6] But to his death, Ralph Lloyd continued to conceive of the Lloyd Center as an extension of the central business district, which in his mind Portland needed more than ever as its population increased by nearly half during the 1940s. In this conception, Lloyd focused on "the big picture," so to speak: the Lloyd Center would improve the competitive position of Portland relative to other metropolises, particularly those on the Pacific Coast.

Over time, the Lloyd Center would become linked to the regional shopping center that bears its name. Before then, it was associated with the Lloyd Corporation's plans for the entire area around Holladay Park. Indeed, the U.S. Department of the Interior Building that provided a home for all of BPA's Portland employees was hailed as the first unit of the Lloyd Center, as it was conceived upon the completion of that project.[7] The Lloyd Center, as real estate consultant Larry Smith explained in his report, "Lloyd's in Portland," which he delivered in January 1951, would encompass office, professional, and apartment buildings, a hotel, an auditorium, and a regional shopping center.[8] Even then, Ralph Lloyd continued to think of the Lloyd Center as an extension of Portland's downtown across the Willamette River.

Notwithstanding the construction of the cluster of office buildings for BPA and other public and private clients west of Holladay Park, a considerable amount of the property that Lloyd had acquired in 1926 remained undeveloped when he retained Smith in March 1950. The real estate consultant was quick to observe that significant development had occurred, and was occurring, in accordance with Lloyd's "fundamentals of a basic plan." As an evolving program, the Lloyd Center presented Smith with a situation unlike any he had encountered. It was neither a downtown redevelopment effort nor a comprehensive undertaking on a greenfield or vacant suburban site. Further, no other major American city had such a large block of land under single ownership so close to its downtown.[9] The development of the Lloyd Center would differ in kind from "the ordinary shopping center development" in that it would include businesses and services not generally associated with such projects.[10] Reflecting a remarkable intersection of factors that had delayed the project for twenty-five years and then propelled it forward in the 1950s, the Lloyd Center was unique in its time.[11]

The Lloyd Center before the Regional Shopping Center

Ralph Lloyd first referred to his Holladay Park project as the Lloyd Center in a letter of 16 June 1937 to Hubbard, Westervelt & Mottelay, commercial real estate brokers in Los Angeles. In discussing possible means of financing the project, Lloyd made clear that his model remained essentially unchanged from a decade earlier, with a major hotel as its centerpiece. It "will be similar to the Ambassador Hotel and shops in Los Angeles, but much more modern," he promised. The hotel would be "the most beautiful and unique project of its kind on the American continent." He insisted that Portland remained "in dire need of a new hotel and particularly one that will give it a center for its social affairs as well as a high-class shopping district." Now, however, Lloyd devoted attention to an unnamed department store and chain stores that would comprise a significant part of the project. The department store soon withdrew its consideration, but as late as the spring of 1938, Lloyd reported that a number of chain stores remained interested in the Lloyd Center. As we have seen, nothing materialized with respect to the project before World War II put hospitality and retail development on hold nationally.[12]

Ralph Lloyd's failure to secure a department store tenant for his Beverly Hills block in the years leading up to the attack on Pearl Harbor prompted him to sell the property. Notwithstanding his failure to develop the block as he intended, Lloyd's efforts to do so and concurrent developments in the immediate vicinity of the property elevated the importance of retail in his conception of the Lloyd Center. Indeed, he now likened his vision for Holladay Park to the Business Triangle district of Beverly Hills, as it had developed over the previous decade. In January 1945, he tried to interest Bullock's executives in his Holladay Park site, making it clear that he was seeking a department store to anchor a "shopping center around the park" that, in tandem with a hotel, would constitute the Lloyd Center.[13]

Ralph Lloyd's vision for Portland's East Side remained unchanged over the next five years, until his bid to recruit Meier & Frank as a department store anchor and the Hilton and Statler groups as hotel operator faltered. The Lloyd Center, in this conception, included a large department store, a hotel, chain stores, a theater, a bank, and a professional office building.[14] Its moderate scale, at least in relation to the regional shopping center that eventually materialized on the site, is suggested by Lloyd's plans to build a complex of two-story apartment buildings on the four blocks that lay immediately to the east of the Holladay School.[15] These blocks eventually would be reserved for the shopping center.

Late in 1947, Lloyd queried Aaron Frank's interest in opening a branch store that would anchor the Lloyd Center, as he conceived it. The move would be unprecedented for Portland. In contrast to Los Angeles, no department store had expanded beyond its downtown flagship location. On the face of it, Frank does not appear to have been a prime candidate as a first mover in this regard. His mammoth, 712,500-square-foot flagship store was the largest department store west of the Mississippi River. But Ralph Lloyd's friendship with Julius L. Meier, president of the company until his death in 1937, dated from at least the mid-1920s, and he continued to have a personal relationship with Frank, who assumed his presidency of the company upon the death of his uncle. Lloyd envisioned a branch store with either three or four stories that, together with a basement, would provide at least 200,000 square feet of retail floor space. Lloyd wooed Frank for nearly two years. Ultimately, negotiations faltered when Frank stood pat on a branch store that Lloyd considered too small to anchor the Lloyd Center. The Lloyd Corporation would open discussions with Allied Stores Corporation, which emerged in the postwar period as one of the largest retail operations nationally under the direction of B. Earl Puckett, when the company indicated a willingness to establish a beachhead in Portland with a store large enough to challenge Meier & Frank's hegemony in the market.[16]

At the same time, Lloyd's efforts to interest the Statler and Hilton hotel chains in his Holladay Park location also faltered. Not for lack of apparent demand: As Lloyd put it, among Pacific Coast cities, Portland "most badly needed" a major hotel. According to real estate consultant Smith, Portland could handle a 50 percent boost in hotel capacity. Both Arthur Douglas, president of the Statler Hotels group, and Hilton's W. R. Irwin expressed an interest in opening a hotel in Portland. Irwin went further, indicating his satisfaction in a possible Holladay Park location. But he insisted that the Lloyd Corporation first develop a master plan for the area, and recommended that it retain Welton Becket and Walter Wurdeman, architects of the recently completed downtown Los Angeles headquarters building for General Petroleum, and Henry Dreyfuss, a New York- and Pasadena-based industrial engineer, to execute it. Irwin promised that Hilton would enter into negotiations for a hotel, once the master plan had been completed.[17]

With the outbreak of hostilities on the Korean Peninsula, Ralph Lloyd shelved his plans. Because of the risks embedded in the business, a hotel could only be operated profitably "under peace conditions." Given wartime restrictions that "apparently are going to be imposed upon us," now was not the time to go ahead with the project. Lloyd advised the director of the East Side Commercial Club and other local business leaders.[18] Articulating a popular belief in the ability of the Kremlin to orchestrate the diplomatic maneuvers and military actions of communist states worldwide, Lloyd expressed the hope that "Joe Stalin will finally decide that it is better for Russia and the world in general that we have peace rather than war, so that we and they can go about our work as usual rather than living in a state of uncertainty as at present."[19] Indeed, "uncertainties as to the future in regard to war and adjustments after war" persuaded the cautious developer to postpone all development of the Lloyd Center.[20]

Yet if the Korean War provided the proximate reason to set aside his plans, Ralph Lloyd continued the pattern of risk-averse behavior that he displayed during the interwar period. He approached the executives of the Meier & Frank department store and the Hilton and Statler hotel chains in much the same way as he approached downtown Los Angeles department store executives regarding the opening of a branch store in Beverly Hills. Without a fully articulated plan for the materialization of his property, Lloyd relied on potential tenants to provide the real estate development solutions that eluded him. As was the case with their interwar counterparts, however, these executives required more assurances and finer details from Lloyd than the small businessman was prepared to offer if they were to commit corporate capital to large-scale projects in an unproven commercial area.

The development of Northgate Shopping Center, which opened in 1950 on a fifty-acre site some five miles north of Seattle, profoundly influenced the reconception of the Lloyd Center. Indeed, the "paradigmatic" project, which offered a total of 400,000 square feet of retail floor area, defined the regional shopping center as a distinct type. Working closely with Rex L. Allison, president of the Bon Marché department store—part of the Allied Stores empire—John Graham Jr.'s design of Northgate represented the intersection of two trends in retail dispersion that found their genesis in 1920s Los Angeles, namely the local shopping center and the branch department store. Graham's wealth of merchandising experience was crucial to realizing Northgate as the remarkably successful project that it became. With demand for architectural services at a low ebb when he completed his training at Yale University in 1931, Graham spent the ensuing six years in department store merchandising management before he returned to his profession of training. He opened a New York office of his father's architectural firm, which counted among its many downtown Seattle commissions the flagship Bon Marché department store. Graham returned to Seattle in 1946, the same year in which Allison was named president of the Bon Marché.[21]

In designing Northgate for its owner, Allied Stores, Graham transplanted the retail configuration of Seattle's principal shopping thoroughfare to a greenfield site, placing a 200,000-square-foot Bon Marché emporium in the center of the project. To maximize foot-traffic, the architect aligned some eighty small stores that offered between 500 square feet and 1,000 square feet of sales area along a narrow, forty-eight-foot-wide, 1,500-foot-long pedestrian mall that constituted "the backbone of the center," as the Community Builders Council of the Urban Land Institute, which reviewed the project at its fall 1951 meeting in Seattle, put it. A second major department store located opposite the Bon Marché and a third department store located at a terminus of the mall served as additional magnets, drawing shoppers into and along the store-lined mall. Supermarkets that served a similar function were located at each end of the mall. To enhance the downtown, "one-stop shopping" experience that he aimed to replicate, Graham incorporated services, such as a bowling alley, restaurants, and a skating rink, into the project. A three-story medical-dental office building and 3,000-seat theatre comprised the major portions of a northwest wing that extended at right angles from the department store at one end of the mall. Graham would apply the principles that were responsible for Northgate's success to the design of the Lloyd Center, whose smaller footprint would require the architect to introduce into the design vertical elements not found in Northgate.[22]

Above all, the competitive advantage of the regional shopping center over downtown retail districts was accessibility, parking in particular. Some 275,000 people lived within a fifteen-minute drive of the Northgate site. As built, the regional shopping center accommodated about 3,000 vehicles; Graham planned for the addition of another 1,000 spaces as the commercial complex expanded. Parking areas surrounded the shopping center, whose nine buildings were separated by arcades that allowed direct access to the pedestrian mall. In addition, both the department stores and all the smaller stores that were not located along inner arcades had entrances that faced both surface parking and the interior pedestrian mall. Once they reached the mall, shoppers did not encounter vehicular traffic, as they did in crossing downtown streets.[23]

Several features had a direct impact on Northgate's efficiency, practicality, and profitability. Borrowing from the Broadway-Crenshaw Center in Los Angeles, a 550,000-square-foot, department store–anchored project that opened in November 1947, Graham incorporated an underground service tunnel that extended the length of the pedestrian mall with access to full basements beneath each store. Eighteen-foot ceiling heights obviated the need to install air conditioning, given Seattle's climate, and allowed tenants to add a mezzanine level, if they so desired. (According to the Community Builders Council of the Urban Land Institute, tenants, who were responsible for finishing their interior spaces, opted to hang ceilings, finding the eighteen-foot ceilings to be too high.) Individual buildings, excepting the Bon Marché store and the medical-dental building, consisted of a single story and were large and rectangular in plan with ninety-six-foot, steel, bow-string trusses that facilitated flexible spatial distribution among stores, which were partitioned by concrete block and plaster curtain walls. Northgate's plain architecture, which the Community Builders Council described as "contemporary," emphasized economy and convenience. At this moment of creation, when Northgate stood alone in the field, Graham deliberately avoided aesthetic features that typically adorned the storefronts of downtown establishments. Simple exteriors served the purpose of framing merchandise displayed in large, plate-glass windows. The underground delivery tunnel and the structural and architectural features of the buildings constituted means by which Graham and Allison realized their idea of Northgate as "suburban shopping made pleasurable."[24]

The success of Northgate also owed to the configuration of control of the project. The Broadway-Crenshaw Center established the precedent of the anchor department store, rather than the real estate developer, initiating the project, planning it, and taking responsibility for attracting and managing tenants. Broadway executives relied on Coldwell, Cornwall & Banker, real estate brokers and consultants, for advice. In the case of Northgate, Allison invested responsibility for site selection and planning, merchandising, leasing, and construction management in John Graham's firm. At the same time, Allied Stores broke new ground in assuming the risk of ownership. In the case of Broadway-Crenshaw Center, Northwestern Mutual Life Insurance Company retained ownership of the buildings and leased them to the Broadway, which subleased space to tenants. In the case of the Lloyd Center, the Lloyd Corporation would defer to architect Graham on the most important aspects of the project. As we shall see, this deference of owner to architect created tensions in the relationship that apparently were absent in the case of Northgate. In the case of the Lloyd Center, ownership would be invested not in the anchor department store, but would remain with the developer.[25]

When W. R. Irwin conditioned Hilton's interest in the Lloyd Center on the development of a master plan, Ralph Lloyd contacted John Graham Jr. They met for the first time in December 1949. On Graham's advice, the Lloyd Corporation retained Larry Smith & Company, which specialized in market analysis, to prepare a demographic and economic study of Portland and analyze the impact of the highway proposed by the State of Oregon for Sullivan's Gulch on the market for East Side hotels, office buildings, retail stores, and apartments. Smith, in turn, retained Howard Bagnall Meek, chair of the Department of Hotel Administration at Cornell University, to assess the viability of the Holladay Park location as the site of a major hotel, and Robert Alexander, chair of the Los Angeles Planning Commission, to prepare an analysis of the local market for multi-unit residential properties. Together with

Graham, Smith had worked with Allison in planning Northgate Shopping Center. After completing his preliminary analysis, Smith retained Graham to develop a site plan. With the delivery of a master plan for the Lloyd Center, the regional shopping center became the centerpiece of Ralph Lloyd's ambitions for his holdings in the vicinity of Holladay Park.[26]

"Lloyd's in Portland"

In his second interim report, which he delivered to Ralph Lloyd in October 1950, Larry Smith focused on the competitive threat posed by the central business district even as he recommended to his client that he proceed with the large-scale development of his Holladay Park properties. For the lack of a response from downtown retail merchants to a 30 percent increase in the population of Multnomah County over the past decade provided the Lloyd Corporation with a market opening to profitably exploit, concluded Smith. To meet the competition of downtown and neighborhood stores, the real estate consultant estimated that the Lloyd Center would need to capture half of the retail demand generated by Multnomah County's increase in population since 1940. This translated into annual retail sales of $65 million, which was equal to 13 percent of the total retail volume for the county in 1948. Smith anticipated that personal and business services offered by the Lloyd Center would generate an additional $10 million in sales. On this scale, the Lloyd Center would reduce by one-third the rate of increase in retail sales that downtown retail merchants had enjoyed since 1940. The Lloyd Corporation was poised to profit handsomely by capturing a significant share of Portland's retail market.[27]

Given its close-in location, it was imperative that the Lloyd Corporation develop the Lloyd Center as a "metropolitan property," offering all the products and services that shoppers found downtown, concluded Smith. At the same time, the Lloyd Corporation had to avoid replicating conditions that were making central business districts obsolete. Portland's downtown retained advantages, such as density and public transportation nodes. Notwithstanding the fact that downtown department store executives were demonstrating their willingness to demolish landmark buildings, such as the Portland Hotel, to clear sites for garages, they could not compete with the Lloyd Center in terms of parking, however. Thus the Lloyd Corporation would have to provide the parking ratios of a suburban shopping center. Smith estimated that the Lloyd Center would need "every last car space that you can get"—at least 6,000, and as many as 9,000, spaces. When Ralph Lloyd wrote to architect Graham, suggesting that the Lloyd Center could provide fewer parking spaces than the typical suburban center, Smith was quick to caution him not to throw away the project's primary source of competitive advantage over Portland's retail core.[28]

In a set of final reports that he delivered in early 1951, Larry Smith reemphasized Ralph Lloyd's view that development around Holladay Park would constitute an extension of a central business district confined by geography. Upon further consideration, Smith now lowered his assessment of the competitive threat level posed by the central business district. The real estate consultant observed that, hampered by narrow streets, short, 200-by-200-foot blocks, and a lack of alleys and off-street parking, downtown development efforts had failed to keep pace with the city's increase in population, 85 percent of which lay to the east of the Willamette River. Leading stores had modernized their properties without adding appreciable retail floor space. Construction of professional office space also had lagged demand. In fact, Smith noted, the cluster of office buildings that the Lloyd Corporation had constructed for BPA employees constituted the only recent, concentrated commercial development completed citywide. Repeated attempts on the part of civic leaders to address downtown congestion associated with the automobile had failed at the ballot box. Notwithstanding the success of a one-way street program that had gone into effect in March 1950, Smith expected downtown traffic congestion to persist. Moreover, downtown development was "almost prohibitively expensive." Private interests were in no position to revitalize the district. Smith concluded that the size of the congestion problem, combined with a lack of municipal leadership to address it, had created a substantial economic opportunity for the Lloyd Corporation to construct a large-scale commercial center. For the only solution to the problems facing Portland's central business district lay in decentralization and the East Side was its logical trajectory.[29]

In keeping with the conception of the Lloyd Center as "metropolitan property," the master plan for the Lloyd Center incorporated both existing and planned office and apartment buildings, a 400- to 600-room hotel, a regional shopping center, and ancillary projects.[30] Fully 89 percent of Portland's residents would enjoy easy access to it—by automobile, of course. Smith did not consider the argument recently advanced by the Portland Planning Commission that streetcars, buses, and arterial street traffic made the central business district more accessible to more residents than Holladay Park.[31] Owing to its proximity, Smith concluded, the Lloyd Center would be "simply an extension of the downtown shopping area," not a new city in Victor Gruen's sense of the term. As such, it would differ in meaning and substance from the regional shopping centers planned for suburban locations. It would be denser, yet not congested. Ralph Lloyd's extensive, undeveloped holdings provided a "unique location" with "unique possibilities" for commercial success. Indeed, Smith's demographic and traffic analyses persuaded him that only Lloyd's holdings around Holladay Park could support a major commercial center in the Portland metropolitan area on the scale that he and architect Graham were proposing.[32]

With master planning completed, the Lloyd Corporation turned its attention to securing a major department store tenant that would constitute the linchpin of the project.

Wooing Allied Stores

As an extension of a business trip to Portland, Richard Von Hagen traveled to Seattle in July 1951 to meet with Bon Marché President Rex L. Allison, who, John Graham had reported, was "anxious to have a proposal" from the Lloyd Corporation. Allison assured Von Hagen that Allied Stores Corporation was interested in establishing a department store on Portland's East Side with some 500,000 square feet of total floor area. With this encouragement, the Lloyd Corporation incorporated negotiations with Allied Stores as a department store anchor for the Lloyd Center into the suite of activities for which it retained Larry Smith & Company as planning consultants, beginning 1 September 1951.[33]

Given Ralph Lloyd's reluctance to press forward with the Lloyd Center, owing to prevailing controls on raw materials associated with the Korean War, negotiations with Rex Allison proceeded haltingly over the ensuing year. Nevertheless, a proposal that the Bon Marché president might bring before Allied Stores President B. Earl Puckett took shape over the summer of 1952. On the basis of a meeting with Allison on 23 May 1952, Larry Smith outlined the terms of a thirty-five-year ground lease on the understanding that the lessee would construct a 448,000-square-foot store at an estimated cost of $6,048,675. The real estate consultant proposed a complicated rental formula that promised to generate $672,075 in annual income for the Lloyd Corporation based on estimated sales. In this reckoning, the department store would generate about 20 percent of the rental income produced by the Lloyd Center. The Lloyd Corporation would contribute land and 20 percent of the cost of prorated items, such as the truck delivery concourse, multi-level parking garage, and common areas, which Smith estimated to equal $4,088,126 in total. In all, the

Lloyd Corporation would invest $2,317,625 in the construction of the department store. Reported Smith: "Allison obviously wants the store and I understand in New York that he has a better chance of getting this approved than many other stores that are in the plan [sic] process."[34] With Smith's proposal in hand, Rex Allison made an apparently successful case to Puckett for a Lloyd Center department store.[35]

In October 1952, Allied Stores issued a letter of intention to occupy a four-floor, 450,000-square-foot department store in the Lloyd Center, on the condition that an additional 400,000 square feet of retail floor area, of which at least 300,000 square feet was located on the ground floor, opened for business at the same time as did the emporia. Moreover, at least 150,000 square feet of this floor area had to be under lease before Allied Stores would break ground on its store. The prospective anchor tenant also expected the Lloyd Center to mirror Northgate in the variety of retail choices that it offered consumers. Under the configuration of the Lloyd Center, as contemplated in the latest iteration of thinking among Graham, Smith, and the Lloyd Corporation, 584,995 square feet of a 1,342,990-square-foot shopping center would constitute ground floor area. Thus the terms of the letter of intention would require the Lloyd Corporation to open some 70 percent of the shopping center's ground floor area at the same time as the department store. With eleven buildings contemplated as two-story structures and given the feasible footprint of the Lloyd Center in the blocks that lay north of Holladay Park, at least 984,500 square feet and as much as 1,060,990 square feet of retail space would have to be built simultaneously. Observed Leland L. Rebber, who served as technical adviser to Ralph Lloyd on the project: "They [Allied Stores] would thus be starting business with almost the complete center functioning about them."[36]

After four months of discussion, Allied Stores issued a proposal that served the purpose of initiating active planning for the development of the Lloyd Center. The project team—initially composed of architect Graham, real estate consultant Smith and his colleague, Robert J. Crabb, and representatives of the Lloyd Corporation, including Ralph Lloyd, Franz Drinker, General Counsel W. Joseph McFarland, Rebber, Von Hagen, and Auditor David W. Yule—set an aggressive schedule, with a construction start in 1955 and an opening in the summer or fall of 1956. Interim milestones included obtaining all zoning changes and street vacations by May 1953; ordering working drawings from the architect, securing financing, and executing the Allied Stores lease by the end of 1953; and completing the leasing of retail space by June 1954. Acquiring properties and demolishing the residences that occupied them would remain ongoing until the site was cleared. Harland Bartholomew & Associates would plan traffic flow.[37]

The linchpin of the tentative work schedule, of course, was the Allied Stores lease. For, as Larry Smith noted, no regional shopping center project had proceeded without a secure commitment from the department store anchor.[38] Ominously, the interim milestone of obtaining a signed lease agreement by the end of 1953 slipped. Nevertheless, on the assumption that Allied Stores would commit to the project in a timely manner, the Lloyd Corporation authorized Graham to construct a model of the Lloyd Center in November and, on 1 December 1953, formally engaged Graham as project architect.[39]

Securing the department store lease seemed to be all but assured when, in February 1954, Allied Stores signed a letter of tentative agreement for the construction of the store. Discussions with Rex Allison, who traveled to Los Angeles in April to review the model of the Lloyd Center on display at the Lloyd Corporation's headquarters, failed to produce a lease agreement, however.[40] Richard Von Hagen nevertheless proceeded with a publicity dinner in May in Portland. With Puckett in attendance, the Lloyd Corporation unveiled Graham's model for local review and Allied Stores and the Lloyd Corporation announced plans for the construction of the Lloyd Center. In the ensuing days, the Lloyd Corporation hired Richard G. Horn as project manager and retained Norris, Beggs & Simpson as leasing agents. A year later, some 400,000 square feet of retail space was under lease, meeting the expectation of Allied Stores, as expressed in its initial letter of intention. Major tenants under lease included J. C. Penney, the shopping center's junior department store, and variety stores J. J. Newberry and Woolworth's. Effused Larry Smith, "This was one of the most effective leasing jobs in a major project from the standpoint of timing that I know."[41]

Ultimately negotiations with Allied Stores broke down. No one issue appeared to be a deal breaker. Yet opening an emporium in Portland was part of an expansion program that met resistance on the part of the board of directors when it became evident that the organization lacked the capability to accomplish it. Regionally, the company was experiencing "serious management problems," as a three-story addition to the flagship Bon Marché store in Seattle "had not proved out" economically. Technically, Allied Stores did not withdraw from the project. In the fall of 1955, the Allied Stores president asked for a year's grace. Engaged in oil field development at Ventura Avenue and on the Oxnard plain of Ventura County, however, the Lloyd Corporation could not afford to wait as Puckett struggled to gain control of an expansion program that was stretching his company's resources.[42]

The Lloyd Corporation turned to Meier & Frank. Just five months after real estate consultant Smith predicted that it would take at least eight months and as many as five years to secure a replacement for Allied Stores, the parties executed a lease for a 300,000-square-foot anchor department store. Meier & Frank reconsidered its earlier interest in the Lloyd Center as a site suitable only for a small branch store—and shelved plans to open a branch store farther afield—because it recognized that it now had no room to expand its flagship store or facilitate easier access to it. As Larry Smith predicted in 1951, traffic congestion and a lack of parking had not been addressed by civic leaders and, together, were deterring or preventing consumers on the East Side from venturing downtown. Meier & Frank executives concluded that a branch store that was more accessible to the majority of Portland's population and offered almost all the goods on offer at their flagship store would improve customer satisfaction and contribute to, rather than cannibalize, overall sales.[43]

Richard Von Hagen was "determined that things move rapidly from now on."[44] In his estimation, courting Allied Stores had delayed the project by as many as fourteen months. The Lloyd Center was scheduled to open in the first or second quarter of 1959, explained Von Hagen to Proctor H. Barnett, executive general manager of the Investments, Mortgage Loan, and Real Estate Investment Department of the Prudential Insurance Company of America, from whom the Lloyd Corporation was seeking construction loans for both the hotel and the shopping center.[45]

Design of the Lloyd Center

With Allied's letter of intention dated 27 January 1953 in hand, the Lloyd Corporation initiated the design phase of the Lloyd Center. With architect Graham in the chair, the project team held regular meetings, beginning on 4 March 1953, on the assumption that concluding a lease with Allied Stores was simply a matter of time.

To the greatest extent possible, Graham intended to adapt his spatial configuration for Northgate to the layout of the Lloyd Center. Dubbed by Leland Rebber as "the Country Club idea," Graham prescribed an integrated shopping center surrounded by parking, with open pedestrian malls interior to the shopping area.[46] From the anchor department store that faced Holladay Park to the south, quality- and popular-priced wings extended eastward to NE 9th Avenue and westward to NE 15th Avenue, respectively. So that the hotel would be located close to women's apparel and other higher-end stores, Graham relocated it to the west of Holladay Park. A separate, third wing, featuring food, furniture, and hardware stores as well as eating establishments, would extend northward from the Allied store. Drawing directly from the design of Northgate, an underground truck concourse would facilitate the separation of delivery and trash collection from pedestrian traffic and customer parking. Graham's preliminary plan provided for covered parking on three levels, both under and around buildings, and additional parking in as many perimeter surface areas as would be needed to increase the number of available spaces to 8,000 or more. To realize this plan, all streets between NE Halsey and NE Multnomah Streets and NE 9th and NE 15th Avenues would have to be vacated and the lots between NE Multnomah and NE Broadway Streets and between NE 9th and NE 15th Avenues would have to be acquired and cleared of the structures that occupied them. They included residences, several apartment buildings, and the Broadway Columbia Market at 1216 NE Broadway Street, which the Lloyd Corporation had completed at the bottom of the Great Depression.[47] As Rebber noted, Graham's "Country Club idea" promoted a "casual, out in the open, free of the crowded city, effect" for shoppers.[48]

At the same time, Ralph Lloyd's technical adviser recommended that the developer also consider a "more compact, more conventional urban plan" for the Lloyd Center. In this configuration, buildings would face streets "in a typical manner similar to the conventional arrangements in uptown business sections." In this configuration, retailers could display their wares to both pedestrians on municipal sidewalks and occupants of passing automobiles. All stores would be two stories in height, with a basement, except for the four-story anchor department store. City streets and sidewalks would function as promenades, with stores retaining their identities. Parking would be confined to a six-level structure hidden from street view. Surface parking, which Rebber considered to be wasted space, would be eliminated. As it required only 54 percent as much land as Graham's "spread-out country club idea," Rebber argued that his scheme would allow the Lloyd Center to expand in line with Portland's growth and adapt to changing tastes, as the Wilshire and Westwood Districts of Los Angeles had done to date. To choose among alternatives, Rebber asked Lloyd to consider his vision for the property and determine what layout would best realize his dreams. The record is silent on what Lloyd, whose health deteriorated during the summer of 1953, may have said in reply.[49]

The death of Ralph Lloyd on 9 September ensured that John Graham's "Country Club idea" for the Lloyd Center prevailed. For, with the oilman's passing, Leland Rebber was dropped from the project team. Project Manager

Richard G. Horn and Lloyd Corporation General Counsel W. Joseph McFarland would note that Lloyd died knowing that his dream for the Lloyd Center would become a reality.[50] For the moment, it was business as usual. At the project meeting held on 24 September, Richard Von Hagen, now president of a corporation owned by his wife and her three sisters, stated that the four women had decided to proceed with the Lloyd Center to honor their father's wishes. Therefore the death of Ralph Lloyd would have no impact on planning. Von Hagen would see the project to completion.[51]

Over the next four years, the Lloyd Corporation worked with municipal officials on site preparation. Ultimately it secured the street setbacks and vacations and zoning changes it needed to clear the site as necessary between NE 9th and NE 15th Avenue and between NE Broadway and NE Multnomah Streets, as architect Graham's "spread-out country club idea" required.[52] The City approved the Lloyd Corporation's petitions to widen the streets around the project and it implemented a system of one-way street couplets to facilitate traffic movement to and from the Lloyd Center. City officials supported the Lloyd Corporation's petitions in the belief that the project would benefit Portland as a whole, generate much-needed tax revenues, promote intracity retail competition, counterbalance decentralizing trends in retail development, and increase demand for East Side residential construction. Thus they backed Graham's blueprint for the suburbanization of urban space.[53]

The Lloyd Corporation also worked with Oregon's highway department to provide access to the Lloyd Center from the Banfield Expressway, which the department was constructing through Sullivan's Gulch. In December 1949, the highway commission estimated that it would need to take 736,500 square feet from twenty of the parcels that Ralph Lloyd had acquired in 1926 from the Balfour-Guthrie Trust Company. Negotiations ensued to establish the fair market value of this property, which included a portion of the golf course. On 16 September 1952, the Lloyd Corporation conveyed seventeen parcels to the State in exchange for $398,000 and other considerations.[54] Subsequently, it cooperated with State highway officials on ramp design, the design and location of a bridge over the Banfield Expressway to a NE 15th and NE 16th Avenue one-way street couplet, and the extension of Lloyd Boulevard into NE 16th Avenue.[55]

During 1954 through 1956, leasing agent George Beggs and Lloyd Corporation representatives, in particular Project Manager Horn, but also Franz Drinker, General Counsel McFarland, and Auditor Yule, visited regional and community shopping centers across the country to assist the Lloyd

Corporation as owner to reach agreement with architect Graham and consultant Smith "on the basic philosophy and basic planning" of the proposed Lloyd Center.[56] Beggs and Lloyd Corporation representatives discussed operating procedures with managers and owners and made personal observations regarding site layout, shopping center design, and tenant selection. They also reviewed brochures and other marketing materials and circulated articles on shopping centers in the trade press amongst themselves. Armed with their conclusions and recommendations, they advised the owner on validating or challenging Graham's design. Finally, they assisted the Lloyd Corporation in financing the construction of the project by reporting on the financial performance of stores—national chains in particular—thereby ensuring that the Lloyd Center conformed to the evolving norms of the regional shopping center as a type.

Beggs, Drinker, Horn, McFarland, and Yule did not lack for properties to visit. For between 1953 and 1956, the number of regional shopping centers tripled. Early in 1954, President Eisenhower signed into law changes to the federal revenue code that Congress intended to stimulate a slowing economy. Foremost among them was an acceleration of the depreciation deduction on new construction. Since 1934, straight-line depreciation of the value of buildings and other income-producing property over a forty-year period had provided an incentive to the developers and owners of buildings that real estate experts agreed was substantial. Nevertheless, Congress prescribed accelerated depreciation as a cure for mild recession. The 1954 Revenue Act offered two formulas, both of which shifted the deduction to the early years of the project without changing the overall the amount of the tax benefit. Accelerated depreciation enabled investors "to recapture the risk capital in a building venture during a brief, tax-protected period," real estate economist Miles L. Colean explained.[57] In doing so, historian Thomas W. Hanchett concludes, the law "radically changed the business of developing buildings."[58]

For their part, Richard Horn and his Lloyd Corporation colleagues visited many of the properties on a list of regional shopping centers with at least 350,000 square feet of store area and constructed between 1950 and 1956 that real estate market analyst Homer Hoyt compiled for *Urban Land*, the flagship publication of the Urban Land Institute (table 3).[59] They also visited a number of community shopping centers.[60] As classified by Hoyt, these centers offered 100,000–300,000 square feet of store area on a twenty-to-fifty-acre site and featured either a junior department store, such as J. C. Penney, or a variety store, such as W. T. Grant or J. J. Newberry, as the largest retail unit.[61]

Horn in particular conveyed his impressions of the shopping centers that he visited at the same time as he catalogued operations for the benefit of Von Hagen. For instance, Evergreen Plaza—a 450,000-square-foot center anchored by The Fair and Carson, Pirie, Scott & Company department stores—was "not impressive," because The Fair was performing poorly as the primary department store.[62] Shopper's World was "the most attractive of any of the smaller centers," with wide and spacious upper level walkways that imparted a "mall level feeling" as one passed by stores.[63] Northland, a colossal regional center developed by the J. L. Hudson Company, was "just as good and even better [in October 1955] than when you [Von Hagen] saw it," with everything "spotlessly clean and beautifully landscaped."[64] Levittown, a 725,000-square-foot center constructed in conjunction with a 17,000-unit residential community, was "one of the more cheaply built centers" with "most ordinary" storefronts. Nevertheless, the property's manager was "gleeful" about the center's return on investment to date—evidence that supported Horn's observation that every center that he visited during October 1955 "seemed to be doing well."[65] Horn's visits convinced him that "anyone who doesn't fight to get into the Lloyd Center just simply doesn't know what goes on."[66] He added: "I wish all our unsigned and wanted tenants could know as I do so well now, how successful these shopping centers are and how essential it is to them that they participate."

Notwithstanding the impressionistic content overall of often-lengthy memoranda, Horn and his colleagues did make concrete recommendations and draw conclusions for the benefit of the owner. Each of the men agreed that incorporating "quality" into the Lloyd Center should be the paramount consideration of the owner. W. Joseph McFarland, for instance, was adamant that attractive storefronts and malls were a necessity and that it would be a mistake to economize on the cost of their construction. It was imperative that, as owner, the Lloyd Corporation retained ultimate design control to ensure that consumers enjoyed a leisurely shopping experience.[67] For his part, Franz Drinker noted that shopping centers easily could look unattractive. Malls and parking areas therefore needed to be "adequately and beautifully landscaped" with sculptures, fountains, drinking fountains, benches, music, and lighting. The Lloyd Center also needed to incorporate a preponderance of "quality" stores that drew "quality" customers. Because the Lloyd Center would be "more complete" than other properties, Drinker recommended the inclusion of recreation facilities.[68] (A 15,000-square-foot outdoor ice skating rink would be constructed at mall level in the middle of the center.) Such recommendations informed discussions that engaged architect, owner, leasing agents, real estate consultants, and municipal planners over more than four years, culminating in agreement on the design of the Lloyd Center late in 1957.

The observations and recommendations of Richard Horn and his colleagues may be read as prescriptions for spatial ordering and aesthetic rendering for the benefit of the white, middle-class female consumer, whose power both within the family and in the public realm increased in line with the emergence of the regional shopping center as a centralized market structure. Historian Lisabeth Cohen aligns the development of such centers with the "feminization of public space" in postwar and increasingly suburban America.[69] Drinker, Horn, and McFarland sought to ensure that the Lloyd Center provided an attractive, comfortable, and safe environment for the Portland-area housewife, enticing her to exercise her power as a mass consumer and thereby consolidate her economic and social gains.

The replacement of the Allied Stores department store with the smaller Meier & Frank emporium precipitated major design changes that involved, for instance, the reconfiguration of parking. Many of the design changes implemented between 1953 and 1957, however, owed to leasing outcomes and a periodic need to "rebalance" the center to satisfy the location and spatial demands of tenants. As a result, both the size of the Lloyd Center and the estimated cost to build it increased.

The plan presented to the Lloyd Corporation by architect Graham in 1953 contemplated a Lloyd Center that comprised fourteen blocks. In consultation with real estate consultants Crabb and Smith and leasing agent George Beggs, Graham allocated Block E to the anchor department store, Blocks A and J to other major, single tenants, and Blocks O and P to individual grocery stores. The remaining blocks would offer consumers goods and services from multiple related stores. For instance, Block D would house purveyors of appliances, furniture, radios and televisions, and wallpaper and flooring; cameras and film processing; and candy, drugs, and soda fountain drinks.[70] As depicted in figure 31, which Smith provided in his final report to the Lloyd Corporation in December 1957, Blocks A, E, J, O, and P (as well as Blocks L and N) indeed were reserved for individual major tenants. The remaining blocks adhered to the principle of assembling stores that offered goods from complementary retail categories under the same roof. At the same time, to meet the desires of tenants, retail categories were shifted among blocks. For instance, appliances and furniture were reassigned to Block F.[71]

As owner, the Lloyd Corporation struggled to ensure that the project "penciled," that is, that it could be built within its budget and operate profitably.

On the one hand, it sought to demonstrate to prospective insurance company lenders that the Lloyd Center would generate rental income sufficient to support the construction loan of at least $24 million that it sought. On the other, it sought to satisfy the desires of tenants even as it limited the total area of the Lloyd Center to a size that it could construct within its budget. Once it agreed on lease terms, a tenant might select a store location not expected by the project's planners or demand additional space. For instance, when Lerner's, a clothing store, Leeds, a shoe store, and Woolworth's requested more space in Block H, architect Graham increased the depth of the building to 170 feet, bringing its south wall in line with Block J, the prospective home of J. C. Penney. Ultimately, Woolworth's needs could not be accommodated in Block H; it leased Block N instead. In accommodating tenant preferences, the owner increased the size of the Lloyd Center. In June 1955, the rentable area of the Lloyd Center reached 1,226,046 square feet, up from 1,179,687 square feet two years earlier. Building a shopping center of this size would strain the resources of the Lloyd Corporation, if not bust its pro forma budget. To address the issue, project planners reduced the number of stores in selected categories and rearranged retail categories by block. At the conclusion of the design phase of the project, the rentable area of the Lloyd Center stood at 1,195,962 square feet.[72]

The professionals who consulted the Lloyd Corporation adhered to a consensus among industry experts that the regional shopping center serve as an integrated "one-stop" retail district, offering at least three competing stores in each retail category to meet the needs of the postwar mass consumer. Through "planned competition," the Lloyd Center would meet the expectations of the middle-class female homemaker. As articulated in gendered terms by developer William Zeckendorf, whose sprawling Roosevelt Field Mall opened on Long Island, New York, in August 1956, the 1950s housewife insisted on comparing the prices and quality of similar items in at least three stores before she decided what to buy. To accommodate the time involved in considering merchandise offered by competing merchants, the Lloyd Center included three restaurants and a cafeteria, three candy stores, two bakeries, two soda fountain shops, and a coffee bar. The inclusion of numerous kiosks and service stores ensured that the Lloyd Center would comprise "a city within a city," as Franz Drinker put it, that enticed the wife as the chief household consumer to both fulfill her desires and meet her family's needs before she departed the property in her automobile.[73]

At the same time, leasing space to nationally or regionally recognized competitors within retail categories provided a means of meeting the expectations

Table 3
Regional Shopping Centers Visited by Lloyd Corporation Representatives

Name of Center	*Location*	*Anchor Department Store(s)*	*Area of Center (sq. ft.)*
Capitol Court	Milwaukee, WI	Schusters	700,000
Cross Country	Yonkers, NY	Wanamaker; Gimbels	1,000,000
Evergreen Plaza	Chicago	The Fair; Carson, Pirie, Scott & Co.	450,000
Gulfgate	Houston, TX	Joske	800,000
Hillsdale	San Mateo, CA	Macy's	600,000
Lakewood	Long Beach, CA	The May Co.	800,000
Levittown	Philadelphia, PA	Pomeroy; Sears, Roebuck & Co.	725,000
Northgate	Seattle, WA	Bon Marché	700,000
Northland	Detroit, MI	J. L. Hudson	1,000,000
Old Orchard	Skokie, IL	Marshall Field; The Fair	1,000,000
Shoppers World	Framingham, MA	Jordan Marsh	500,000
Southdale	Minneapolis, MN	Dayton; Donaldson	800,000
Stanford	Palo Alto, CA	Emporium	536,000
Stonestown	San Francisco	Emporium; Butler Bros.	650,000
Westgate	Cleveland, OH	Halle Bros.	600,000

Source: Homer Hoyt, "Impact of Suburban Shopping Centers in September, 1956," Urban Land 15 (September 1956): tables on 4, 5.

of the insurance companies that the Lloyd Corporation approached for financing. Long-term leases provided the security needed to secure a loan. Potential lenders required as much as 80 percent of the property to be under lease before they would consider extending a construction loan, which at the time had a median amortization period of twenty-five years. Further, insurance companies expected at least half of the space in a property not including the anchor department store be leased to national chains with excellent credit ratings. George Beggs did lease space to local businesses. At the same time, figure 31 indicates the extent to which chains dominated list of tenants who committed to occupying the Lloyd Center upon its completion.[74]

John Graham's final design and working drawings also detailed the features and structures that addressed the ancillary needs of shoppers, above all the female consumer and her children. They included parking for 8,500 vehicles, as the lender demanded, 5,000 of which the architect located within two- and three-level aboveground structures. Three levels of parking provided access to stores located in the core of the project centered on the Meier & Frank store. Two-level, tilted-deck structures with direct access to municipal streets served Blocks K, L, O, and P (figure 31). Escalators conveyed shoppers between levels. Three fifty-foot-wide malls connected the major buildings and converged in a central plaza that featured the aforementioned ice skating rink, where mothers could leave their children while they shopped. Covered walks, arcades, and balconies would protect female shoppers from inclement weather. Shaded benches would provide moments of respite en route between stores. A center-wide sound system would play soothing music throughout the day. All stores would provide air conditioning.[75]

The Lloyd Corporation, as owner, had little financial margin for error as the Lloyd Center moved into the construction phase. Total projected cash flow before taxes from total annual rent of $3,085,599 was now just under $700,000, which represented a decrease of more than half from the expected cash surplus of $1,503,855 as of December 1953. The expected cash surplus before taxes on guaranteed rents stood at just $44,210. The reduction in expected cash flow owed to increases in taxes, vacancy allowances, and mortgage charges. Smith observed that the projected cash surplus at the guaranteed annual rent level of $2.33 million was "very thin"; it easily could be eliminated by vacancies of just 20,000 square feet above the present allowance of 25,000 square feet at average rentals. Moreover, expected construction costs had increased from $22.2 million to $28 million.[76] The financial data provided support for a 5.5 percent, thirty-year, $24 million construction loan from Prudential. Even though it may have been "the largest single mortgage loan ever disbursed by the company" (as of 1961), it was imperative that the owner exert control over the cost of construction. This was easier said than done, given the relationships among the members of the building team that governed the construction of the Lloyd Center.[77]

Construction of the Lloyd Center

Rather than use the "design-negotiate-build" project delivery method preferred by Ralph Lloyd, Richard Von Hagen chose to realize the Lloyd Center

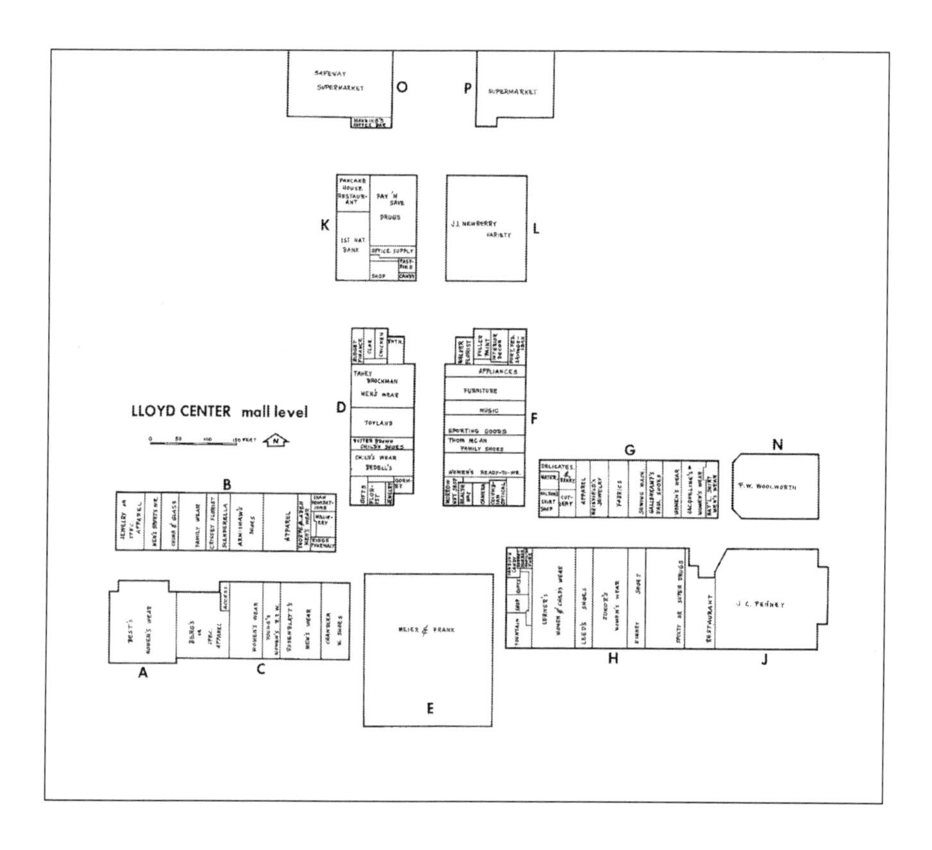

Figure 31. Lloyd Center, mall level. Note the blocks identified by letter. (Larry Smith & Co., "Lloyd Center—Terminal Report," 2 December 1957. The Huntington Library, San Marino, California, LC drawer 12, box 2.)

through "design-bid-build." Under this approach, the owner was responsible for securing the land and necessary permits as well as conceiving and planning the project. The architect, as the owner's agent, designed the project, often in consultation with a structural engineer. (In the case of large design firms, such as John Graham and Company and Welton Becket, architect and engineer worked under a common corporate roof.) Once the owner approved plans and specifications developed by the architect, general contractors, who had no involvement with the design process, bid on them. The bidding process was either invited, whereby the architect restricted the number of contractors who could bid, or open to any firm that responded to the request for proposals. The owner and successful bidder established a contractual relationship that was wholly separate from the one that defined owner-architect relations. The architect supervised construction, monitoring adherence on the part of the general contractor to approved working drawings and other construction documents and managing the process by which modifications to the design—so-called change orders—were proposed to, and approved by, the owner. The general contractor was responsible for procuring materials and coordinating operations, but typically subcontracted the building of the project to various specialized firms.[78]

Design-bid-build emerged as prevailing method of project delivery in the commercial segment of the building industry early in the twentieth century. With the development of tall and complex commercial structures, responsibility for designing, engineering, and erecting buildings devolved to specialized professionals, displacing the master builder, who both designed the building and became intimately involved in its construction. At the same time, a system of assigning responsibility for construction through competitive bidding developed in the context of an interrelated web of legal, managerial, and professional practice. By World War I, the separation of design and construction in commercial building was well established.[79]

Historically, architects, structural engineers, and builders found common ground in criticizing the approach. Often, once construction on a design-bid-build project commenced, problems manifested themselves as errors or ambiguities in design documents, gaps in plans, or incompatibilities between design and construction methods. Contractually obligated to proceed according to the working drawings and related documents, the builder issued change orders to address these discrepancies, requiring the owner to supply additional capital to the project. Slips in schedules, owing to change orders, increased a project's interim financing costs too. And since project management was fragmented under design-bid-build, the members of the building team had a financial incentive to adopt adversarial positions. The owner thus assumed the project's execution risk; designer and builder acted to protect their margins by attempting to show that problems associated with the project were not their fault. Nevertheless, design-bid-build resisted both reform and innovation. It remained the preferred method of commercial project delivery when the four daughters of Ralph and Lulu Lloyd moved ahead with the Lloyd Center in the wake of their father's death. The deployment of design-bid-build to deliver the project saddled the owner with additional costs that it funded out of diminishing surpluses generated from petroleum extraction.[80]

The regional shopping center provided the architect an "almost unparalleled" opportunity to orchestrate the design and construction of a complex commercial project. Where it materialized on a greenfield site, in particular, the regional shopping center as a "suburban retail district" offered "both a design freedom and a planning responsibility of staggering proportions." The architect's newfound position of leadership on retail project teams stood in contrast to his or her allegedly subordinate role only two decades earlier. As *Architectural Forum* framed the transformation, "The modern architect who opened up the front of the downtown store in the Thirties has come a long way."[81] John Graham Jr. may not have enjoyed the notoriety of Victor Gruen. Nevertheless, his peers recognized him as a pioneer in the development of the regional shopping center. He relished the leading role that he and his firm played in realizing these complex projects. Said Graham, "We can and do handle everything involved in a particular project from area and site studies right down to selection of ashtrays that'll go inside the completed building."[82] By 1956, John Graham and Company was either designing or constructing eleven regional shopping centers.[83] Of course, the Lloyd Center was one of them.

In a memorandum dated 10 September 1954 in which he proposed to construct the Lloyd Center in six phases, John Graham asserted his bona fides as a master builder and assured the Lloyd Corporation that he could realize the project "within a reasonable cost." "Extensive study by a strong team of experienced technicians [with] actual experience gained in several other projects located in widely separated parts of the country," the architect advised, had made possible his "scientific planning" of the project. It remained only to explicate a construction program that satisfied the requirements of tenants, who, Graham argued, sought to encumber the owner with as much of the cost as possible of finishing the retail store. The architect was in the best position, according to Graham, to mediate this contested space "to produce the most building for the least cost." On this project, the architect proposed to control the cost of tenant's work out of his office, reviewing the construction clauses in as-yet-to-be-signed leases and working with the leasing agent to modify them, as necessary, to limit the cost of construction.[84]

Research conducted by architect Victor Gruen and real estate consultant Larry Smith, who previously had collaborated on a study for *Progressive Architecture* that purported to inform all aspects of shopping center planning, design, construction, and operations, showed that constructing storefronts and finishing interior retail spaces accounted for at least 25 percent of the owner's capital investment in a project. In short, Gruen and Smith concluded, tenant improvements, as they were known, constituted "the crux of shopping center finances." It was imperative that the owner strike an appropriate balance in sharing the cost of these improvements and convey the division of cost responsibility to the tenant before quoting a rent. Otherwise the tenant would assume that the rent included a "turnkey," or finished, retail space. Absorbing too much responsibility for tenant improvements could jeopardize the viability of the project before it opened, Gruen and Smith advised, as owners had found it difficult, if not impossible, to secure rent premiums sufficient to justify the additional investment. At the same time, absorbing too little responsibility for finishing retail stores invariably reduced the amount of guaranteed rent that the owner might obtain from tenants, given the substantial investment required on the part of the latter to finish their stores. Moreover, the overall quality and attractiveness of the project might suffer as tenants maneuvered to reduce their cash outlay.[85]

With these research findings in hand, Smith and his colleague Robert J. Crabb scrutinized Graham's memorandum for the benefit of the owner. The real estate consultants agreed that Graham's phased approach to construction on a bid basis was sound. At the same time, Smith disagreed wholeheartedly with Graham's assertion that the architect was well placed to control tenant work without a fixed budget. Indeed, in most cases, including projects in which Graham participated, owners had established control over these expenditures. In the case of Capitol Court in Milwaukee, for instance, Schusters, the owner of both the project and its anchor department store, "absolutely refused to go along [with the approach] in the face of Jack's insistence that it is the only way to do it." Schusters expected to "make substantial savings by setting up a budget and checking their progress against it as they go along."[86] Crabb concurred: proceeding with construction without a predetermined total cost estimate based on a specific leasing program "would be unacceptable to any owners with whose work we are familiar."[87] Under a system of owner control, project managers or leasing agents, rather than the architect, would enjoy discretion to negotiate concessions to the budget. It remained for the Lloyd Corporation to demarcate the point at which the owner's responsibility for completing finish work beyond the shell of the building ended and the tenant's responsibility began.

Crabb elaborated the division of responsibilities under "the standard historical approach," involving "the great majority of premises" in regional shopping centers. The owner absorbed the cost of standard storefronts and entrance doors, plastered ceilings, flooring finished to receive the tenant's floor covering, the priming coat of paint for interior trim, electrical wiring, utilities delivered to the store, and duct work and other heating equipment. For its part, the tenant supplied and installed interior partitions, electrical and plumbing fixtures, floor coverings, light coves and other non-standard ceilings, finish painting, display windows, fixtures and furnishings, air conditioning, signs, and mechanical equipment, such as elevators and escalators. Crabb noted, however, that many lease negotiations in which he had been involved had produced agreement on the part of the owner to pay the cost of items under the tenant's column of responsibilities. Over the next five years the Lloyd Corporation would work with Crabb, Graham, and Smith to settle the boundaries of this contested space to limit project costs.[88]

Descriptions of landlord's and tenant's work, prepared in the offices of John Graham and Company and appended as "Exhibit B" of the lease, accumulated a growing number of items under "Tenant's Work in Tenant's Premises at Tenant's Expense." An early version of the exhibit, submitted by Graham to the project team in May 1955, included eleven items, all of which had appeared on Crabb's list a year earlier.[89] Two years later, the tenant's responsibilities had grown to twenty items. For its part, the Lloyd Corporation agreed to supply air conditioning, but won concessions on ten of the new items.[90] The ability to transfer costs associated with finishing stores to tenants indicated the growing attractiveness of the Lloyd Center as the project moved from planning to design to construction. The willingness of tenants to invest capital in developing attractive retail spaces also validated the competitive model adopted by the owner. The inclusion of multiple purveyors of similar goods within the shopping center redounded to the benefit of the Lloyd Corporation. Not only did retailers compete for the opportunity to sell their goods in the Lloyd Center, they recognized the need to distinguish their stores from competitors who also leased space in the shopping center.

The cost to the Lloyd Corporation of providing its tenants with sanitary facilities; a service concourse; heating, ventilation, and air conditioning (HVAC); electric wiring; and standard interior finishing was not insignificant, however. Moreover, given that tenants often did not finalize their designs until after the Lloyd Corporation had let the contracts for the construction of the Lloyd Center, many tenant improvements were implemented as change orders to said contracts. The owner would finance these change orders, estimated at the start of construction to equal five percent of the cost of the project as bid, out of free cash flow generated largely from crude oil production at Ventura Avenue, which was declining from its 1953 peak.[91]

John Graham promised that the competitive bid system that he proposed for the project would eliminate "excessive" construction costs and contractor profits. By 1956, the estimated average cost of building the Lloyd Center exceeded the average cost other projects in Graham's portfolio, such as Capitol Court and Gulfgate, by as much as nine dollars per square foot. The differences owed to the need to construct multi-level parking structures on the site's more confined, urban location and to the delay associated with securing the anchor department store. But in breaking up the project into four units that were constructed under separate contracts, Graham added to the cost and length of the project.[92]

Architect Graham rejected outright negotiating a fee with a large general contractor for the project, as Ralph Lloyd had done three decades earlier with the Austin Company for his proposed Walnut Park building.[93] In a proposal to Richard Von Hagen, A. E. Holt, a senior vice president with Guy F. Atkinson Company, argued the advantages of investing responsibility for the execution of the project in a single firm operating under a fee contract. Holt estimated that the Lloyd Corporation would save four to six percent, or $800,000 to $1.2 million, in overhead, materials purchasing, subcontracting, labor, change orders, and profit margin. For instance, under a fee contract, the Lloyd Corporation would absorb only the actual construction cost of change orders. Under the bid system, Holt noted, a minimum 10 percent profit would accrue to the contractor, as a modification to the basic contract. Further, any construction savings achieved by the contractor operating under a fee contract would benefit the owner. Efficiencies gained in project coordination and scheduling would minimize the project's time to completion. Guy F. Atkinson's résumé, including a distribution building for the General Electric Company; The Dalles dam, locks, and powerhouse; and "all the building construction" for Richfield Oil Company, demonstrated the company's capacity to deliver a regional shopping center under the design-build method that Holt proposed. Whether Von Hagen brought the proposal to Graham's attention is unclear. Just weeks after Holt submitted it, Graham's office produced the bidding procedure under which the Lloyd Center was constructed.[94]

In November 1957, Graham's office extended bid invitations to approved contractors. It divided the work into four lump-sum construction contracts and a fifth contract for site grading. (Contracts for earth removal and the boiler plant would also be let.) Three contracts pertained to the main sections of the Lloyd Center. Section 1 comprised Blocks D, E, and F; Section 2 was composed of Blocks G, H, J, and N; and Section 3 consisted of Blocks A, B, C, K, and L (figure 31). Each of these sections involved approximately equal amounts of work. Section 4 consisted of a Safeway supermarket together with its associated parking structures and truck tunnel. Contractors could bid on any or all of the construction contracts, which would be bid in successive weeks, beginning in late November. Contractors could win multiple construction contracts, which would be awarded separately on 31 December 1957 (Section 1) and 15 February 1958 (Sections 2, 3, and 4). In lieu of a bonus clause for completing work ahead of schedule and a strict penalty clause for a failure to complete work on schedule, Graham adopted a "time is of the essence" clause without a stipulated dollar amount. Bidders would bid the completion date, which would be established in the contract. Separate bids would be accepted for electrical, HVAC, mechanical, and plumbing work, elevators, and escalators for each construction contract. The low bidders for each section would become subcontractors of the general contractor. The general contractor would receive a management fee equal to a percentage of the work completed by the subcontractor. Graham estimated that the Lloyd Center would open on 1 March 1960. Owing to change order implementation; statewide labor actions on the part of laborers, operating engineers, and the Teamsters; project management issues; and the decision of Meier & Frank to switch architects for the design of its department store, the Lloyd Center would open exactly five months later than Graham predicted.[95]

As owner, the Lloyd Corporation chose to rely on Graham's office for project supervision. In May 1957, Larry Smith recommended to Von Hagen that he retain an exclusive representative to coordinate construction activities for the purposes of cost control and timely project completion. The individual would oversee a team with direct responsibility to the owner: three or four tenant coordinators, project engineers who would address and resolve construction issues associated with each of the contracts, and a project cost accountant.[96] In essence, Smith described a program of construction management that would gain popularity a decade later when, in the context of accelerating inflation, owners sought ways of configuring work to realize expected returns on their investments. In doing so, owners would adopt a proactive role in the building team that, in dismissing Smith's proposal, the Lloyd Corporation declined to do.[97]

John Graham's office provided project supervision through a field team composed of a chief inspector, three architectural inspectors, an electrical inspector, a mechanical inspector, and an administrative assistant. In addition, representatives from Graham's office, such as architects Alfred H. Fast and Robert Kammer, stood ready to visit the site as needed. Together, the members of Graham's team comprised "the greatest inspection organization I have heard of in connection with any one project." Nevertheless, a relationship between owner and architect that already was strained by delays in the delivery of working drawings would fray as construction costs mounted in line with change orders and as schedules slipped while the architect awaited drawings and other information from tenants. Graham's field team performed myriad tasks. It interpreted drawings and specifications, reviewed shop drawings, selected materials, approved samples, inspected work, managed construction schedules, issued instructions to contractors, prepared change orders, and issued payment certificates. The extent of its work, Graham would later concede, increased "many times" by his decision to execute the project under five contracts. As a result, his team would provide supervisory services "far in excess of that contemplated at the time the contract for architect-engineer service was entered into."[98]

On 11 April 1958, the Lloyd Center broke ground with the daughters of Ralph and Lulu Lloyd, Richard Von Hagen, and Aaron M. Frank in attendance.[99] After Henry M. Mason Company, a local contractor, graded the site, contractors Donald M. Drake Company of Portland and Max J. Kuney of Spokane, Washington, would construct the main sections of the shopping center under contracts awarded by the Lloyd Corporation (table 4). Mason would build the Safeway store and associated parking structures. To the roughly $15.5 million in work awarded through bidding, Graham's office added an estimated $4.75 million in additional work that would be required to complete the project. Ultimately, it would cost some $22 million to construct the Lloyd Center under these contracts. An additional $6 million in costs were incurred in the implementation of 309 change orders, which were approved through a formal procedure designed to limit the financial exposure of the owner (table 5).

Table 4
Lloyd Center Construction Contracts

No.	*Contractor*	*Basic Contract Bid*	*Basic Contract Disbursements*	*Amended Contract Disbursements*
1	Donald M. Drake Co.	$4,890,000	$7,746,252	$9,786,671
2	Max J. Kuney	$4,000,000	$5,357,181	$6,495,303
3	Donald M. Drake Co.	$5,305,000	$7,643,311	$10,056,573
4	Henry M. Mason	$439,206	$682,653	$707,608
5	Henry M. Mason	n/a	$561,088	$694,510
6	n/a	$207,160		
7	Howard Brewton	$274,920	$45,175	$309,470
			$22,035,660	$28,005,135

Source: David W. Yule, Report 24: Final Phase I Disbursement, 4 August 1960, LC drawer 10, box 6; Lloyd Center, original contract bids, LC drawer 6, box 3.

Table 5
Lloyd Center Change Orders

No.	*Contractor*	*Requests for Proposal*	*Change Orders*
1	Donald M. Drake Co.	120	81
2	Max J. Kuney	122	92
3	Donald M. Drake Co.	116	93
4	Henry M. Mason	9	9
5	Henry M. Mason	22	20
6	not available	4	1
7	Howard Brewton	24	13
		417	309

Source: Lloyd Center, Change Orders, LC drawer 6 box 4.

Changes to the construction contract, Graham's office advised, "may become necessary due to construction conditions in the field, omissions or errors in the Working Drawings and Specifications, or changes that the Owner may deem advisable." Requests for such changes originated from both members of the building team and tenants. All requests for changes to the contract required Graham's office to produce a request for proposal and revised drawings, which it submitted to the contractor after consultation with the owner's representatives. The contractor prepared an estimate, which experts retained by Graham's office vetted. Requests for proposals required the approval of the owner, engineering consultants retained by the Prudential Life Insurance Company, and Prudential's local office before they formally became change orders that amended the basic construction contract.[100]

Change orders addressed all aspects of construction, from overtime wages to individual items, such as fire doors and escalator safety switches, to the finish of individual tenant retail spaces to an all-encompassing category of "general revisions."[101] Implementing change orders required the Lloyd Corporation to supply additional capital to complete the construction phase of the project. For the Prudential Life Insurance Company disbursed only $22.3 million under its loan at the time of the Lloyd Center's opening on 1 August 1960. The Lloyd Corporation funded the balance of the $28,956,176 total cost of construction.[102] Despite diminishing during the second half of the 1950s, capital generated from petroleum extraction enabled Ralph Lloyd to conceive and develop the Lloyd Center until his death and enabled the Lloyd Corporation to finance its planning and design, as well as a significant portion of the cost of construction.

Once it opened, the Lloyd Center operated as a centralized node of automobile-centered consumption wholly dependent on a robust regional petroleum product market based in Southern California. Shoppers arrived in vehicles fueled by gasoline refined in Los Angeles from crude oil extracted from the fields of the Coastal Region, Los Angeles Basin, and San Joaquin Valley and shipped by tanker to bulk petroleum terminals that lined the Willamette River north of downtown.[103] Thus some of the crude oil pumped from the Lloyd Corporation's Ventura Avenue wells may have been ultimately consumed by Portland-area shoppers in traveling from their homes to the company's regional shopping center. The parking structure that enveloped the Lloyd Center's pedestrian malls and retail blocks and provided a temporary home for hundreds of automobiles, in particular, expressed the close association between the Southern California oil industry and the suburbanization of Portland's East Side.

"Suburban Convenience in a Downtown Location"

Ralph Lloyd closely linked the development of his East Side holdings to downtown expansion in keeping with a city leveraging its location and natural resources to become one of America's major metropolitan areas. Yet the imperative of marketing the Lloyd Center to ensure its profitability persuaded Richard Von Hagen and his managers to establish an identity for the property as a market structure in competition with Portland's central business district.

To ensure the financial success of a project that represented total investment exceeding $30 million, the Lloyd Corporation and the professionals whose services it retained positioned the Lloyd Center as a "shopping hub, not only for Portland and its suburbs, but for the city's primary area, comprising all sections of Oregon west of the Cascade Mountains, and southwest Washington." By juxtaposing the Lloyd Center's "suburban setting of beauty, convenience, and efficiency" and a congested central business district that was confined by the Willamette River and the West Hills, Lloyd Corporation executives and managers began to conceive of the Lloyd Center as a "city within a city" that offered "all the facilities of a modern business district."[104] In October 1954, only one year after Larry Smith confirmed that the Lloyd Center had been conceived "more as adjunct to the downtown district than as a suburban shopping center," Franz Drinker urged Richard Von Hagen to think of the Lloyd Center in these exact terms.[105] Two years later, W. Joseph McFarland illustrated the extent to which outside observers influenced the thinking of Lloyd Corporation executives and managers when he routed to Von Hagen an article published in *Printers' Ink*, a trade publication, that described the Lloyd Center as "literally a new downtown center."[106] Fast forward to September 1962: Richard G. Horn confirmed the shift away from thinking of the Lloyd Center as an extension of Portland's downtown when he defined the property as a "complete city" that was "more than a place to buy merchandise, more than a pleasant place to shop, and more than a place where it is convenient to park."[107] It made business sense to market the Lloyd Center as a competitor of Portland's downtown rather as than a complementary extension of it.

In thinking of the Lloyd Center as a "complete city," Lloyd Corporation executives and managers did not displace entirely the oilman's belief that the development of his East Side holdings constituted a necessary extension of

Portland's central business district. Indeed, the same marketing brochure that touted the Lloyd Center as "a city within a city" also called readers' attention to the "logical" extension of Portland's "compressed and fixed" downtown business area to the East Side. In language that recalled the boosterism deployed by Ralph Lloyd in exhorting local business and civic leaders to improve Portland's competitiveness in the "tournament of cities," such an extension was as "inevitable as is the city's continued population growth and commercial progress."[108] In offering potential tenants "suburban convenience in a downtown location," however, the Lloyd Corporation "did little to disguise their ambition to have the project become the preferred retail [and business] center for the metropolis," as Richard Longstreth writes.[109] Competition rather than complementarity would frame Portland's early postwar trajectory as a regional metropolis.[110]

Conclusion

The Lloyd Corporation succeeded as a small business in both the upstream segment of the oil industry and commercial real estate. In contrast to many of the small businesses that have drawn the attention of scholars, the Lloyd Corporation competed on a sustained basis at the center of these markets. Under the direction of Ralph Bramel Lloyd and then Richard R. Von Hagen, his son-in-law, the company operated as a leading independent in California in competition against and cooperation with the integrated companies that dominated the industry. It invested surplus earnings from petroleum extraction in real estate, a "vast and fragmented" industry that proved to be open to entry to any well-capitalized investor. Its development efforts helped to shape the character of several urban districts, on the East Side of Portland in particular. Ralph Lloyd was fortunate that his father acquired ranch lands that lay atop one of America's largest onshore oil fields—though it remained for him to lead the search for oil on this property in partnership with Joseph B. Dabney and E. J. Miley. The royalty income that ultimately flowed into the coffers of the Lloyd Corporation as a result of their establishing the presence of substantial petroleum reserves in the area of the former Lloyd ranch ensured that the company would be well-funded and have the financial resources to weather difficult times, unlike most American small businesses.[1] As we have seen, success in both petroleum extraction and commercial real estate was contingent on the company's ability to address issues at the level of the firm and, at the same time, navigate turbulent macroeconomic conditions. This study adds to our understanding of the contribution of small business to two important sectors; the evolving relationships among owner, architect, and contractor within the building team; the importance of public capital in mobilizing real estate development in depression and war; and the direct and indirect links between oil and urban development on the Pacific Coast. It also shows how one small business handled ownership and management succession upon the death of its founder.

Ralph Lloyd's role in the development of the Ventura Avenue field supports Amherst University Professor Willard L. Thorp's contemporaneous assertion that individual ability "seems to be the outstanding cause for difference among business enterprises." Lloyd benefited Associated and Shell as an entrepreneur acting on his conviction that he could mobilize their "static" power to mutual advantage. As a lessor and operator, he cooperated with both majors, which had the resources and expertise to tackle the formidable geology of the field. Subsequently, he assumed the risk of testing its unproven area with the drill. As a result, the Lloyd Corporation, a small business by any definition, became one of California's leading producers of crude oil. Yet, even as he competed against the majors, Lloyd recognized that maximizing his profits required ongoing cooperation with them. Such symbiotic relations constitute an important factor in the oil business and one that is buried under an avalanche of academic and popular literature that pits independents against majors in matters of business and politics.

Statistics on exploratory wells drilled explain why many an independent operator made its name in the search for oil. Yet nonintegrated firms played an important role in producing oil. Production statistics confirm the growing participation of independents in California during the first half of the twentieth century, despite the early dominance of three majors. At the same time, the numbers conceal the role of independents in developing extractive regions in cooperation with other firms. After proving the presence of substantial reserves along the Ventura anticline, Lloyd, Dabney, and Miley participated in the development of a gigantic field in ways that do not show up in statistics. This approach to the oil business is underplayed, if not generally overlooked, in the literature. It is entirely absent in the literature on the California oil industry.

The decisions of Associated and Shell not to extend the proven area of the field illustrate why majors could not simply exclude new entrants through lease acquisition. Even as they cooperated with the two majors, Lloyd and Dabney—having bought out Miley—retained their option of reentering the field. Following Dabney's death in 1932, Ralph Lloyd, acting through the vehicle of the Lloyd Corporation, exercised this option when the two majors chose not to invest in its eastward extension. During the Great Depression, the capital budgets of Associated and Shell were limited. Their managers had to justify risk taking on the basis of expected return. As integrated firms, they acted more conservatively than independents during a period of oversupply. The development of the Ventura Avenue field also illustrates how individual circumstances also mattered in the decision-making of independent operators. As a closely held family firm, the Lloyd Corporation invested in oil development to limit its exposure to the democratic tax regime of the New Deal. It also did so because that same tax regime privileged oil exploration over commercial real estate development.

Ralph Lloyd's relations with Associated and Shell illustrates why the oil business during the "gusher age" was in large part "shaped by risk," to use the words of Dalit Baranoff.[2] Since the organizational capacities of the "first movers" in the industry were not decisive in the search for oil, companies had an incentive to cooperate in petroleum extraction. Even as market structures of capital-intensive manufacturing industries were evolving toward oligopoly, extractive industries attracted new entrants, many of whom thrived over the long-term strictly as exploration and production companies.

Ralph Lloyd's approach to the oil business could have been applied wherever drilling was an expensive proposition. His long-term interest in directing the development of the Ventura Avenue field, however, is atypical of the symbiotic relations detailed in the literature. In the histories of the Texas oil industry cited in this book, for instance, independents thrived by obtaining either cheap acreage in areas ignored by majors or farmouts of acreage held by them. That is, the interest of majors in sharing risk drove cooperative relationships. Moreover, the interest of independents in field development was generally short-term. They managed risk by selling out even before payout, to obtain capital for the next prospect. Roger Olien and Diana Davids Olien detail one exception: lacking the capital to invest in a Permian Basin (West Texas) play, Haymon Krupp and Frank Pickerell tried to interest majors, among them Humble Oil and Refining Company. When Humble declined the prospect, Krupp and Pickerell turned to Michael Benedum, a well-capitalized, large independent. In 1923, Benedum discovered the Big Lake field and stayed in for decades.[3] Additional research may uncover cases similar to Lloyd's, in California and elsewhere. Until then, Ralph Lloyd's approach of working with majors across decades without relinquishing control of operations stands out as distinctive in the literature.

Industries such as oil and base-metal mining may be distinguished from both oligopolistic and competitive manufacturing industries, in terms of the dominance of national firms.[4] In the automobile industry, "typically cited as the exemplar of mass production," as a recent study put it, Chrysler, Ford, and General Motors exerted market power nationally by the end of the interwar period. That is, in any given market, the presence of other firms was limited, except, perhaps, in the immediate proximity of factory operations.[5] A competitive industry, such as paint and varnishes, was one in which there was "no evidence of concentration or of a controlling combination," as Professor Thorp put it, owing to low barriers to entry and high unit transportation costs relative to the value of the product.[6] During the 1910s and 1920s, half-a-dozen firms established national markets in consumer paints.[7] Yet they did so largely through the acquisition of regionally prominent companies. A national presence required operations in numerous geographically dispersed locations, as there were no economies of scale in distribution. Local and regional firms proliferated, producing and marketing paints at low cost, selling products through independent dealer networks, and specializing in coatings by, for instance, adapting exterior house paints to local climatic conditions.[8] The industry expanded throughout the first two-thirds of the twentieth century, peaking at 1,579 firms in 1963. During the 1960s and 1970s, mergers and acquisitions gradually diminished the size of the industry. Yet, even then, it remained highly competitive. As of 1967, Sherwin-Williams, the leading firm, held only 15 percent of the national market.[9] Often, local or regional firms were market leaders. For instance, homeowners in many California markets preferred the products of San Francisco-based W. P. Fuller to any other.[10] In Milwaukee, Wisconsin, two local firms led the market from the 1930s to the 1960s.[11] The leading national firm was the only one with a local factory, which it had acquired.[12]

In the oil business, the dominance of integrated firms was regional, not national. As of 1940, for example, the top five producers held 43 percent and 50 percent of the California and Texas markets, respectively. Yet there was no overlap of market leaders between the two regions.[13]

The configuration of the oil industry varied by region, because operations were closely linked in terms of place. The leading firms in California ranked high nationally. At the same time, their market power did not extend into other regions, as it did for the so-called Big Three automakers. California's oil majors became large, multi-divisional organizations by leveraging their access to substantial reserves through integrating transportation, refining, distribution, and marketing functions. Yet entry by independent exploration and production firms remained a defining characteristic of the business. During the first four decades of the twentieth century, which were characterized by the increasing dominance of big businesses across swathes of the American economy, these small businesses expanded their presence in a critical segment of a strategic industry. These firms were not peripheral purveyors of specialty products. Rather they used their comparative market advantages to compete against and cooperate with their major counterparts in the center of the business.

Perhaps taking his cue from his father, Ralph Lloyd speculated in real estate before capital generated from petroleum extraction enabled him to become a significant investor in selected metropolitan markets. At a general level, his direction of surplus capital into real estate was tied to a notion, if not a well-conceived strategy, of investing capital derived from depleting crude oil reserves into assets that would provide a sustained and stable, if not increasing, income stream. But during his lifetime, the oilman's real estate program depended on infusions of energy-generated capital. Only with the completion of the Lloyd Center seven years after his death did the real estate segment of the Lloyd Corporation's business become self-sustaining. Indeed, the completion of the regional shopping center marked the beginning of the evolution of the Lloyd Corporation into a property management organization under the direction of Richard R. Von Hagen, a process that was completed with its exit from the Ventura Avenue field as an operator in the 1980s. For by the early 1960s petroleum exploration and development in Ventura County had largely played out.[14] In contrast to planning and executing his oil field business within a coherent strategy of maximizing long-term production in one of America's largest onshore fields, Lloyd's approach to real estate acquisition was often ad hoc and speculative. Oil-generated wealth, however, enabled him to hold property through the worst of economic times and allowed him to plan its eventual development. Still, given the performance of his real estate portfolio during his lifetime, it is clear that factors other than financial return enticed Ralph Lloyd to invest so much of his accumulated capital in undeveloped lots in selected markets. Ultimately, he identified himself as not simply an oilman but a builder and booster of cities and of the region in which they were located.

Ralph Lloyd's career may be located at the intersection of petroleum extraction and urbanization on the Pacific Coast. As a speculator in commercial real estate, he reinvested capital that he accumulated from the exploitation of reserves in Southern California. In doing so, he contributed to an oil-driven transformation of a regional economy and the region's urban spaces. Through the development of his commercial real estate portfolio, Lloyd also demonstrated how oil indirectly extended the reach of Los Angeles as an "energy capital." He never fully realized his commercial real estate ambitions in the City of Angels. Nevertheless, the metropolis's development during the first three decades of the twentieth century informed his thinking on the development of his holdings on Portland's East Side and provided a blueprint for action. To advocate on behalf of public infrastructure projects in support of his commercial real estate program, Lloyd drew directly from projects in which he was involved or with which he was familiar. Further, buildings constructed in Los Angeles inspired his Portland projects. They included signature buildings, such as the Ambassador and Biltmore Hotels, and myriad examples of vernacular architecture in Hollywood and along Wilshire Boulevard. Lloyd's entrepreneurship as both oilman and real estate developer illustrates the dependence of urbanization on fossil fuel production not only in Southern California, where the link has been well documented, but also in the Pacific Northwest, where the connection has remained largely unexplored.

In depression and war, public capital provision catalyzed the development of Ralph Lloyd's real estate portfolio. The materialization of a portion of Portland's East Side as a cluster of office buildings for Bonneville Power Administration employees in particular demonstrated the impact of public investment associated with New Deal programs. Federal expenditure to accomplish the economic and social objectives of public power advocates enabled Lloyd to mobilize private capital that otherwise he likely would have held as cash. His contemporaneous failure to develop his Beverly Hills Business Triangle block for private clients throws his successful business relationship with BPA Administrator Paul Raver into sharp relief.

In developing real property, Ralph Lloyd and the Lloyd Corporation deployed various methodologies to deliver their buildings. In selecting the Austin Company to design and construct selected buildings, Ralph Lloyd utilized an early version of design-build, whereby the engineering firm both designed and constructed the structure. The cases presented in this book demonstrate the advantages of the approach. In the planning, design, and construction of the Lloyd Center, the Lloyd Corporation relied on design-bid-build, the predominant delivery method of the twentieth century. Indeed, the case of the Lloyd Center provides a useful study that illustrates why, as I elaborate elsewhere, owners have increasingly utilized design-build in recent decades to control project costs and schedules.[15] For the majority of his projects, Ralph Lloyd relied on design-negotiate-build, a hybrid approach that is not discussed elsewhere in the literature. Under this method, Lloyd as owner continued to rely on architects to design the project initially without input from the building contractor but provided selected contractors with the opportunity to comment on the plans and specifications before he negotiated the terms of a contract with one of them. By adopting this approach, Lloyd exerted a measure of owner control over his projects. More research is needed on the alternative (that is, non-design-bid-build) means by which commercial projects were delivered before the sector ground nearly to a halt during the Great Depression. At the same time, in utilizing the voluminous real estate files of the Lloyd Corporation Archive, this study has shed much needed light on project delivery methods and relationships among the members of the building team on both signature and vernacular projects from 1920 to 1960. It also has broken new ground in analyzing the means by which the Lloyd Corporation procured and delivered commercial projects between 1933 and 1945. However depressed the building industry may have been during this period, projects moved forward. In the case of the Lloyd Corporation, public capital was a crucial factor in making its projects "pencil."

As a "citizen outside the government," Ralph Lloyd acted within a variety of urban milieus on behalf of his business objectives. As a party interested in maximizing the productivity of oil field operations at Ventura Avenue, he acted as intermediary on behalf of outside oil companies whose primary interest lay in extracting petroleum with minimal interference from the City of Ventura. In Portland, he established personal relationships with civic leaders and worked with them individually and through interest groups to accomplish the eventual development of his Holladay's Addition and Walnut Park properties. At the same time, he complemented his advocacy with the investment of a portion of his accumulated capital in public infrastructure construction. In Los Angeles, where his potential individual influence was diluted by the participation of hundreds of developers with similar aims, Lloyd acted within various "improvement" associations to secure passage of municipal infrastructure projects that increased the value of his real estate portfolio.

Ralph Lloyd was a leading voice for the independent operator as the industry contended with a crisis of overproduction from the late 1920s through the Great Depression. In this context, he championed cooperation among California's oil producers, in keeping with the associationalist vision offered by Herbert Hoover as commerce secretary and president. He advocated on behalf of price support, both directly and through production control, as president of the Oil Producers Sales Agency of California and as a member of other boards and organizations. He also helped to develop the production code for the California oil industry under the National Industrial Recovery Act. With his opposition to tax policy as it evolved at the federal, state, and local levels during the 1930s, Lloyd adopted an increasingly libertarian perspective on the role of government in business. But as a practical businessman, he never dismissed entirely the role of government in business affairs at any level and continued to work with municipal officials in support of his projects. And he happily met the needs of the Bonneville Power Administration for office space without directly questioning its legitimacy or its reform objectives.

The editors of *Energy Capitals: Local Impact, Global Influence* note that the "strong and complex connections at the intersection of energy-led development [and] urban growth . . . are intuitively obvious. Yet they are largely absent in the existing literature."[16] In explicating and interrogating the career of an individual who operated successfully as independent oil operator and commercial real estate developer, this book elaborates the interchangeability of energy and capital and contributes to our understanding of the direct and indirect impacts of energy-capital conversion on urbanization. In layering mid-century mass consumption at the base of the pyramid that Gray Brechin has deployed to visualize Lewis Mumford's Mega-Machine, the study further illustrates the extent to which fossil fuel energy constitutes, as human ecologist Alf Hornberg asserts, "the vital essence that flows through human societies."[17]

Archival Collections

Bancroft Library, University of California, Berkeley
　Associated Oil Company, *The Record*
California Institute of Technology Archives
　Historical File of Joseph B. Dabney
California State University, Long Beach, Library, Virtual Oral/Aural History Archive
　Interview of James E. Herley
　Interview of Thomas M. Rowan, Jr.
City of Los Angeles, Archives and Records Center
　Meeting Minutes and Files, City of Los Angeles Council
City of Portland, Oregon, Archives and Records Center
　Documents, City of Portland Council
　Meeting Minutes, City of Portland Council
　Meeting Minutes, City of Portland Planning Commission
　City Engineer Subject Files, Department of Public Works
City of San Buenaventura, California
　Minute Book of the Board of Trustees
Henry E. Huntington Library, San Marino, California
　Ralph Arnold Collection
　John Anson Ford Papers
　Homer D. Crotty Papers and Addenda
　Lloyd Corporation Archive
Library of Congress
　William Gibbs McAdoo Papers
Long Beach, California, Public Library
　Long Beach History Collection
Milwaukee County Historical Society
Museum of Ventura County
　Biographical Files
　Oral History Collection
　　Interview of F. W. Hertel
　　Interview of Richard Maulhardt
　　Interview of Joseph D. McGrath, Jr.
　　Interview of Verne Patmore

Interview of Benton Turner
Interview of Richard Willett

Oregon Historical Society
Lloyd Center, vertical file

Stanford University, Special Collections and University Archives
Paul Shoup Papers

University of California, Los Angeles, Special Collections
Shannon Crandall Papers

University of California, Southern Regional Library Facility, Los Angeles
Eberle Economic Service, Weekly Letters and Supplements

University of California, Santa Barbara, Special Collections
Lawrence B. Romaine Trade Catalog Collection

University of Oregon Libraries, Special Collections and University Archives
John C. Ainsworth Papers

NOTES

INTRODUCTION: THE ROLE OF THE INDEPENDENT OIL OPERATOR

1. Mansel G. Blackford, *A History of Small Business in America*, 2d ed. (Chapel Hill: University of North Carolina Press, 2003).
2. William Issel, "'Citizens Outside the Government': Business and Urban Policy in San Francisco and Los Angeles, 1900–1932," *Pacific Historical Review* 56 (1987): 117–45.
3. See, for example, Lloyd to Norton, 2 December 1927, Port drawer 1, box 4 (hereafter in the form: drawer number–box number, that is, 1–4).
4. Roger M. Olien and Diana Davids Olien, *Wildcatters: Texas Independent Oilmen* (Austin: Texas Monthly Press, 1984). See, also, idem, *Oil in Texas: The Gusher Age, 1895–1945* (Austin: University of Texas Press, 2002).
5. Gerald D. Nash, "Oil in the West: Reflections on the Historiography of an Unexplored Field," *Pacific Historical Review* 39 (1970): 193–204.
6. Thomas M. Rowan Jr., interviewed by Ann Andriesse, 12 April 1982, Long Beach Area History: Petroleum Entrepreneurs, VOAHA.
7. Geoffrey G. Snow and Brian W. MacKenzie, "The Environment of Exploration: Economic, Organizational, and Social Constraints," in *Economic Geology: 75th Anniversary Volume*, ed. Brain J. Skinner (Lancaster, PA: Economic Geology Publishing, 1981), 871–96.
8. "Small Operators," *California Oil World*, 1 May 1930.
9. Roger M. Olien and Diana Davids Hinton, *Wildcatters: Texas Independent Oilmen* (College Station: Texas A&M University Press, 2007), xiii.
10. James W. McKie, "Market Structure and Uncertainty in Oil and Gas Exploration," *Quarterly Journal of Economics* 74 (November 1960): 550–1; Olien and Olien, *Wildcatters*, 2–3.
11. *Oil and Gas Prospect Wells Drilled in California through 1980* (Sacramento: California Division of Oil and Gas, 1982) (compiled by the author).
12. Quoted in Martin R. Ansell, *Oil Baron of the Southwest: Edward L. Doheny and the Development of the Petroleum Industry in California and Mexico* (Columbus: Ohio State University Press, 1998), 91. The Bureau of Corporations, which was established in 1903 as an agency within the new Department of Commerce and Labor, was investigating the Standard Oil Trust's activities in the San Joaquin Valley.

13. W. J. Lovingfoss, interview with the author, Carpinteria, California, 3 June 1998.
14. Olien and Olien, *Oil in Texas*, 220–1.
15. Michael P. Barnard, "Brief History of the Santa Ana Valley and the Barnard Family," 1980, Biographical File B-Ba, VCM; Yvonne G. Bodle, "The McGrath Story: 100 Years of Ranching on the Oxnard Plain," *Ventura County Historical Society Quarterly* 22 (Summer 1977): 1–25; Robert Glass Cleland, *The Cattle on a Thousand Hills: Southern California, 1850–1870* (San Marino, CA: Huntington Library Press, 1941), 135–83; W. H. Hutchinson, *Oil, Land, and Politics: The California Career of Thomas Robert Bard*, vol. 1 (Norman: University of Oklahoma Press, 1965), 45–75, 103–32, 235–48; Richard Maulhardt, interviewed by George Appel, 15 June 1988, VCM; Joseph D. McGrath, Jr., interviewed by Joyce Pederson, 12 April 1991, VCM; Gerald T. White, *Formative Years in the Far West: A History of Standard Oil Company of California and Predecessors Through 1919* (New York: Appleton-Century-Crofts, 1962), 4–19.
16. James Willard Hurst, *Law and the Conditions of Freedom in the Nineteenth-Century United States* (Madison: University of Wisconsin, 1956).
17. A gigantic field has expected or realized production of 100 million barrels of crude oil or natural gas equivalent (BOE).
18. Ralph G. Frame, "Santa Maria Valley Oil Field," *Annual Report of the State Oil and Gas Supervisor of California* 24 (October-November-December 1938): 37–43; William W. Porter II, "Santa Maria Valley—Another Great Field," *Petroleum World* (July 1937): 24–30; "Union Oil Buys Up Moore Lease on Sugar Lands," *Santa Maria Daily Times*, 28 October 1940; Earl M. Welty and Frank J. Taylor, *The Black Bonanza*, 2d rev. ed. (New York: McGraw-Hill, 1958), 237–42; CCCOP, *Annual Review of California Crude Oil Production, 1940* (Los Angeles, CCCOP, 1941) (compiled by the author).
19. J. C. Gilbert and J. H. Siemon, "Marine Drilling in California," *Petroleum World* (April 1933): 15–18; K. P. Goble, "Pacific Western's Big Well and G.P.'s More No. 2 Spotlight Elwood-Goleta," *Petroleum World* (April 1930): 85–86; K. P. Goble, "Interest in Santa Barbara Centers on Goleta Test," *Petroleum World* (July 1930): 96; "Possibilities of the Sespe at Elwood," *Petroleum World* (November 1931): 30–31; S. G. Dolman, "Elwood Oil Field," *Annual Report of the State Oil and Gas Supervisor* 16 (January-February-March 1931): 6–7; Sarah S. Elkind, "Oil in the City: The Fall and Rise of Oil Drilling in Los Angeles," *Journal of American History* 99 (June 2012): 82–90; Charles S. Jones, *From the Rio Grande to the Arctic: The Story of the Richfield Oil Corporation*

(Norman: University of Oklahoma Press, 1972), 37–51; James T. Lima, "The Politics of Offshore Energy Development," Ph.D. diss., University of California, Santa Barbara, 1994, 150–68; Paul Sabin, "Beaches versus Oil in Greater Los Angeles," in *Land of Sunshine: An Environmental History of Metropolitan Los Angeles*, ed. William Deverell and Greg Hise (Pittsburgh, PA: University of Pittsburgh Press, 2005), 95–114; Nancy Quam-Wickham, "'Cities Sacrificed on the Altar of Oil': Popular Opposition to Oil Development in 1920s Los Angeles," *Environmental History* 6 (April 1998): 189–209.

20. James E. Herley, interviewed by Ann Andriesse, 18 March 1982, Long Beach Area History: Petroleum Entrepreneurs, VOAHA.
21. Richard C. Schwarzman, "The Pinal Dome Oil Company: An Adventure in Business, 1901–1917," Ph.D. diss., University of California, Los Angeles, 1967, 25–44, 82–153.
22. Walker A. Tompkins, *Little Giant of Signal Hill: An Adventure in American Enterprise* (Englewood Cliffs, NJ: Prentice-Hall, 1964), 73–80.
23. Walter H. Case, ed., *Long Beach Community Book: Part 2: Biographical* (Long Beach, CA: A. H. Cawston, 1948), 269–70.
24. On earlier methods deployed in support of the search for oil, see Paul Lucier, *Scientists and Swindlers: Consulting on Coal and Oil in America, 1820–1890* (Baltimore, MD: Johns Hopkins University Press, 2008), 244–71.
25. Taylor to McAdoo, 28 April 1922, McAdoo Papers, box 263.
26. William Rintoul, *Spudding In: Recollections of Pioneer Days in the California Oil Fields* (Fresno, CA: Valley Publishers, 1978), 57.
27. Olien and Olien, *Oil in Texas*, 129–33.
28. McKie, "Market Structure and Uncertainty in Oil and Gas Exploration," 544–5.
29. Snow and MacKenzie, "The Environment of Exploration," 875.
30. McKie, "Market Structure and Uncertainty in Oil and Gas Exploration," 563.
31. Jones, *From the Rio Grande to the Arctic*, 40–7; Paul W. Prutzman, "Petroleum in Southern California," *California State Mining Bureau Bulletin* 63 (San Francisco, 1913): 396–401.
32. "California Petroleum Situation," *Oil, Paint, and Drug Reporter*, 10 June 1901. A wildcatter drills wells in unproven areas. That is, he or she engages in exploration.
33. Ralph Arnold and Robert Anderson, "Geology and Oil Resources of the Santa Maria Oil District, Santa Barbara County, Cal.," Bulletin 322, U.S. Geological Survey (Washington, DC, 1907), 91.

34. McKie, "Market Structure and Uncertainty in Oil and Gas Exploration," 550–2.
35. *Annual Report for the Associated Oil Company*, fiscal years ending 31 December 1918–1923 (compiled by the author).
36. FTC, *Report of the Federal Trade Commission on the Pacific Coast Petroleum Industry*, pt. 1 (Washington, DC: GPO, 1921), 44–45; Hugh A. Matier, "This Company's Oil Conservation Policy Is Building Up Reserves for the Future," *Petroleum World* (October 1927): 60; Barbara L. Pederson, *A Century of Spirit: Unocal, 1890–1990* (Los Angeles: Unocal Corporation, 1990), 34–55.
37. FTC, *Report of the Federal Trade Commission on the Pacific Coast Petroleum Industry*, pt. 1 (Washington, DC: GPO, 1921), 44–45; Moody's Investor Service, *Moody's Manual of Industrial Securities, 1940* (New York: Moody's, 1941) (compiled by the author); Olien and Olien, *Oil in Texas*, 153–5.
38. Kendall Beaton, *Enterprise in Oil: A History of Shell in the United States* (New York: Appleton-Century-Crofts, 1957), 174–8; Robert H. Wheeler and Maurine Whited, *Oil from Prospect to Pipeline*, 4th ed. (Houston, TX: Gulf Publishing, 1981), 71–79.
39. Beaton, *Enterprise in Oil*, 373–85.
40. Humphery to Byles, memos of 31 October 1932 and 2 December 1932, box 7; "Associated Oil Company: Explanation of Changes in 1933 Capital Budget," 8 February 1933, box 5; Humphery to Shoup, 4 January 1935, box 1 (all found in Shoup Papers).
41. Donald T. Critchlow, *Studebaker: The Life and Death of an American Corporation* (Bloomington: University of Indiana Press, 1996).
42. Willard L. Thorp, "The Merger Movement and the Paint and Varnish Industry," an address delivered at the annual meeting of the American Paint and Varnish Manufacturers' Association, Washington, DC, 15 October 1929, 14.
43. Ralph Andreano, "The Structure of the California Petroleum Industry, 1895–1911," *Pacific Historical Review* 39 (1970): 171–184, 185 (table IV), 189 (table VII), 191 (table VIII); Ansell, *Oil Baron of the Southwest*, 23–51, 92–95; H. L. Barber & Co., *Facts and Figures about California Oil* (Chicago: the company, 1910), Lawrence B. Romaine Trade Catalog Collection, Department of Special Collections, Davidson Library, University of California, Santa Barbara, series I, sub-series A, folder: "Oil," box 1; "The Associated Oil Company," *Mining and Oil Bulletin* (November 1922), 646–50; Gray Brechin, *Imperial San Francisco: Urban Power, Earthly Ruin* (Berkeley: University of California Press, 1999), 256–60; Carl J. Mayer and George A. Riley, *Public Domain, Private*

Dominion: A History of Public Mineral Policy in America (San Francisco: Sierra Club, 1985), 155–69; Pederson, *A Century of Spirit*, 15–47; White, *Formative Years in the Far West*, 208–361, 576 (table III).

44. Beaton, *Enterprise in Oil*, 67–104; FTC, *Report of the Federal Trade Commission on the Pacific Coast Petroleum Industry*, pt. 1, 61–3, 64 (table 15), 65, 89–116; Olien and Olien, *Wildcatters*, 13–14; Welty and Taylor, *The Black Bonanza; White, Formative Years in the Far West*; Harold F. Williamson, Ralph L. Andreano, Arnold R. Daum, and Gilbert C. Klose, *The American Petroleum Industry: The Age of Energy, 1899–1959* (Evanston, IL: Northwestern University Press, 1963), 536–40. Alfred D. Chandler Jr. argues that the process of integration at the national level was complete by 1920 (*The Visible Hand: The Managerial Revolution in American Business* [Cambridge, MA: Belknap Press of Harvard University Press, 1977], 350–53). For national rankings of US oil companies by assets, see Ansell, *Oil Baron of the Southwest*, 252 (appendix D).
45. Beaton, *Enterprise in Oil*, 174–88; Pederson, *A Century of Spirit*, 59–78; Rintoul, *Spudding In*, 146–61; Jules Tygiel, *The Great Los Angeles Swindle: Oil, Stocks, and Scandal During the Roaring Twenties* (New York: Oxford University Press, 1994), 3–16.
46. Paul Sabin, *Crude Politics: The California Oil Market, 1900–1940* (Berkeley: University of California Press, 2005), 64–78, 113.
47. James E. Herley, interviewed by Ann Andriesse, 18 March 1982, Long Beach Area History: Petroleum Entrepreneurs, VOAHA.
48. Joe S. Bain, *The Economics of the Pacific Coast Petroleum Industry, Part I: Market Structure* (Berkeley: University of California Press, 1944), 45 (table).
49. FTC, *Report of the Federal Trade Commission on the Pacific Coast Petroleum Industry*, pt. 1, 64 (table 15); *Petroleum World, Annual Review, California Oil Industry* (Los Angeles, 1934) (compiled by the author).
50. Bain, *The Economics of the Pacific Coast Petroleum Industry, Part I*, 42; CCCOP, *Annual Review of California Crude Oil Production, 1940* (compiled by the author).
51. CCCOP, *Annual Review of California Crude Oil Production, 1940* (compiled by the author).
52. Blackford, *A History of Small Business in America*, 51–70, 102–13.
53. Blackford, *A History of Small Business in America*, 51–70, 102–13.
54. Christopher J. Castaneda and Joseph A. Pratt, *From Texas to the East: A Strategic History of Texas Eastern Corporation* (College Station: Texas A & M University Press, 1993).

55. See, for example, Bryan Burrough, *The Big Rich: The Rise and Fall of the Greatest Texas Oil Fortunes* (New York: Penguin, 2009); Mike Cochran, *Claytie: The Roller-Coaster Life of a Texas Wildcatter* (College Station: Texas A & M University Press, 2007).
56. Martin V. Melosi, "Houston: Energy Capital," *New Geographies* 2 (2009): 99, quoted in Stephanie LeMenager, *Living Oil: Petroleum Culture in the American Century* (New York: Oxford University Press, 2014), 13.
57. Sarah S. Elkind, "Los Angeles: The Energy Capital of Southern California," in *Energy Capitals: Local Input, Global Influence*, ed. Joseph A. Pratt, Martin V. Melosi, and Kathleen A. Brosnan (Pittsburgh, PA: University of Pittsburgh Press, 2014), 77–93.
58. LeMenager, *Living Oil*, 71–81.
59. Richard Longstreth, *The Drive-In, The Supermarket, and the Transformation of Commercial Space in Los Angeles, 1914–1941* (Cambridge, MA: MIT Press 1999); Richard Longstreth, "The Forgotten Arterial Landscape: Photographic Documentation of Commercial Development Along Los Angeles Boulevards During the Interwar Years," *Journal of Urban History* 23 (1997): 437–59. On the development of Hollywood as a commercial district, see Richard Longstreth, *City Center to Regional Mall: Architecture, the Automobile, and Retailing in Los Angeles, 1920–1950* (Cambridge, MA: MIT Press, 1997), 81–101.
60. Michael R. Adamson, *A Better Way to Build: A History of the Pankow Companies* (West Lafayette, IN: Purdue University Press, 2013).
61. In the public sector, design-build projects are often procured through competitions that typically award points in three categories, namely, qualifications, design, and price. For a discussion, see, for instance, Adamson, *A Better Way to Build*, 266–77.
62. Longstreth, *City Center to Regional Mall*, 103–41.
63. Robert Higgs, "Regime Uncertainty: Why the Great Depression Lasted So Long and Why Prosperity Resumed after the War," *Independent Review* 1 (Spring 1997): 561–90.
64. Brechin, *Imperial San Francisco*, 19.
65. William Leach, *Land of Desire: Merchants, Power, and the Rise of a New American Culture* (New York: Vintage, 1993), 173–6.

CHAPTER 1. DEVELOPING THE VENTURA AVENUE FIELD

1. Lloyd to King and McLaughlin, 28 October 1925, RBL 4–1.
2. Homer D. Crotty, "Biography of Ralph B. Lloyd," TS, 2nd version, Crotty Papers, box 113, 2–7.

3. Crotty, “Biography of Ralph Lloyd,” 7–12; “Lewis Marshall Lloyd,” *California and Californians*, ed. Rockwell D. Hunt, vol. 3 (Chicago: Lewis Publishing, 1926), 122.
4. *Ventura Free Press*, 8 July 1887, quoted in Glenn S. Dumke, *The Boom of the Eighties in Southern California* (San Marino, CA: Huntington Library Press, 1944), 172.
5. Crotty, “Biography of Ralph B. Lloyd,” 16–26; Dumke, *The Boom of the Eighties in Southern California*, 168–72; Richard J. Orsi, *Sunset Limited: The Southern Pacific Railroad and the Development of the American West, 1850–1930* (Berkeley: University of California Press, 2005), 18–21; Rintoul, *Spudding In*, 137–8; “Ventura Avenue,” *The Record* 7 (September 1926): 3–8; White, *Formative Years in the Far West*, 106–31. The completion of the Santa Fe Railroad to Los Angeles, in 1887, also fueled the regional boom, and broke the Southern Pacific Railroad’s monopoly in California. *The Record* was the company magazine of the Associated Oil Company.
6. Crotty, “Biography of Ralph B. Lloyd,” 25–9; Dumke, *The Boom of the Eighties in Southern California*, 259–67.
7. Crotty, “Biography of Ralph B. Lloyd,” 27–9; Willoughby Rodman, *History of the Bench and Bar of Southern California* (Los Angeles: William J. Porter, 1909), 191.
8. Ralph B. Lloyd, “Father Saves Life in Brush Fire by Leap into Canyon,” *California Oil World*, 21 January 1926, 14; F. W. Hertel, “Ventura Is One of California’s Greatest Fields,” *The Oil Weekly*, 4 March 1927, 47–58; Benton Turner, oral history, VCM, 30 January 1986, 2–3; Rintoul, *Spudding In*, 138–40.
9. Crotty, “Biography of Ralph B. Lloyd,” 30–4, 38–40.
10. Richard White, *Railroaded: The Transcontinentals and the Making of Modern America* (New York: Norton, 2011), 466–71.
11. Crotty, “Biography of Ralph B. Lloyd,” 38–40.
12. Crotty, “Biography of Ralph B. Lloyd,” 34–5; Lloyd, “Father Saves Life in Brush Fire by Leap into Canyon.”
13. Hudson quoted in Crotty, “Biography of Ralph B. Lloyd,” 42.
14. “Lewis Marshall Lloyd,” 122; Charles Montville, Benjamin Brooks, and Edwin M. Sheridan, “Mariano Erburu,” *History of Santa Barbara, San Luis Obispo and Ventura Counties*, vol. 2 (Chicago: Lewis Publishing, 1917), 713; Crotty, “Biography of Ralph B. Lloyd,” 49–50.
15. Crotty, “Biography of Ralph B. Lloyd,” 51–6.
16. Crotty, “Biography of Ralph B. Lloyd,” 56–9; “William E. Hampton,” *Notables of the West*, vol. 2 (New York: International News Service, 1915), 203; “Ralph

Bramel Lloyd," *California and Californians,* ed. Rockwell D. Hunt, vol. 5 (Chicago: Lewis Publishing, 1926), 73.

17. Crotty, "Biography of Ralph B. Lloyd," 58.
18. For instance, "Irregular heart action," related to the condition, but apparently not identified as such, was the basis for the denial of a policy for Ralph Lloyd at the age of forty-two by the Union Central Life Insurance Company (Folsom to Lloyd, 11 September 1917, Shannon case #21, LCA).
19. "Ralph B. Lloyd," timeline, Photos 1–3; "William E. Hampton," 203; Crotty, "Biography of Ralph B. Lloyd," 58–9.
20. Lloyd to Hampton, 12 May 1909, Shannon case #24, LCA; Lloyd to Hampton, 23 June 1909, Shannon case #24, LCA; "Ralph B. Lloyd," timeline, Photos 1–3; Crotty, "Biography of Ralph B. Lloyd," 60–1.
21. Lloyd to Hampton, 6 August 1909; Lloyd to Hampton, Kirsch, and Sackett, 7 August 1909; Hampton to Kirsch, 10 August 1909; Lloyd to Kirsch and Sackett, 16 August 1909; Lloyd to Kirsch, 1 September 1909; Lloyd to Hampton, 18 December 1909; Hampton to Lloyd, 30 December 1909; Lloyd to Hampton, 3 January 1910 (all found in Shannon case #24, LCA); "General Machinery News," *Modern Machinery* 24 (January 1910): 216; Ralph B. Lloyd, submission to Appleton's *Cyclopedia of American Biography,* 11 May 1922, RBL 2–1; "Ralph Bramel Lloyd," 73–4; Pacific Tank and Pipe Company, advertisement, *The Timberman* (September 1910), 26 (quoted).
22. Hampton to Lloyd, 19 September 1910, Shannon case #5, LCA; Lloyd to Hampton, 8 February 1911, Shannon case #5, LCA; "William E. Hampton," 203; Crotty, "Biography of Ralph B. Lloyd," 62–3. Talk of a merger to address the "very fierce" competitive conditions in the industry dated from the Panic of 1907 (Hampton to Whitney, 16 October 1907; National Wood Pipe Company to Kirsch, 9 March 1909 [quoted] [all found in Shannon case #24, LCA]). On the negotiations, see the correspondence, dated March–April 1911, in Shannon case #5, LCA.
23. Mortgage, 20 April 1910, LCR 1–1; Gates to Lloyd, 25 July 1925, LCR 2–7; Lloyd to Gates, 31 July 1925, LCR 2–7; Ralph B. Lloyd, submission to Appleton's *Cyclopedia of American Biography,* 11 May 1922, RBL 2–1; "Lloyd Outlines Project Dreams," *Oregonian,* 4 April 1929, I:1; "Guide to the Larrabee Family Papers, 1840–2004," Center for Pacific Northwest Studies, Western Washington University, 2004; E. Kimbark MacColl, *The Growth of a City: Power and Politics in Portland, Oregon, 1915–1950* (Portland, OR: Georgian Press, 1979), 3, 325 (quoted). On the career of the notorious Ben

Holladay, see Jewel Lansing, *Portland: People, Politics, and Power, 1851–2001* (Corvallis: Oregon State University Press, 2005), 130–63.

24. Lloyd to California Farmland Company, 5 February 1912, Shannon case #7, LCA.
25. Lewis Lloyd held 1,300 shares in VL&W; Sarah Lloyd held 450 shares; each of the children held 250 shares. With the distribution of Sarah Lloyd's shares, the children could outvote their father by 150 shares (VL&W, Agreement between Stockholders, 15 March 1909, VLW 1–2).
26. Crotty, "Biography of Ralph B. Lloyd," 63–4.
27. Crotty, "Biography of Ralph B. Lloyd," 64; "Warren Estelle Lloyd," *Men of Nineteen-Fourteen* (Chicago: American Publishers Association, 1915), 467. In 1908, Warren E. Lloyd published *Psychology, Normal and Abnormal: A Study of the Processes of Nature from the Inner Aspect.* He and his wife, Caroline Alma Goodman, who was a noted sculptor, shared with their friends "a fashionable interest in the occult" (Tom Hiney, *Raymond Chandler: A Biography* [New York: Grove Press, 1997], 38–9). On Warren Lloyd and his family, see, also, Frank MacShane, *The Life of Raymond Chandler* (New York: Penguin, 1976), 24–7; Tom Williams, *Raymond Chandler: A Life* (London: Aurum Press, 2012), 51–8.
28. VL&W, Agreement between Stockholders, 15 March 1909, VLW 1–2.
29. "Ralph B. Lloyd," timeline, Photos 1–3; Crotty, "Biography of Ralph B. Lloyd," 64.
30. "Crude Oil Production in California from the Industry's Inception to January 1, 1925," *Petroleum World* (January 1925): 70–1; Hutchinson, *Oil, Land, and Politics*, vol. 1, 65–75, 235–43, 271–4; White, *Formative Years in the Far West*, 6–10, 11 (quoted), 12–19, 23–57; Gerald T. White, "California's Other Mineral," *Pacific Historical Review* 39 (May 1970): 145–50.
31. Daniel Kevles, "Joseph B. Dabney," May 1991, box Z9, Historical Files, Archives, California Institute of Technology.
32. "Emmor Jerome Miley," *Notables of the West*, vol. 2, 427.
33. "Emmor Jerome Miley," *Notables of the West*, vol. 2, 427; "Directory of California Oil Operators," *Annual Report of the State Oil and Gas Supervisor of California* 1 (1917): 251.
34. "Emmor Jerome Miley," *Notables of the West*, vol. 2, 427; "Directory of California Oil Operators," *Annual Report of the State Oil and Gas Supervisor of California* 1 (1917): 267.
35. As Martin R. Ansell notes, however, crude oil was apparently relatively easy to find in the San Joaquin Valley at this time. From 1907 to 1912, fully 92

percent of all wells drilled in California (most of them found in Kern County) discovered crude oil reserves in commercial quantities, compared to 73 percent nationally (*Oil Baron of the Southwest*, 95–7).

36. "Directory of California Oil Operators," *Annual Report of the State Oil and Gas Supervisor of California* 1 (1917): 251, 267; "Joseph B. Dabney," obituary, *LAT*, 12 September 1932; "Miley Drills 7592 Feet to Get New Well," *LAE*, 25 December 1925, III:21 (quoted); State Consolidated Oil Co. to Lloyd, 26 February 1914, with attached, A. M. Buley, letter to stockholders, Shannon case #11, LCA; "Twenty Five Years of Successful Oil Development," annual midwinter number, *LAT*, 1 January 1925, pt. 2, 28.
37. Lloyd to Bank of California, Portland branch, 22 January 1912, Shannon case #6, LCA (quoted); "History of the Discovery of Oil on the Lloyd Ranch, Ventura County, California," undated, enclosed with Lloyd to First National Bank, 22 January 1916, Shannon case #16, LCA; Crotty, "Biography of Ralph B. Lloyd," 75–6.
38. Grant to Lloyd, 4 December 1912, Shannon case #8, LCA; Lloyd to Baylis, 11 January 1913, Shannon case #9. LCA; Lloyd to Buckalow, 16 January 1913, Shannon case #9, LCA; Lloyd to Doheny, 24 February 1914, Shannon case #11, LCA; Crotty, "Biography of Ralph B. Lloyd," 72–3.
39. "Assignment of One-Third Interest in Oil Lease," 3 June 1913, Shannon case #9, LCA; Lloyd to Dabney, 13 August 1913, Shannon case #10, LCA; Lloyd to Black, 9 September 1913, Shannon case #10, LCA; "New Company in Simi Field," *Oil Age*, 26 September 1913, 1; Crotty, "Biography of Ralph B. Lloyd," 73–5. Hidalgo Oil Company operated out of an office in the Union Oil Building in downtown Los Angeles that Ralph Lloyd had leased in June at a rate of $25 per month.
40. Lloyd to Dabney, 26 August 1913; Lloyd to Boss, 26 August 1913 (quoted) (all found in Shannon case #10, LCA). Gravity is an industry measure of the specific gravity of petroleum relative to water: the higher the number, the lower the density and viscosity of the oil, and the higher its value. Oil with a gravity of more than ten floats on water. In 1916, the National Bureau of Standards adopted the Baumé scale that had been developed in France to measure the specific gravity of liquids less dense than water. In 1921, the newly established American Petroleum Institute adopted the standard.
41. Lloyd to Warren E. Lloyd, 8 December 1913, Shannon case #11, LCA.
42. Lloyd to Sinsheimer, 23 December 1913, Shannon case #11, LCA.
43. Lloyd to Dabney, 13 August 1913; Lloyd to Oxnard, 27 September 1913 (all found in Shannon case #10, LCA).

44. Lloyd to Neosho Oil Co., 24 November 1914, Shannon case #13, LCA; Beaton, *Enterprise in Oil*, 109 (quoted).
45. Lloyd to Havens, 25 August 1914, Shannon case #12, LCA.
46. Dabney to Lloyd, 25 March 1914; Agreement between Lloyd and Chapin, 3 April 1914; Assignment of Half Interest in Oil Lease (5), 4 April 1914; Lloyd, Cheney & Geibel to Security Trust and Savings Bank, 4 April 1914; Lloyd to Chapin, 7 April 1914; Trial Balance, Hidalgo Oil Co., 1 May 1914; Assignment and Surrender of Oil Lease, 22 June 1914 (all found in Shannon case #23, LCA); Dabney and Miley to Lloyd, 1 April 1914; "Assignment of Oil Lease (Patterson Ranch Co.)," 4 April 1914; "Assignment of Oil Lease (Pierre Lapeyere)," 4 April 1914; Lloyd to Dabney, 25 May 1914; Hidalgo Oil Co. to Patterson Ranch Co., 9 June 1914 (all found in Shannon case #11, LCA); Lloyd to Havens, 25 August 1914, Shannon case #12, LCA (quoted); Lloyd to Neosho Oil Co., 24 November 1914, Shannon case #13, LCA, Lloyd to Mrs. Hugh T. (Roberta Lloyd) Dobbins, 12 August 1916, Shannon case #17, LCA; Crotty, "Biography of Ralph B. Lloyd," 75.
47. During the 1910s, E. L. Doheny posed a serious challenge to California's majors. As of 1919, his California Petroleum Corporation ranked fourth statewide in terms of output and proven acreage (Ansell, *Oil Baron of the Southwest*, 130–46, 236–7; FTC, *Report of the Federal Trade Commission on the Pacific Coast Petroleum Industry*, pt. 1, 64 [table 15]).
48. Lloyd to Doheny, 24 February 1914; Lloyd to Doheny, 27 March 1914 (all found in Shannon case #11, LCA); Lloyd to Sherman, 20 June 1916, Shannon case #17, LCA; Lloyd and Chapin to Doheny, 26 December 1916, Shannon case #18, LCA; Lloyd to Doheny, 4 December 1916; Lloyd to Doheny, 26 December 1916; Lloyd to Doheny, 22 January 1917; Doheny to Lloyd, 22 January 1917 (quoted); Lloyd to Doheny, 23 January 1917; Lloyd to Doheny, 7 February 1917 (all found in Shannon case #19, LCA); Lloyd to Blalock, 26 August 1926, LCL 1–1; "Directory of California Oil Operators," *Annual Report of the State Oil and Gas Supervisor of California* 7 (August 1921): 118; William S. W. Kew, "Structure and Oil Resources of the Simi Valley," *Contributions to Economic Geology, 1918, Part II,* Bulletin 691, U.S. Geological Survey (Washington, DC: GPO, 1919), 340–4; William S. W. Kew, *Geology and Oil Resources of a Part of Los Angeles and Counties, California*, Bulletin 753, U.S. Geological Survey (Washington, DC: GPO, 1924), 182–3.
49. Ralph B. Lloyd, submission to Appleton's *Cyclopedia of American Biography*, 11 May 1922, RBL 2–1 (quoted); Lloyd to Dabney, 27 October 1913, Shannon case #10, LCA; "History of the Discovery of Oil on the Lloyd Ranch, Ventura County,

California," undated, enclosed with Lloyd to First National Bank, 22 January 1916, Shannon case #16, LCA (quoted); Lloyd to Argobrite, 4 May 1916, Shannon case #16, LCA; Lloyd to Roberta Lloyd Dobbins, 12 August 1916, Shannon case #17, LCA; Lloyd Corporation, "Oil Documents," June 1935, LCL 9–5.

50. Humphery to Lloyd Corp., South Basin Oil, and VL&W, 6 March 1936, LCL 10–1; Crotty, "Biography of Ralph B. Lloyd," 78.
51. Lloyd to "Bertie" (Roberta Lloyd Dobbins), 2 November 1915, Shannon case #15, LCA.
52. Lloyd to Morgan, Fliedner & Boyce, 22 November 1915, Shannon case #16, LCA.
53. Lawrence Vander Leck, "Report on the Ventura Oil Field, Ventura County, California," *Annual Report of the State Oil and Gas Supervisor* 5 (February 1920): 13.
54. Lloyd to First National Bank, Los Angeles, 22 February 1916 (quoted); Lloyd to Adams, 25 January 1916 (all found in Shannon case #16, LCA). On the accomplishments of William W. Orcutt as Union's chief geologist and land department manager, see Pederson, *A Century of Spirit*, 30–1; "Union Oil Co. Is Proud of Its Great Past, Looks Ahead to a Greater Future," *Petroleum World* (May 1940): 50–2; Welty and Taylor, *The Black Bonanza*. A casing string consists of sections of steel pipe, successively narrower in diameter, that are connected by threaded steel couplings, lowered into the well bore, and cemented in place. Drilling teams run casing into a well to seal off or protect formations that lay adjacent to the well bore.
55. Lloyd to McLaughlin, 24 November 1922, RBL 2–1 (quoted).
56. Lloyd to First National Bank, 22 February 1916, Shannon box #16, LCA (quoted); "History of the Discovery of Oil on the Lloyd Ranch, Ventura County, California," undated, enclosed with Lloyd to First National Bank, 22 January 1916, Shannon case #16, LCA; Lloyd to Argobrite, 4 May 1916, Shannon case #16, LCA; Lloyd to Ventura Abstract Co., 6 June 1916, Shannon case #17, LCA.
57. Lloyd to Hartman, 27 November 1916, Shannon case #18, LCA.
58. Lloyd, "Father Saves Life."
59. Brechin, *Imperial San Francisco*, 261, 349n15.
60. Lloyd to Doheny, 20 March 1916, Shannon case #16, LCA (quoted); Lloyd to Van der Linden, 23 August 1921, RBL 2–4 (quoted); Ansell, *Oil Baron of the Southwest*, 130–46.

61. Lloyd to Van der Linden, 3 April 1916, Shannon case #16, LCA; Lloyd to Shell Co. of California, 2 October 1916, Shannon case #18, LCA; Beaton, *Enterprise in Oil*, 108–9.
62. Lloyd to Shell Co. of California, 24 July 1916, Shannon case #17, LCA (quoted); Lloyd to Roberta Lloyd Dobbins, 12 August 1916, Shannon case #17, LCA; Vander Leck, "Report on the Ventura Oil Field, Ventura County, California," 13.
63. Lloyd to Shell Co. of California, 28 August 1916, Shannon case #17, LCA (quoted).
64. Dabney, Lloyd, Miley, and Buley to Shell Co. of California, 16 October 1916, Shannon case #18, LCA.
65. Lloyd and Dabney to Lewis M. Lloyd and Warren Lloyd, 22 September 1916, Shannon case #18, LCA.
66. Lloyd to Shell Co. of California, 28 August 1916, Shannon case #17, LCA.
67. Lloyd to Lewis M. Lloyd, 19 May 1917, Shannon case #20, LCA (quoted); Vander Leck, "Report on the Ventura Oil Field, Ventura County, California," 13–18. On the Barnard and other "inferior" properties that Ralph Lloyd and Joseph Dabney chose not to lease, see Lloyd to Van der Linden, 10 February 1917, Shannon case #19, LCA.
68. Vander Leck, "Report on the Ventura Oil Field, Ventura County, California," 13–15.
69. Lloyd quoted in Beaton, *Enterprise in Oil*, 108; Kevles, "Joseph B. Dabney."
70. Lloyd to Hobson, 23 September 1916, Shannon case #21, LCA.
71. Lloyd to Katherine Hartman, 23 September 1916, Shannon case #18, LCA; McLaughlin to Lloyd, 25 July 1923, RBL 3–3; Vander Leck, "Report on the Ventura Oil Field, Ventura County, California," 17–20; "The Associated Oil Company," *Mining and Oil Bulletin*, 650, 668; Beaton, *Enterprise in Oil*, 109–11; Rintoul, *Spudding In*, 140–3. In 1908, A. C. McLaughlin was appointed superintendent of exploration of Kern Trading and Oil Company, the oil subsidiary of the railroad. On 23 July 1918, he was named assistant to Paul Shoup, who was Associated's newly appointed president. A few months later, McLaughlin became vice president and executive head of operations.
72. A basic rotary rig included a hoist, or draw works, that raised and lowered the drilling tools; a system of blocks that improved on the lifting power of cable rigs; a motor-driven table that rotated the drill bit; and a pump and circulating system that used drilling fluids, or muds, to clean the hole as drilling proceeded. Advances in rotary drilling technology and fluids enabled operators to open up

major fields in California during World War I and the 1920s (Williamson et al., *The American Petroleum Industry: The Age of Energy*, 29–34).

73. R. P. McLaughlin, "General Statement and Review of Departmental Work," *Annual Report of the State Oil and Gas Supervisor* 1 (1917): 3 (quoted); Vander Leck, "Report on the Ventura Oil Field, Ventura County, California," 6–10, 11 (quoted), 12 (quoted).
74. Lloyd to McLaughlin, 23 February 1920, RBL 2–4 (quoted); Lloyd to McLaughlin, 18 March 1922, RBL 2–1; Lloyd to Van der Linden, 29 March 1922, RBL 2–2; Lloyd to Hartman, 4 March 1924; Lloyd to Hartman, 15 September 1924; Lloyd to Hartman, 24 November 1924; Lloyd to Hartman, 8 December 1924; Lloyd to Hartman, 30 December 1924 (all found in RBL 3–1); McDuffie to Lloyd, 4 December 1925, RBL 4–4 (quoted); *Ira Gosnell and Lena Bowyer v. Ralph B. Lloyd and Lloyd Corporation*, 215 Cal. 244 (1932); Beaton, *Enterprise in Oil*, 110–12. The State Mining Bureau later recognized Gosnell No. 1, which was spudded in February 1917, as the field's discovery well. By redrilling the well, Shell was able to produce some 250,000 barrels from it before abandoning it in 1951. Van der Linden would rise through the ranks to become Royal Dutch/Shell's chief of production worldwide. Shell retained its interest in the two wells that it had drilled on the Hartman lease until October 1925 (*Hartman Ranch Co. v. Associated Oil Co.*, 10 Cal. 232 [1937]).
75. Lloyd to Humphery, 8 October 1928, LCL 3–1; Beaton, *Enterprise in Oil*, 177–8, 223.
76. Lloyd to Van der Linden, 3 April 1916, Shannon case #16, LCA; Lloyd to Van der Linden, 24 March 1919; Lloyd to Shell, 31 October 1919; Lloyd to Shell, 14 November 1919 (all found in RBL 2–6). William Meischke-Smith served as president from 5 August 1914 to 11 October 1915 and again from 13 January 1916 to 30 April 1919. J. C. Van Eck succeeded him on both occasions.
77. Van der Linden to Lloyd, 3 April 1919; Van Eck to Lloyd, 8 December 1919; Van Eck to Lloyd, 2 February 1920 (all found in RBL 2–6).
78. Lloyd to Van der Linden to Lloyd, 24 March 1919, RBL 2–6.
79. Dabney to Lloyd, 3 October 1916, Shannon case #18, LCA (quoted); Van Eck to Lloyd, 2 February 1920, RBL 2–6 (quoted).
80. Lloyd to Shell Co. of California, 5 February 1920, RBL 2–6.
81. Crotty, "Biography of Ralph B. Lloyd," 88–9.
82. "Tentative Agreement between Mssrs. Lloyd and Dabney on one hand and A. C. McLaughlin, representing Associated Oil Company, on the other," 19 February 1920, RBL 2–4; Humphery to Lloyd Corp., South Basin Oil, and VL&W, 6 March 1936, LCL 10–1.

83. Lloyd to McLaughlin, 23 February 1920, RBL 2–4 (quoted).
84. Lloyd to Van Eck, 23 August 1920 (quoted); Van Eck to Lloyd, 7 September 1920 (quoted) (all found in RBL 2–6); Beaton, *Enterprise in Oil*, 762.
85. "Doheny Job Explained by McAdoo," *LAT*, 2 February 1924, I:1; "M'Adoo Is Quizzed," *LAT*, 12 February 1924, I:1; Ansell, *Oil Baron of the Southwest*, 213–21; Linda B. Hall, *Oil, Banks, and Politics: The United States and Postrevolutionary Mexico, 1917–1924* (Austin: University of Texas Press, 1995), 36–51; Jessica M. Kim, "Oil Men and Cactus Rustlers: Metropolis, Empire, and Revolution in the Los Angeles-Mexico Borderlands, 1890–1940" Ph.D. diss., University of Southern California, 2012, 222–9; Jordan A. Schwarz, *The New Dealers: Power Politics in the Age of Roosevelt* (New York: Knopf, 1993), 21–31.
86. "McAdoo to Practice Law Here," *LAT*, 2 March 1922, II:1.
87. McAdoo to Milton, 20 March 1922, McAdoo papers, box 262. See, also, McAdoo to Ward, 15 March 1922; McAdoo to Bollings, 22 March 1922 (all found in McAdoo papers, box 262).
88. Schwarz, *The New Dealers*, 26.
89. McAdoo to Baruch, 24 March 1922, McAdoo papers, box 262.
90. McAdoo to McAdoo Jr., letters of 22 and 28 March 1922, McAdoo papers, box 262; McAdoo to Doheny, 18 April 1922, McAdoo Papers, box 263; Lloyd to McLaughlin, 24 November 1922, RBL 2–1; Chapin Hald, "Doheny Home from Mexico," *LAT*, 9 May 1922, II:1; John N. Blackburn, "M'Adoo Punches Cattle for Day," *LAT*, 15 May 1922, II:1; "Doheny Job Explained by McAdoo," *LAT*.
91. McAdoo to Doherty, 22 February 1923, McAdoo papers, box 275.
92. McAdoo to Carden, 12 July 1922, McAdoo papers, box 265.
93. McAdoo to Williams, 15 November 1922, McAdoo papers, box 271. Jameson Petroleum Company incorporated in California on 7 June 1922 with a capitalization of $300,000 ("Directory of California Oil Operators," *Annual Report of the State Oil and Gas Supervisor* 10 [August 1924], 136). As William Gibbs McAdoo reported to Daniel Roper, the Wilson Administration's wartime Commissioner of Internal Revenue, J. W. Jameson was "a successful oil man of the highest quality and character." Through McAdoo, Roper invested $5,000 in the venture (McAdoo to Williams, 17 June 1922; McAdoo to Roper, 3 July 1922 [quoted] [all found in McAdoo papers, box 265]).
94. Williams to McAdoo, 12 October 1925, McAdoo papers, box 319. On the progress of the Jameson company's drilling campaign, see McAdoo to Roper, 14 February 1923, box 275; McAdoo to Roper, 7 April 1923, box 276; McAdoo to Baruch, 21 April 1923, box 277; McAdoo to Howard, 11 May 1923, box 277;

McAdoo to Williams, 27 June 1923, box 279 (all found in McAdoo papers); "Oil Production Record Seen," *LAT*, 4 September 1922, II:11; "Santa Fe Field Area Enlarged," *LAT*, 16 April 1923, I:11; "Oil Peak Not Yet Reached," *LAT*, 2 July 1923, II:9; Beaton, *Enterprise in Oil*, 174–88. In September 1926, Richfield Oil Company, an emerging California major, acquired the Jameson concern. At the time, its assets included producing wells in several Los Angeles Basin fields and a 10,000-barrel refinery at Vernon (Richfield Oil Company of California, *1926 Annual Report* [Los Angeles: the company, 1927]; "Richfield Adds to Its Holdings," *LAT*, 29 September 1926, I:14).

95. McLaughlin to Lloyd, 21 November 1922, RBL 2–1; Lloyd to McLaughlin, 24 November 1922, RBL 2–1 (quoted). Contemporaneously, Henry L. Doherty, president and chief engineer of Bartlesville, Oklahoma–based Cities Service Oil Company, through his Doherty Research Company, was promoting unitization, which invested control of both development and production in a single entity in the belief that it would operate the field on behalf of its members in the most efficient and equitable manner possible (D. R. Knowlton, "Unitization—Its Progress and Future," *Drilling and Production Practice*, American Petroleum Institute Report 39 [1939]: 630–5). On California's experiment with unitization, see Sabin, *Crude Politics*, 123–7.

96. "Oil Documents," June 1935, LCL 9–5. In April 1925, Ralph Lloyd and Joseph Dabney subleased the Joseph and William Sexton ranch properties to Milham Exploration, a company in which renowned mining engineer John Hays Hammond was interested. (Hammond would make the cover of *Time* magazine's issue of 3 May 1926.) The transactions completed their subleasing of their leases in the field. Associated controlled some 5,000 acres; Shell, a little more than 4,000 acres; and Milham Exploration, 4,000 acres (Lloyd to King, 24 April 1925, RBL 4–1). Milham Exploration, however, enjoyed no success in testing the easternmost properties on the Ventura anticline and surrendered the two subleases back to Lloyd and Dabney in January 1928.

97. Lloyd to Van der Linden, 20 October 1922, RBL 2–2 (quoted); "Ventura Avenue," *The Record* 7 (September 1926): 3–8; "Engineering at Ventura," *The Record* 9 (December 1928): 12–13; Benton Turner, oral history, 6–7.

98. Lloyd to Rosemary Dobbins Lloyd, 1 July 1932, not sent, LCL 7–3; Humphery to Lloyd Corp., South Basin Oil, and VL&W, 6 March 1936, LCL 10–1; Exhibit "A," Modification of Lease, 1 January 1938, Crotty Papers, box 3; Crotty, "Biography of Ralph B. Lloyd," 90–1, 106–8. Exhibit "A" reproduced the 8 August 1922 agreement between VL&W, on the one hand, and Ralph Lloyd and Joseph Dabney, on the other.

99. Lloyd to McLaughlin, 5 September 1925, RBL 4–1 (quoted); Lloyd to McLaughlin, 30 December 1925, RBL 4–1 (quoted); "'25 in the Light of '24," *The Record* 6 (January 1925): 8–9; "Ventura Draws a Pair of Aces," *The Record* 6 (March 1925): 7; "Gusher Wells Double Output of Ventura Avenue Field," *Petroleum World* (March 1925): 53; "Ventura Avenue," *The Record* 7 (September 1926): 3–8; F. W. Hertel, oral history, VCM, 1 September 1979, 2–8.
100. "Years More of Drilling, Says Lloyd," *VCS*, 31 March 1926; Carl P. Miller, "Ventura Oil Field Praised," *LAT*, 25 June 1926, I:12; Associated Oil Company, "Statement of Oil Royalty, Lloyd Lease, Ventura County, September 1926," 13 October 1926, LCL 1–1; Lloyd to Roberta Lloyd Dobbins, 16 December 1926, LCL 1–2 (quoted).
101. "Associated Lloyd Is Doing 4500," *California Oil World*, 12 February 1925; "General Petroleum Planning Big Campaign," *VCS*, 12 December 1925; Joseph Jensen, "Petroleum Development in California during 1926," *Oil Bulletin* (March 1927): 257–69; "Ventura's Ultimate Output 250,000,000 Barrels of Oil," *Petroleum World* (September 1928): 53; Beaton, *Enterprise in Oil*, 189–90. The Ventura Avenue field, now known simply as the Ventura field, has produced more than one billion barrels of crude oil to date.
102. "Roll Call," undated [February 1924], VLW 1–2; Lloyd to Rosemary Dobbins Lloyd, 15 December 1925, RBL 4–2; Lloyd to Rosemary Dobbins Lloyd, 1 July 1932, not sent, LCL 7–3; Crotty, "Biography of Ralph B. Lloyd," 94–5. Lewis Lloyd held 1,507 of 2,424 outstanding shares in VL&W (Annual Report to Stockholders, VL&W, for the year ended 31 December 1917, 11 December 1917, VLW 1–2). On the sudden death of Warren Lloyd, see Williams, *Raymond Chandler*, 81–2.
103. Lloyd, Dabney, Miley, Lloyd Corp., and South Basin Oil Co. to Associated Oil Company, 15 September 1926, LCL 1–1; Agreement between Lloyd Corp., South Basin Oil Co., and E. J. Miley and Associated Oil Company, 31 December 1926, LCL 1–1; Lloyd to Bank of America, "Statement as to Financial Standing of Lloyd Corporation, Ltd.," 13 September 1934, LCL 9–1; Lloyd to Security-First National Bank of Los Angeles, 2 June 1936, LCL 10–4; *Ira Gosnell and Lena Bowyer v. Ralph B. Lloyd and Lloyd Corporation*, L. A. No. 11554 (1932); "Associated Oil Tax Reduction Plea Is Denied," *VCS*, 20 July 1931. As of November 1928, Ralph Lloyd held 8,198 of 10,000 shares in the Lloyd Corporation. His wife and four daughters held all but one of the remaining shares (Lloyd Corporation, List of Stockholders, 23 November 1928, LCL 3–4). For his part, E. J. Miley sold his overriding royalty interest in

the properties subleased to Associated to various firms and individuals (Crotty, "Biography of Ralph B. Lloyd," 110–11).

104. Lloyd to McLaughlin, 23 February 1920, RBL 2–4.

105. Lloyd to McLaughlin, 26 February 1920, RBL 2–4; King to Lloyd, 19 June 1922, RBL 2–1; Lloyd to King, 20 June 1922, RBL 2–1; Lloyd to King, 7 April 1925; Lloyd to King and McLaughlin, 28 October 1925 (quoted); Lloyd to Taff, 15 December 1925 (all found in RBL 4–1).

CHAPTER 2. LOCAL ELITES, OUTSIDE COMPANIES, AND VENTURA'S OIL BOOM

1. "The Best Kind of Growth," *VFP*, 5 June 1925, 2.
2. On comparisons between Santa Barbara and the cities of the Los Angeles Basin, see Kevin Starr, *Material Dreams: Southern California Through the 1920s* (New York: Oxford University Press, 1990). For a study of a California city that pursued qualitative growth in more recent times, see Ryan M. Kray, "The Path to Paradise: Expropriation, Exodus, and Exclusion in the Making of Palm Springs," *Pacific Historical Review* 73 (February 2004): 85–126.
3. Rowan Miranda and Donald Rosdil, "From Boosterism to Qualitative Growth: Classifying Economic Development Strategies," *Urban Affairs Review* 30 (1995): 870.
4. The terms "qualitative" and "quantitative" growth may be found in the taxonomy of Rowan Miranda and Donald Rosdil, which also includes historical preservation, environmentally harmful, and redistributive growth ("From Boosterism to Qualitative Growth," 868–79).
5. Kevin R. Cox and Andrew Mair, "Locality and Community in the Politics of Local Economic Development," *Annals of the Association of American Geographers* 78 (1988): 307–25; John R. Logan and Harvey L. Molotch, *Urban Fortunes: The Political Economy of Place* (Berkeley: University of California Press, 1987), 50–98; Harvey L. Molotch, "The City as a Growth Machine" *American Journal of Sociology* 82 (1976): 309–31; Harvey L. Molotch, "Growth Machine Links: Up, Down, and Across," in *The Urban Growth Machine: Critical Perspectives Two Decades Later*, ed. Andrew E. G. Jones and David Wilson (Albany: State University of New York Press, 1999), 248–9 (quoted). For a recent assessment of growth machine and civic boosterism theory, see Mark Boyle, "Growth Machines and Propaganda Projects: A Review of Readings of the Role of Civic Boosterism in the Politics of Local Economic Development," in *The Urban Growth Machine*, 55–70.

6. Hutchinson, *Oil, Land, and Politics*, vol. 1, 101–2; Roy Pinkerton, *The County Star: My Buena Ventura* (Ventura, CA: s.n., 1962), 42; Gertrude L. Reith, "Ventura: Life Story of a City," Ph.D. diss., Clark University, 1963, 107–20.
7. Hutchinson, *Oil, Land, and Politics*, vol. 1, 271–4, 328–31; Reith, "Ventura," 102–4; Welty and Taylor, *The Black Bonanza*, 71–74, 90; White, "California's Other Mineral," 145–50; White, *Formative Years in the Far West*, 6–19, 23–57.
8. Lima, "The Politics of Offshore Energy Development," 143–9; Harvey Molotch and William Freudenburg, eds., *Santa Barbara County: Two Paths*, OCS Study MMS 96–0036 (Camarillo, CA: U.S. Minerals Management Service, 1996), 10–11; *SBMP*, quoted in Robert Sollen, *An Ocean of Oil: A Century of Political Struggle Over Petroleum Off the California Coast* (Juneau, AK: Denali Press, 1998), 11.
9. Molotch and Freudenburg, *Santa Barbara County*, 11.
10. Starr, *Material Dreams*, 263–302.
11. Indeed, in 1947, voters defeated by a three-to-one margin a measure to amend the City's charter to prohibit drilling for oil within city limits (Molotch and William Freudenburg, *Santa Barbara County*, 46).
12. E. Baden Powell, "Geology of Santa Barbara Mesa Field," *Petroleum World* (November 1934): 22–23, 30, 44; Harvey Molotch, William Freudenburg, and Krista E. Paulsen, "History Repeats Itself, But How? City Character, Urban Tradition, and the Accomplishment of Place," *American Sociological Review* 65 (2000): 804–8.
13. In 1953, the Santa Barbara City Council banned drilling for oil within city limits. In 1965, voters approved a measure to amend the City's charter to include the ordinance (Molotch and William Freudenburg, *Santa Barbara County*, 46).
14. Beaton, *Enterprise in Oil*, 174–88; Rintoul, *Spudding In*, 146–61; Sabin, *Crude Politics*, 64–78, 113; Quam-Wickham, "'Cities Sacrificed on the Altar of Oil'"; Tygiel, *The Great Los Angeles Swindle*, 3–16. Albert W. Atwood wrote the classic account of the Los Angeles oil boom ("When the Oil Flood Is On," *Saturday Evening Post*, 7 July 1923, 3–4, 86–101; "Mad from Oil," *Saturday Evening Post*, 14 July 1923, 10–11, 92–105).
15. Roger W. Olien and Diana Davids Olien, *Oil Booms: Social Change in Five Texas Towns* (Lincoln: University of Nebraska Press, 1982).
16. "Industries of Ventura Have Pay Rolls Running into Thousands Monthly," *VFP*, 21 January 1925, 1; "Oil Field Resembles Big Beehive," *VFP*, 7 February 1925, 1; E. E. Wiker, "Business Fine in Ventura; Oil Wells, Building Cause Activity," *VCS*, 10 September 1925, 4, 6; "City Grows 25 Pct. in Year," *VCS*, 31

December 1925, 1; "'1926 Will See Record Set for Ventura Oil Development'—Lloyd," *VWP*, 26 March 1926, 1; "Editorial Comment," *VWP*, 7 May 1926, 4; "Phenomenal Progress Made in All Lines, Figures Show," *VCS*, 31 December 1926, 1; Los Angeles Directory Co., *Ventura County Directory, 1930* (Los Angeles, 1929).

17. Douglas G. McPhee, "Solution of Deep Drilling Problem in the 'Avenue Field' Rejuvenates a Pioneer Oil District," *Petroleum World* (April 1925): 33 (quoted); "City Grows 25 Pct. in Year."
18. Quoted in McPhee, "Solution of Deep Drilling Problem in the 'Avenue Field' Rejuvenates a Pioneer Oil District," 96.
19. "Nearly $200,000 Is Spent for Homes in First Three Months," *VFP*, 2 April 1925, 1; "Will Build Homes in City on Lots Owned by Wage Earners," *VFP*, 2 June 1925, 1; "Derth [*sic*] of Housing Service Now Drives Many to Camp," *VFP*, 11 June 1925, 1 (quoted); "Our Greatest Need Today," *VCS*, 13 July 1925, 4; "Yes, We Need That Hotel," *VCS*, 16 July 1925, 4; "18 Oil Families Come in Group," *VCS*, 10 August 1925, 1; Wiker, "Business Fine in Ventura"; "Squatters Are Told to Move," *VCS*, 9 February 1926, 1; Pinkerton, *The County Star*, 4–9, 7 (quoted).
20. "Oil Strike Draws Investors Here," *VFP*, 4 February 1925, 1.
21. "Thirty Acres Will Be Sub-Divided on Avenue," *VFP*, 10 March 1925, 1; "Simpson Property on Ventura Ave. Sold to Group," *VFP*, 18 April 1925, 5; "Ventura Avenue's Lots Sell Easily," *VFP*, 18 April 1925, 5; Wiker, "Business Fine in Ventura"; McPhee, "Solution of Deep Drilling Problem in the 'Avenue Field' Rejuvenates a Pioneer Oil District," 33 (quoted); Lynn Weitzel, "Historic Simpson Street Dedication," *Ventura County and Coast Reporter*, 17 May 1990.
22. "Keep Rentals Fair," *VCS*, 5 May 1926, 4.
23. "City Over Vast Oil Pool, Claim," *VFP*, 11 February 1925, 1.
24. "January Eventful in City's History," *VFP*, 31 January 1925, 1.
25. "Oil Field Resembles Big Beehive."
26. "City Has State's Largest Oil Well," *VFP*, 29 January 1925, 1 (quoted); "Well Is Producing 4450 Barrels Daily," *VFP*, 30 January 1925; "New Oil Well Is Due In Tonight," *VFP*, 5 February 1925, 1; "City Over Vast Oil Pool, Claim"; "Gusher Wells Double Output of Ventura Avenue Field," *Petroleum World* (March 1925): 53. On speculation and promotion at Santa Fe Springs, for instance, see Tygiel, *The Great Los Angeles Swindle*.
27. Verne Patmore, oral history, VCM, 26 October 1979, 5–7, 30–31, 7 (quoted); Richard Willett, oral history, VCM, 4 June 1987, 9 (quoted); "City Grows 25 Pct. in Year."

28. "'1926 Will See Record Set for Ventura Oil Development'—Lloyd," 1 (quoted); "Years More of Drilling, Says Lloyd," *VCS,* 31 March 1926, 1 (quoted); "Lloyd's Good News," *VCS,* 16 December 1926, 4; "Lloyd Makes Public Late Find in Oil," *VCS*, 20 December 1926; "11-Acre Lloyd Lease Tract in Motion, Can't Be Drilled," *VCS*, 30 April 1928, 1.
29. "Oil Strike Draws Investors Here."
30. "Oil Field Resembles Big Beehive."
31. "Let Ventura Rejoice," *VFP,* 13 April 1925, 2.
32. "Editorial Comment," *VWP,* 7 May 1926, 4.
33. "January Eventful in City's History"; Wiker, "Business Fine in Ventura"; "The Best Kind of Growth"; "Johnson Forecasts City's Largest Year in Building," *VCS,* 7 December 1926, 2 (quoted); Hoover quoted in "Biggest Year Looms Ahead," *VCS*, 31 December 1926, 1.
34. Quoted in McPhee, "Solution of Deep Drilling Problem in the 'Avenue Field' Rejuvenates a Pioneer Oil District," 96.
35. McPhee, "Solution of Deep Drilling Problem in the 'Avenue Field' Rejuvenates a Pioneer Oil District," 33.
36. "Ventura Grows," *California Oil World*, 8 May 1930, 6.
37. Brief biographies of Austad, Ferro, Randall, Rea, and Sheridan may be found in Sol N. Sheridan, *History of Ventura County, California*, vol. 2 (Chicago: S. J. Clarke Publishing, 1926), 42, 109, 131, 171, 554–5.
38. Austad to Lloyd, 15 February 1926, LCL 1–5.
39. "Here's Way to Aid Ventura to Grow," *VFP,* 10 February 1925, 1; "Movement To Get Industries Here," *VFP,* 13 February 1925, 1; "The Best Kind of Growth" *VFP* (quoted); "Get Yourself a 'Home, Sweet, Home,'" *VFP,* 15 June 1925, 2 (quoted); "Campaign To Sell Ventura Approved," *VFP,* 23 June 1925; Ventura Chamber of Commerce, advertisement, *VFP,* 29 June 1925 (quoted); "Financing New Building," *VCS,* 17 July 1925, 4; "What Santa Barbara Plans; What Ventura Might Also Do," *VCS,* 30 July 1925; "A Beach Hotel," *VCS,* 24 February 1926, 4. On the earthquake that struck Santa Barbara on 29 June 1925 and the city's response, see Starr, *Material Dreams,* 287–92.
40. "Editorial Comment," *VWP,* 7 May 1926, 4; "Editorial Comment," *VWP,* 14 May 1926, 4; "Avenue Is Mecca of Realtors, Investors," *VFP,* 31 January 1925, 1; Hoover quoted in "Biggest Year Looms Ahead."
41. Wiker, "Business Fine in Ventura"; "Fortunate Ventura," *VCS,* 7 November 1925, 4; "City Grows 25 Pct. In Year"; "Ventura Keeps Up in Building," *VCS,* 7 December 1926, 2; "Joe Daley Has Much Faith in Ventura County Future," *VCS,* 17 December 1926, 4; "Leggett Drug Company Is Another New Avenue

Store," *VCS*, 17 December 1926, 11; "Rotarians Hear of Rosy Future," *VCS*, 29 December 1926; Hoover quoted in "Biggest Year Looms Ahead"; "Record Here in Building," *VCS*, 31 December 1926, 8; "Avenue Has Big Changes," *VCS*, 31 December 1926, 2.

42. Los Angeles Directory Co., *Ventura County Directory, 1930*; Reith, "Ventura," 137–42; Judith P. Triem, *Ventura County: Land of Good Fortune: An Illustrated History* (Northridge, CA: Windsor Publications, 1985), 118–21; "Avenue Before Oil Was Idyllic Site to Live," *VS-FP*, 3 May 1987, 18; Reed Fuji, "Industry Helped Make City Thriving Place It Is Today," *VS-FP*, 20 August 1989, 7–8.
43. Olien and Olien, *Oil Booms*, 18 (quoted).
44. Austad and Sheridan to Lloyd, 20 October 1925, LCL 1–5; "January Eventful in City's History" (quoted); "Only 28 Miles Between Ventura and a Greatly Added Importance," *VCS*, 28 July 1925, 1 (quoted); "Breakwater Is Real City Need," *VCS*, 17 August 1925, 1; "All Lined Up to Push Road to The Finish," *VCS*, 6 November 1925, 1.
45. "All Ventura County Will Win, or Else Will Lose, Together," *VCS*, 24 August 1925, 1; Sol N. Sheridan, "Maricopa Road Cost Will Be Small, Usefulness Great," *VCS*, 10 September 1925, 1–2; Butcher quoted in "Vote Maricopa Road District Is Supervisors' Decision," *VCS*, 26 September 1925, 1; Sol N. Sheridan, *History of Ventura County, California*, vol. 1 (Chicago: S. J. Clarke Publishing, 1926), 275–94; John R. Wallace, "Ventura," in *Ventura County Directory, 1930*, 8–9; Pinkerton, *The County Star*, 42–43.
46. Pinkerton, *The County Star*, 42–43.
47. "Success Comes Twice to Ventura Tool Co.," *VCS-FP*, 19 October 1949; "Ventura Tool Co. Had Modest Start," *VCS-FP*, 14 October 1953; Jack Smalley, "Completing the Maricopa Road," *VCS-FP*, magazine section, 10 October 1964, 12–13; Sid Hayes, "Fritz Huntsinger: A Man for All Mankind," *Pacific Oil World* 71 (September 1978): 20–25; Krista Paulsen, Harvey Molotch, Perry Shapiro, and Randolph Bergstrom, *Petroleum Extraction in Ventura County, California*, OCS Study MMS 98–0047 (Camarillo, CA: U.S. Minerals Management Service, 1998), section 3–1.
48. F. W. Hertel, "Ventura's Submarine Pipe Line," *The Record* 5 (November 1924): 10–12; Hutchinson, *Oil, Land, and Politics*, vol. 1, 328–31; Reith, "Ventura," 124–6.
49. Sheridan, *History of Ventura County*, vol. 1, 377–81.
50. Pinkerton, *My County Star*, 43–44; W. H. Hutchinson, *Oil, Land, and Politics: The California Career of Thomas Robert Bard*, vol. 2 (Norman: University of

Oklahoma Press, 1965), 162–204; Charles C. Teague, *Fifty Years a Rancher* (Los Angeles: Ward Ritchie Press, 1944).

51. Pinkerton, *My County Star*, 44–46; "All This Marching Up and Down the Hill Would Be Funny If the Public Didn't Pay the Bill," *VCS*, 19 September 1929, 1; "Now the County Supervisors to Have a Word!" *VCS*, 15 October 1929, 1; Triem, *Ventura County*, 134–6; Powell Greenland, *Port Hueneme: A History* (Oxnard, CA: Ventura County Maritime Museum, 1994).
52. Lloyd to Associated, Shell, and General Petroleum, letters of 22 and 25 January 1926, LCL 1–1; Legh-Jones to Lloyd, 27 January 1926, LCL 1–4; Lloyd to Associated, Lloyd Corp., Shell, General Petroleum, and Pacific Western, 17 September 1929, LCL 4–2; Jurs and Gallagher to Bard, 18 September 1929, LCL 4–6 (quoted); Lloyd to Reinhardt, 7 October 1929, LCL 4–6; Lloyd to Bard, 22 May 1934, LCL 9–1; Sheridan, *History of Ventura County*, vol. 1, 381; Pinkerton, *The County Star*, 46–47.
53. Meetings of the Board of Trustees, 24 May 1926, 1 June 1926, and 8 June 1926, City of San Buenaventura, Minute Book of the Board of Trustees, no. 8, 347–71; "Trustees Should Say No!" *VCS*, 2 June 1926, 4; "That's That!" *VCS*, 9 June 1926, 1.
54. Molotch et al., "History Repeats Itself, But How?" 798–804.
55. "Irvine Not to Press for Decision Today on Pipeline," *SBMP*, 26 February 1931, 1; "E. W. Alexander Attacks Plan of Oil Interests for Building Pipeline," *Santa Barbara Daily News*, 24 February 1931, 1 (quoted); "Kill the Pipe Line" *SBMP*, 23 February 1931 (quoted); "Now Stay Firm," *SBMP*, 28 February 1931 (quoted).
56. "Biggest City Undertaking Is Outlined," *VCS*, 4 August 1925, 1; "Bond Win! People Vote 633–74 For High School," *VCS*, 28 November 1925, 1; "Your Need and Mine," *VCS*, 10 August 1926, 4 (quoted); "Growth! Progress!" *VCS*, 18 December 1926, 1; "If You Want Ventura to Grow," *VCS*, 14 December 1926, 1; "People Will Vote on Two Bond Issues," *VCS*, 14 December 1926, 1; "Here Is a Catechism on That Water Bond Issue," *VCS*, 14 December 1926, 1; Lloyd to Ventura Chamber of Commerce, quoted in "Lloyd's Good News," 4; "Both Bond Issues Fail of Two-Thirds Majority," *VCS*, 16 December 1926, 1 (quoted); "Ferro Charges Hobson Selfish," *VCS*, 16 December 1926, 1; Austad to Lloyd, 17 December 1926, LCL 1–1; Austad to Lloyd, 15 February 1926, LCL 1–5 (quoted).
57. Edmund F. Ball, *From Fruit Jars to Satellites: The Story of Ball Brothers Company Incorporated* (New York: Newcomen Society of North America, 1960); Frank Clayton Ball, *Memoirs* (Muncie, IN: p.p., 1937); David L.

Good, *Orvie: The Dictator of Dearborn: The Rise and Fall of Orville L. Hubbard* (Detroit, MI: Wayne State University Press, 1989), 35–42; David Karjanen, "The Wal-Mart Effect and the New Face of Capitalism: Labor Market and Community Impacts of the Megaretailer," in *Wal-Mart: The Face of Twenty-First-Century Capitalism*, ed. Nelson Lichtenstein (New York: New Press, 2006), 144–6; Richard Lingeman, *Small Town America: A Narrative History, 1620–The Present* (New York: Putnam, 1980), 425–30; June C. Nash, *From Tank Town to High Tech: The Clash of Community and Industrial Cycles* (Albany: State University of New York Press, 1989).

58. Fred W. Viehe, "Black Gold Suburbs: The Influence of the Extractive Industry on the Suburbanization of Los Angeles, 1890–1930," *Journal of Urban History* 8 (November 1981): 10–13.

59. G. Ferro, "Annexation of Avenue Next Important Step," *VCS*, 19 December 1925, 1; "Big 3 Will Not Oppose Inclusion in the City," *VWP*, 16 April 1926; Lloyd to Associated, Shell, General Petroleum, and Petroleum Securities Co., 26 March 1927, LCL 2–1; "Annexation of Big Area Proposed," *VCS*, 20 April 1928, 1 (quoted).

60. Frank C. Renfrew, "Value of Oil to Long Beach," *Mining and Oil Bulletin* (April 1923): 289–90; "City Acquires Fortune in Oil," *Union Pacific Magazine* (January 1925): 12–13; Terry Carpenter, "Production of Petroleum Pays for Public Projects," *Long Beach Press-Telegram*, 31 December 1925; Richard DeAtley, *Long Beach: The Golden Shore* (Houston, TX: Pioneer Publications, 1988), 68–70. As of 31 December 1939, the City of Long Beach had collected $11.96 million in royalties (City of Long Beach, "City's Oil, Long Beach-Signal Hill Field: Income and Expenditures," Long Beach History Collection, Long Beach Public Library). The City benefited enormously from post–World War II oil development on lands chartered by the State of California for harbor development. In 1950, the City ranked fifth among operators statewide (*Thirty-Sixth Annual Review of California Crude Oil Producers* [Los Angeles, CCCOP, 1951]).

61. DeAtley, *Long Beach*, 71; City Council of Signal Hill, *Never by Chance: A Brief History of Signal Hill* (Signal Hill, CA, 1974), 57–59.

62. Austad and Sheridan to Lloyd, 20 October 1925; Lloyd to West Side Chamber of Commerce, 27 October 1925; Lloyd to Ventura Chamber of Commerce, 22 January 1926; Austad to Lloyd, 15 February 1926 (all found in LCL 1–5); Lloyd to Associated, Shell, and General Petroleum, letters of 22 and 25 January 1926, and 17 February 1926 (all found in LCL 1–1).

63. Lloyd to Sheridan, 24 June 1926, LCL 1–5. On the situation in Long Beach to which Ralph Lloyd pointed, see, for instance, "L. B. Row Simmers," *Petroleum World*, 24 June 1926, 1, 12.
64. Lloyd to Lagomarsino, 7 December 1925, RBL 4–2.
65. Lloyd to Sheridan, 24 June 1926, LCL 1–5.
66. Lloyd to Associated, Shell, and General Petroleum, 17 February 1926, LCL 1–1.
67. Legh-Jones to Lloyd, 27 January 1926, LCL 1–4.
68. Lloyd to Davis, 15 April 1926, LCL 1–1; "Editorial Comment," *Ventura Daily Post*, n.d. [April 1926], copy found in LCL 1–5 (quoted); Sheridan to Lloyd, 26 June 1926, LCL 1–5; Lloyd to Shell, 7 July 1926, LCL 1–4 (quoted); Lloyd to Dabney, 7 December 1926, LCL 1–2.
69. Lagomarsino to Lloyd, 23 September 1925, RBL 4–2.
70. Lloyd to Associated, Shell, and General Petroleum, 26 May 1926, LCL 1–1.
71. Lloyd to Associated, Shell, and General Petroleum, 11 June 1926 LCL 1–1; Lloyd to Associated, Shell, General Petroleum, and Petroleum Securities Co., 26 March 1927, LCL 2–1; Lloyd to Associated, Shell, General Petroleum, and Petroleum Securities Co., 28 February 1928, LCL 3–1; Lloyd to Associated, Shell, General Petroleum, Pacific Western, and Standard Oil Co. of California, 21 December 1929, LCL 4–2; Lloyd to Associated and Shell, 3 March 1930; Lloyd to Humphery and Legh-Jones, 19 March 1930; Lloyd to Humphery and Legh-Jones, 25 March 1930 (all found in LCL 5–1); Lloyd to Associated, Shell, General Petroleum, Pacific Western, and Standard Oil Co. of California, 13 May 1931, LCL 6–1; "Is the Chamber of Commerce an Efficient Organization?" *VCS*, 10 November 1925, 1; "Big Companies Give $1250," *VCS*, 17 November 1925, 1; "For Everybody," *VCS*, 18 November 1925, 4; "Your Need and Mine," *VCS*, 10 August 1926, 4; "Asks People to Back C. C.," *VCS*, 18 January 1927, 1.
72. Lloyd to Associated, Shell, General Petroleum, and Petroleum Securities Co., 26 March 1927, LCL 2–1.
73. McLaughlin to Lloyd, 25 July 1923, RBL 3–3 (quoted); Lloyd to Associated, Shell, General Petroleum, and Petroleum Securities Co., 28 February 1928, LCL 3–1; Lloyd to Associated, Shell, General Petroleum, Pacific Western, and Standard Oil Co. of California, 14 March 1929, LCL 4–6; Lloyd to Associated, Shell, General Petroleum, Pacific Western, and Standard Oil Co. of California, 13 May 1931, LCL 6–1; Wallace to Lloyd, 22 July 1931, LCL 6–6 (quoted); Lloyd to Wallace, 23 July 1931, LCL 6–6 (quoted); Lloyd to Associated, Shell, General Petroleum, Pacific Western, and Standard Oil Co. of California, 23 February 1932, LCL 7–1; "County to Lose Big Sum Due to Oil Shut-Down,"

VCS, 2 May 1931, 1; "Associated Oil Tax Reduction Plea Is Denied," *VCS*, 20 July 1931, 1; "Chamber Will Take Part in Oil Tax Meet," *VFP*, 20 July 1931, 2; "County Facing Possibility of Big Tax Suits," *VCS*, 21 July 1931, 1; "Reese Justifies His Assessment of Oil Holdings," *VCS*, 22 July 1931, 1; "Future of Avenue Field Is Brighter," *VCS*, 25 July 1931, 1. Each county in California used a different method of assessing the value of petroleum properties, none of which satisfied operators. See S. A. Guiberson, "Operators Envolve [*sic*] Equitable Formula for the Taxation of Oil Lands," *Oil Age* (November 1925): 13.

74. Ventura Chamber of Commerce, resolution, 27 Feb. 1928, LCL 4–2; Sabin, *Crude Politics*, 56–62.
75. Lloyd to Associated, Shell, General Petroleum, Pacific Western, and Standard Oil Co. of California, letters of 13 May 1931, LCL 6–1 (quoted) and 23 February 1932, LCL 7–1 (quoted).
76. Lloyd to Associated and Shell, 30 December 1926, LCL 1–1 (quoted); Lloyd to Associated and Shell, 13 June 1927, LCL 2–1; Lloyd to Humphery, 22 June 1927, LCL 2–1 (quoted).
77. Associated to Lloyd, 6 January 1927, LCL 2–1; Humphery to Lloyd, 20 June 1927, LCL 2–1; Shell to Lloyd, 14 June 1927, LCL 2–5. Deposits at Ventura's three leading banks increased from $3.75 million in 1924 to $11 million in 1929.
78. Wallace, "Ventura"; "City Water Supply Double Soon by New Shell Well," 19 June 1925; Meeting of the Board of Trustees, 30 April 1928, City of San Buenaventura, Minute Book of the Board of Trustees, no. 9, 312.
79. Los Angeles Directory Co., *Ventura County Directory, 1930;* idem, *Ventura County Directory, 1948–1949* (Los Angeles, 1949); Reith, "Ventura," 141–7.
80. Olien and Olien, *Oil Booms.*
81. Contributions of the Ball family included $450,000 toward the construction of facilities for the YMCA and YWCA, land and $1.75 million for what is now Ball State University, land and $100,000 for Ball Memorial Hospital, and $195,000 for a public auditorium (Ball, *Memoirs*, 121–46; Robert T. Brodhead, *Ball Memorial Hospital: A Commitment to Community* [New York: Newcomen Society of North America, 1996]). For its part, Ford donated more than 1,000 acres of land to the City of Dearborn for public purposes. In addition, the company's share of Dearborn's assessed property tax base reached 75 percent in the 1940s (Good, *Orvie*, 39–40, 53).
82. Molotch et al. "History Repeats Itself, But How?"

CHAPTER 3. MAKING PORTLAND A WONDERFUL CITY

1. "East Side Sees Dawn of New Era," *Sunday Oregonian*, 1 March 1926, I:1.
2. Lowe to Lloyd, 1 March 1926, LCL 1–4. Lowe's firm, Ritter, Lowe & Co., represented the Balfour-Guthrie Trust Company.
3. In 1913, Charles X. Larrabee sold the lots that comprised both blocks to the Anglo-Pacific Realty Company, an English syndicate, in what constituted the largest real estate transaction on the Pacific Coast for the year. With the outbreak of World War I, the syndicate was unable to transfer its funds out of Great Britain. Ultimately, the Oregon Real Estate Company foreclosed on the property. By then, the Anglo-Pacific Realty Company had obtained title to about 30% of the 858 lots and other property outside Portland involved in the original sale. This property was now held by the Balfour-Guthrie Trust Company (Gates to Lloyd, 25 July 1925; Gates to Lloyd, 25 August 1925; Gates to Lloyd, 16 January 1926 (all found in LCR 2-7); "Great Activity in Oregon," *Coast Banker* 12 (February 1914): 161; *Coopey v. Keady*, 73 Or. 66, 144 Pac. 99 (1914).
4. Gates to Lloyd, 25 July 1925; Lloyd to Gates, 31 July 1925; Gates to Lloyd, 10 August 1925; Lloyd to Gates, 17 August 1925; Gates to Lloyd, 25 August 1925; Lloyd to Gates, 23 September 1925; Gates to Lloyd, 5 October 1925; Lloyd to Gates, 15 October 1925; Gates to Lloyd, 28 October 1925; Lloyd to Gates, 3 November 1925; Gates to Lloyd, 9 November 1925; Gates to Lloyd, 30 January 1926; Lloyd to Gates, cable, 6 February 1926; Gates to Lloyd, cable, 7 February 1926; Lloyd to Gates, cable, 19 February 1926; Contract of Purchase between the Oregon Real Estate Company and Ralph B. Lloyd, 26 February 1926; Deed of Oregon Real Estate Co. to Lloyd Corp., 3 February 1927 (all found in LCR 2–7); "R. B. Lloyd to Buy Sullivan's Gulch," *Oregonian*, 28 September 1926, I:1, 6. Each block contained eight lots. Ralph Lloyd's acquisitions included blocks, half blocks, quarter blocks, and individual lots.
5. Killingsworth to Lloyd, 3 March 1926, LCL 1–3.
6. MacColl, *The Growth of a City*, 326.
7. Lloyd to Gates, 21 January 1926, LCR 2–7.
8. Lloyd to Macrae, 18 December 1911, Shannon case #6, LCA; Lloyd to Boss, 26 August 1913, Shannon case #10, LCA; Gates to Lloyd, 25 July 1925, LCR 2–7; Lloyd to Gates, 31 July 1925, LCR 2–7; "Lloyd Outlines Project Dreams," *Oregonian*, 4 April 1929, I:1, 12. During 1925–1926, Ralph Lloyd negotiated for Larrabee's former property with Cyrus Gates, who represented Charles Larrabee's wife and four children. Larrabee had hired Gates as his private secretary in 1890.

9. Lloyd to Bank of Italy 21 January 1925, RBL 4–1; Lloyd to Gates, 15 October 1925, LCR 2–7; Lloyd to Gates, 19 November 1926, LCR 2–7.
10. Lloyd to Norton, 14 July 1926, LCL 1–4; Lloyd to Cutts, 1 August 1927, Port 1–3.
11. Lloyd to Macrae, 18 December 1911, Shannon case #6, LCA (quoted); Watson to Lloyd, 24 February 1912, Shannon case #7, LCA; Lloyd to Killingsworth, 5 September 1914, Shannon case #12, LCA. On the growth of the East Side and the Bennett Plan, see Carl Abbott, *Portland: Planning, Politics, and Growth in a Twentieth-Century City* (Lincoln: University of Nebraska Press, 1983), 49–70. The Steel, Broadway, and Interstate Bridges were completed in 1912, 1913, and 1917, respectively. The Bennett Plan was adopted by the City Council, but it was not implemented. Many of the street and other improvements that Bennett recommended were constructed in piecemeal fashion over time, as funds permitted. On the fate of the plan, see Mansel G. Blackford, *The Lost Dream: Businessmen and City Planning on the Pacific Coast, 1890–1920* (Columbus: Ohio State University Press, 1993), 128–50.
12. Killingsworth to Lloyd, 28 February 1917, Shannon case #19 (quoted); Abbott, *Portland*, 33–35, 61–62; Blackford, *The Lost Dream*, 136–8.
13. Roger W. Lotchin, "The Darwinian City: The Politics of Urbanization in San Francisco Between the Wars," *Pacific Historical Review* 48 (1979): 357–81.
14. Lloyd to Michels, 19 January 1917, Shannon case #21, LCA.
15. See, for example, Abbott, *Portland*, 33–37.
16. Lloyd to Killingsworth, 5 September 1914, Shannon case #12, LCA; Lloyd to Killingsworth, 19 November 1914; Killingsworth to Lloyd, 7 December 1914; Killingsworth to Lloyd, 14 January 1915 (all found in Shannon case #13, LCA); Killingsworth to Lloyd, 9 December 1916, LCR 1–4; Lloyd to Killingsworth, 10 December 1916, LCR 1–4 (quoted); Killingsworth to Lloyd, 28 February 1917, Shannon case #19, LCA; Lloyd to Killingsworth, 3 March 1917, Shannon case #19, LCA; Lloyd to Warren, 18 July 1923, LCR 2–2; Lloyd to Wagner, 7 February 1923, RBL 3–5 (quoted).
17. Lloyd to Killingsworth, 5 September 1914, Shannon case #12, LCA.
18. Lloyd to Killingsworth, 19 November 1914, Shannon case #13, LCA; Lloyd to Killingsworth, 10 December 1916, LCR 1–4 (quoted); Lloyd to Portland City Treasurer, 3 March 1917, Shannon case #19, LCA; Lloyd to Eisman, 30 October 1917, Shannon case #21, LCA; Lloyd to Street, 2 February 1920, RBL 2–6; Lloyd to McKenna, 12 June 1922, RBL 2–1.
19. Tom Sitton and William Deverell, eds. *Metropolis in the Making: Los Angeles in the 1920s* (Berkeley: University of California Press, 2001); Robert M. Fogelson,

The Fragmented Metropolis: Los Angeles, 1850–1930 (Cambridge, MA: Harvard University Press, 1967), esp. 112–82. Building permit data found in Guy E. Marion, "Watching Los Angeles Grow," annual midwinter number, *LAT,* 1 January 1925, pt. 1, 19; Eberle Economic Service, weekly letter, 11 February 1929, 31–32.

20. Alison Isenberg's analysis of the professionals who drove the materialization of America's central business districts applies equally to the making of "close-in" districts, such as Hollywood, which often were not greenfield sites (*Downtown America: A History of the Place and the People Who Made It* [Chicago: University of Chicago Press, 2004], esp. 123–65).
21. Lloyd to Title Guarantee & Trust, 25 January 1917, LCR 1–3; Lloyd to Warren, 25 March 1920, LCR 1–3; Lloyd to Title Guarantee & Trust, 19 October 1920, LCR 2–1; Agreement for Sale of Real Estate, 15 June 1921, LCR 2–2.
22. Longstreth, *City Center to Regional Mall,* 63–71, 81–96.
23. Lease agreement with E. J. Le Fon, 17 October 1919, LCR 1–3; Lloyd to Bank of Italy, 28 October 1919, LCR 1–4; Longstreth, "The Forgotten Arterial Landscape."
24. A ground lease is a long-term agreement—typically 99 years—that permits a tenant to develop the property during the period of the leasehold, after which the land and all improvements on it revert to the owner.
25. Lease, 4 May 1921, LCR 1–3; "Bank Will Erect New Building," *LAT,* 10 April 1921, V:6; "Large Sum for Ground Lease in Hollywood," *LAT,* 15 May 1921, V:1 (quoted); "New Buildings to Rise Soon," *LAT,* 3 July 1921, V:1.
26. Agreement, 17 May 1923, LCR 1–3; "$2,500,000 Involved in 13 Realty Deals," *LAE,* 23 May 1923, I:10.
27. Morrison to Giannini, 18 December 1924, LCR 1–3; "Hollywood & Vermont—S. E. Corner," schedule, undated, LCR 1–3; Petition to the Bank of Italy and Others Interested, undated, LCR 1–3 (quoted).
28. City Planning Commission (Portland), *Major Traffic Street Plan, Boulevard, and Park System for Portland, Oregon,* Bulletin No. 7 (January 1921); Abbott, *Portland,* 96. In 1920, Cheney produced a street plan for the Portland Planning Commission that the City promptly shelved.
29. Scott L. Bottles, *Los Angeles and the Automobile: The Making of a Modern City* (Berkeley: University of California Press, 1987), 92–121; F. C. Spayde, "Eighty Million Dollars for City's Streets," annual midwinter number, *LAT,* 1 January 1925, pt. 1, 24; "Survey Shows Progress in Development of Major Highway System," *LAT,* 12 August 1926, II:12; Charles C. Cohan, "Road-Building Program Calls for $15,000,000," *LAT,* 15 July 1928, V:1. The city was also

spending millions of dollars on street improvements before the November 1924 election. See, for example, "Los Angeles Spends Millions on Streets," *LAT*, 27 September 1923. For a partial list of Ralph Lloyd's associations and club memberships, see Hunt, *California and Californians*, 74.

30. Wheeler to Lloyd, 19 May 1924; Wheeler to Giannini and Sevier, 19 May 1924; Hollywood-Vermont Association & Vermont Beverly Association to Vermont Ave. Property Owners, 11 July 1924; Vermont Property Owners Association, "Property Owners on Vermont, Let Us Reason One with Another," undated (1924); Hollywood-Vermont Association & Vermont Beverly Association to Vermont Ave. Property Owners, undated (1924); Culver to Vermont Property Owners, undated (1924); Vogel to Vermont Property Owners, undated (1924); Ryerson to Vermont Property Owners, 9 September 1924; Lloyd to Vermont-Beverly Association, 22 October 1924; Lloyd to Wheeler, 30 October 1924; Lloyd to Board of Public Works, City Engineer, and City Council, 4 November 1924 (all found in RBL 3–2); Lloyd to Miller, 25 June 1926, LCL 1–3; Lloyd to Haas & Dunnigan, 8 October 1926, LCL 1–2; "Hearty Support Is Forthcoming In Street Plan," *Hollywood News*, 22 May 1924; "Vermont Fill Gets Impetus," *LAT*, 14 October 1928, V:3; "Gala Celebration Planned for Completion of Vermont Avenue," 10 November 1929, *LAT*, V:2; "Hollywood Closes Year Marked by Great Building Progress," *LAT*, 5 January 1930, V:3.
31. See, for example, Lloyd to May Company, 14 November 1924, RBL 3–2.
32. Lloyd to Norton, 2 December 1927, Port 1–4.
33. Lloyd to Kendall, 22 August 1929, LCL 4–4.
34. Abbott, *Portland*, 97–98. On 2 September 1931, Portland renamed and renumbered its streets. The City Council divided Portland into five districts. All of Ralph Lloyd's East Side properties became part of the Northeast (NE) district. The City Council also established a system of street addresses numbered in the hundreds, corresponding to a block: hence, the 100 block, 200 block, and so on. All east-west streets became "streets." All north-south streets became "avenues." Thus, East Seventh Street became NE Seventh Avenue. So, for instance, the address of an apartment building that Lloyd acquired in 1926 changed from 241 East Eleventh Street to 1111 NE Eleventh Avenue. On the renaming convention, see Lansing, *Portland*, 322.
35. Killingsworth to Lloyd, 11 September 1922, RBL 2–1; Lloyd to Killingsworth, 12 October 1922, RBL 2–1; Killingsworth to Lloyd, 1 February 1923, RBL 3–3; Lloyd to Killingsworth, 16 March 1923, RBL 3–3; Lloyd to Warren, 18 July 1923, LCR 2–2.
36. Lloyd to Killingsworth, 5 March 1923, RBL 3–3.

37. Lloyd to Killingsworth, 4 February 1925, LCR 2–3; Lloyd to Killingsworth, 30 March 1925; Lloyd to Goldstaub, 5 May 1925; Lloyd to Killingsworth, 18 May 1925 (all found in RBL 4–2); "Notes Payable—Portland Real Estate," undated (1925), LCR 2–4.
38. Killingsworth to Lloyd, 19 March 1923, RBL 3–3 (quoted); Hossack to Lloyd, 14 July 1924, RBL 3–1; Lloyd to Hossack, 27 August 1924, RBL 3–1; Lloyd to Killingsworth, 6 April 1925, RBL 4–2; Lloyd to Killingsworth, 19 August 1925, RBL 4–2. For examples of projects in Los Angeles that may have provided sources of inspiration for the Walnut Park mixed-use building, see "Theater and Store Building," *LAT*, 28 August 1921, V:4; "Center of Big Development," *LAT*, 15 January 1922, V:1 (MacQuarrie Building); "What Neighborhood Business Districts Are Doing in Building Line," *LAT*, 3 September 1922, V:1; "Apartments and Stores Rise in North Vermont," *LAT*, 24 September 1922, V:1.
39. Killingsworth to Lloyd, 16 March 1925, RBL 4–2; Killingsworth to Lloyd, 3 March 1926; Lloyd to Killingsworth, 9 March 1926 (all found in LCL 1–3). William Killingsworth brokered Ralph Lloyd's Walnut Park acquisitions on behalf of the estate of Francis M. Warren Sr. Warren was a prominent figure in Portland's business community. His father had formed the Warren Packing Company, an early salmon cannery. The family owned real estate on both sides of the Willamette River. Warren was an ill-fated passenger aboard the Titanic. His son was president of the Port of Portland Commission and one of the city's wealthiest individuals (MacColl, *The Growth of a City*, 232).
40. Lloyd to Gates, 17 August 1925, LCR 2–7.
41. MacColl, *The Growth of a City*, 325; "Lloyd Outlines Project Dreams."
42. Lloyd to Ainsworth, 12 February 1927, JCA, box 1; "Addition to New Business Center," *Oregon Journal*, 16 October 1927 (quoted).
43. "Things Are Coming Oregon's Way, Says Ralph B. Lloyd," *Oregonian*, 27 April 1926, I:7.
44. Todd D. Gish, "Building Los Angeles: Urban Housing in the Suburban Metropolis," Ph.D. diss., University of Southern California, 2007, 228 (quoted).
45. Lloyd to Hollinshead, 19 March 1926, LCL 1–2; "R. B. Lloyd to Buy Sullivan's Gulch"; Lloyd to Cox, 3 February 1928, Port 1–3 (quoted); "Lloyd Outlines Project Dreams," I:12 (quoted).
46. Lloyd to Killingsworth, 3 March 1917, Shannon case #19, LCA; Lloyd to Gates, 23 September 1925, LCR 2–7 (quoted); Lloyd to Killingsworth, 27 February 1929, LCL 4–4.

47. Lloyd to *Oregonian*, 12 October 1926, LCL 1–3.
48. Lloyd to Chandler, 9 July 1926, LCL 1–1.
49. Lloyd to Norton, 29 March 1927, Port 1–4; "R. B. Lloyd to Buy Sullivan's Gulch."
50. Lloyd to *Oregonian*, 17 March 1926, LCL 1–3 (quoted); Lloyd to DeLong, 13 September 1926, LCL 1–2; Lloyd to Portland Chamber of Commerce, 10 January 1928, Port 1–3; Mike Davis, *City of Quartz: Excavating the Future in Los Angeles* (New York: Verso, 1990), 25 (quoted).
51. Lloyd to Loeb, 28 January 1929, LCL 4–4.
52. Lloyd to Killingsworth, cable, 30 March 1925, RBL 4–2; Abbott, *Portland*, 21.
53. Lloyd to Ainsworth, 9 December 1926, LCL 1–5.
54. Lloyd to Wilcox, 8 October 1926, Port 1–3.
55. Lloyd to Boss, 9 April 1926, LCL 1–1; "R. B. Lloyd to Buy Sullivan's Gulch," I:6 (quoted).
56. Ralph B. Lloyd, Statement to the People of the State of Oregon, 6 July 1926, LCL 1–2, published as, "Investor Decries Pending State Tax," *Oregonian*, 10 September 1926, I:10; "Things Are Coming Oregon's Way, Says Ralph B. Lloyd"; MacColl, *The Growth of a City*, 35–9. Voters had approved a state income tax in a 1923 special election but repealed it in the following general election. Governor Isaac Lee Patterson (1927–1929) led efforts to reestablish the state income tax, which took effect in 1929 (Statement of the Greater Oregon Association, 4 June 1927, LCL 2–3; "Isaac Lee Patterson," *Biographical Dictionary of the Governors of the United States, 1789–1978*, vol. 3, ed. Robert Sobel and John Raimo [Westport, CT: Meckler Books, 1978], 1277–8).
57. Abbott, *Portland*, 94–109; MacColl, *The Growth of a City*, 257–9, 360–1, 655–9.
58. Abbott, *Portland*, 73–4, 80–90, 119–21; Longstreth, "The Forgotten Arterial Landscape," 441–2; MacColl, *The Growth of a City*, 298–303.
59. Greg Hise, *Magnetic Los Angeles: Planning the Twentieth-Century Metropolis* (Baltimore, MD: Johns Hopkins University Press, 1999), esp. 14–55. On Los Angeles as a dystopian urban model, see, for example, Davis, *City of Quartz*; Mike Davis, *Ecology of Fear: Los Angeles and the Imagination of Disaster* (New York: Vintage, 1999); William Fulton, *The Reluctant Metropolis: The Politics of Urban Growth in Los Angeles* (Baltimore, MD: Johns Hopkins University Press, 2001).
60. Lloyd to Gates, 4 December 1925; Lloyd to Gates, 6 January 1926; Lloyd to Gates, 21 January 1926 (all found in LCR 2–7); Lloyd to Kirsch, 17 March 1926, LCL 1–3 (quoted); Lloyd to Ainsworth, 17 May 1926, LCL 1–1; "East Side Growth, Aim," *Oregonian*, 2 July 1926; Norton to Lloyd, cable, 17 July

1926, LCL 1–4; Norton to Lloyd, 16 October 1926, LCL 1–4; "Things Are Coming Oregon's Way, Says Ralph B. Lloyd."

61. See, for example, Norton to Lloyd, 26 June 1926, LCL 1–4.
62. Osborne and Jessup to Lloyd, 5 April 1926, LCL 1–3; Lloyd to Bales, 6 April 1926, LCL 1–4; Lloyd to Wilcox, 10 May 1926, LCL 1–4.
63. Lloyd to Osborne and Jessup, 19 April 1926, LCL 1–3; Lloyd to Ainsworth, 17 May 1926, LCL 1–1; Norton to Lloyd, 11 June 1926, LCL 1–4 (quoted); Blackford, *The Lost Dream*, 131–8; MacColl, *The Growth of a City*, 271–3, 284 (Laurgaard quoted).
64. McClure to Ainsworth, 17 June 1926, LCL 1–4; Calendar for Meeting of the City Planning Commission to be Held, undated [1926], CPA; Portland Planning Commission, Minutes of Meeting, 8 July 1926, CPA. The planning commission published a revised traffic plan in December 1927. As was the case with the earlier plan, the City Council refused to adopt it.
65. Lloyd to Osborne and Jessup, 19 April 1926, LCL 1–3; Lloyd to Norton, 17 June 1926, LCL 1–4.
66. Norton to Lloyd, 26 June 1926, LCL 1–4; "East Side Growth, Aim."
67. Lloyd to Norton, 1 July 1926, LCL 1–4.
68. Lepper quoted in "East Side Sees Dawn of New Era," I:1.
69. Clark to Lloyd, 2 March 1926, LCL 1–1.
70. East Side Commercial Club to Mayor and City Council, 7 May 1928, Public Works, City Engineer Subject Files, box 32, CPA.
71. Lloyd to Norton, 17 June 1926, LCL 1–4; Norton to Lloyd, 10 July 1926, LCL 1–4; East Side Business Men's Club to City Council, 22 July 1926, attached to Lepper to Norton, 24 July 1926, Port 1–1; "Set Back Line Is Urged for New Buildings," *Portland Bulletin*, undated [July 1926], Port 1–1; Lansing, *Portland*, 306.
72. Lloyd quoted in "Lloyd Outlines Project Dreams"; Lloyd quoted in "R. B. Lloyd to Buy Sullivan's Gulch"; Lloyd to Wilcox, 10 May 1926, LCL 1–4; Lloyd to Ainsworth, 9 December 1926, Port 1–3. A $2 million sewer and harbor project, featuring a 32-foot concrete wall that stretched for twenty blocks along Portland's waterfront, was completed in 1929 (Lansing, *Portland*, 316–7).
73. Lloyd to Norton, 7 December 1926, LCL 1–4; Lloyd to Ainsworth, 9 December 1926, Port 1–3; Lloyd to Laurgaard, 9 December 1926, Port 1–3; Lloyd to Norton, 11 December 1926, LCL 1–4; Ainsworth to Lloyd, 20 December 1926, LCL 1–5 (quoted); Lloyd to Norton, 29 December 1926, Port 1–4 (quoted); Portland Planning Commission, Minutes of Meeting, 13 January 1927, CPA.

74. Lloyd to Norton, 7 December 1926, LCL 1–4; Lloyd to Fletcher, 30 December 1926, LCL 1–2 (quoted); Lloyd to Norton, 1 February 1927, Port 1–4; Lloyd to Lowry, 25 May 1927, LCL 2–3. This section is adapted in part from, Michael R. Adamson, *A Better Way to Build*, 10–14. Courtesy of Purdue University Press. All rights reserved.
75. Lloyd to United States National Bank (Sammons), 12 February 1927, LCL 2–6.
76. Lloyd to Norton, 4 March 1927, Port 1–4 (quoted); Lloyd to Norton, 9 March 1927, Port 1–4; Lloyd to Wright and Gentry, 19 May 1927, LCL 2–6; Joe Fitzgibbon, "Bagdad Theater," *The Oregon Encyclopedia* (Portland, OR: Portland State University and the Oregon Historical Society, 2015), www.oregonencyclopedia.org/articles/bagdad_theater; Bart King, *An Architectural Guidebook to Portland*, 2d ed. (Corvallis: Oregon State University Press, 2007), 234–5, 261–2.
77. Norton to State Superintendent of Banks (Schramm), 5 November 1926, Port 1–3; Lloyd to Norton, 29 March 1927, Port 1–4.
78. Lloyd to Drinker, 23 November 1935, LCR 7–2.
79. On Henry Ford, see, for example, Olivier Zunz, *Making America Corporate, 1870–1920* (Chicago: University of Chicago Press, 1990), 79–90.
80. "Austin Guarantees," advertisement, annual midwinter number, *LAT*, 1 January 1925, pt. 1, 22 (quoted); "Austin Pushes More Ideas for Construction," *Engineering News-Record*, 27 November 1952, 48; Martin Grief, *The New Industrial Landscape: The Story of the Austin Company* (Clinton, NJ: Main Street Press, 1978), 15–23, 34–7.
81. Grief, *The New Industrial Landscape*, 59–60, 65–6, 84–5; Shirk, *The Austin Company*, 12.
82. Lloyd to Austin Co., 19 May 1927, Port 1–3; Lloyd to Wright and Gentry, 19 May 1927, LCL 2–6; "Addition to New Business Center." The design for Ralph Lloyd's Walnut Park mixed-use building incorporated "Spanish Renaissance" and "Moorish" elements similar to those employed by Morgan, Walls & Clements in their contemporary McKinley Building on Wilshire Boulevard in Los Angeles (Donald E. Marquis, "The Spanish Stores of Morgan, Walls & Clements," *AF* 50 [June 1929]: 902–9). The final design, as depicted in the *Oregon Journal*, was strikingly similar to a two-block store and office building designed and built by the Austin Company for the Advanced Property Company at the corner of Pruess Road and Wilshire Boulevard in Los Angeles ("Store Building for Wilshire," *LAE*, 26 December 1926, IV:3). On contemporary mixed-use buildings, see, for example, Longstreth, *The Drive-In, The Supermarket, and the Transformation of Commercial Space in Los Angeles*.

83. Lloyd to Wright and Gentry, 19 May 1927, LCL 2–6; Lloyd to Robert W. Hunt Co., 17 September 1927, Port 1–3 (quoted); "Plans Completed for New Automotive Structure," *LAT*, 4 February 1923, V:14.
84. Norton to Superintendent of Banks (Schramm), 5 November 1927, Port 1–3; Lloyd to Superintendent of Banks (Schramm), 9 November 1927, LCL 2–6; Lloyd to Union State Bank, 25 November 1927, LCL 2–6 (quoted); "Addition to New Business Center." Ralph Lloyd bought the lots associated with the site for the bank in December 1926 for $21,000. In fact, the bank would not operate from that location until the end of 1930 (Agreement, 3 December 1926, LCR 3–2; Lease, 24 December 1930, LCR 3–2).
85. Lloyd to Norton, 9 March 1927, Port 1–4. Ralph Lloyd moved his downtown Los Angeles office from the Union Oil Building to the twelfth floor of the Bank of Italy Building on the corner of Seventh and Olive Streets when the Morgan, Walls & Clements–designed structure opened to great acclaim in March 1923 ("Bank Plans Are Completed," *LAT*, 3 February 1922, III:12; "Bank of Italy Occupies Palatial New Home—Award of Merit," *LAT*, 16 March 1923, II:1).
86. See, for example, Charles W. Ertz, "Estimate of Automotive Building for Ralph B. Lloyd," 8 April 1930, LCR 3–1; Builder's Contract, 5 May 1930, LCL 5–1; Ertz to Lloyd, 19 February 1932, LCL 7–2; Builder's Contract, 18 April 1932, LCL 7–2; Lloyd to Ertz-Burns & Co., 11 December 1937, LCR 8–2; Builder's Contract, 10 January 1938, LCR 8–2; Drinker to Von Hagen, 19 October 1945, LCR 10–4; Von Hagen to Lloyd, 25 March 1946, LCL 19–1; Builder's Contract, 1 July 1946, LCL 19–1. On the careers of Thomas Burns and Charles Ertz, see Richard Ellison Ritz, *Architects of Oregon: A Biographical Dictionary of Architects Deceased—19th and 20th Centuries* (Portland, OR: Lair Hill Press, 2002), 59–60, 124–5.
87. See, for example, Lloyd to L. H. Hoffman, 7 February 1928, Port 1–3; Builder's Contract, 2 March 1928, LCR 3–4; Hoffman to Lloyd, 13 July 1928, LCR 3–4; Lloyd to Hoffman, 20 November 1928, Port 1–3; Lloyd to Hoffman, 19 December 1928, Port 1–3; C. L. Peck to Lloyd, 16 September 1941, BI 3–3; Lloyd to Title Insurance and Trust, 5 October 1944, LCR 4–4; C. L. Peck to Beelman, 21 September 1948, LCL 20–3; Claud Beelman, invoice, 15 November 1948, LCL 21–1; Von Hagen to C. L. Peck, 1 April 1950, LCL 23–3.
88. Norton to Lloyd, 8 December 1926, LCL 1–4; Lloyd to Norton, 10 September 1927, Port 1–4; Barbur to City Council, 2 December 1927, LCL 2–5; Portland Planning Commission, Minutes of Meeting, 8 November 1927, CPA; Portland City Council, Minutes of Meeting, 7 December 1927, CPA; "Big Property

Development Started by Ralph B. Lloyd," *Sunday Oregonian*, 1 January 1928, VII:2.

CHAPTER 4. FALSE START: RALPH LLOYD'S EAST SIDE DREAM FALLS SHORT

1. "Big Property Development Started by Ralph B. Lloyd," *Sunday Oregonian*, 1 January 1928, VII:2.
2. Lloyd to Cox, 3 February 1928, Port 1–3 (quoted); Lloyd to Hoffman, 7 February 1928, Port 1–3.
3. Lloyd to Hoffman, 7 February 1928, Port 1–3; Contract with L. H. Hoffman, 2 March 1928, LCR 3–4.
4. Lloyd to Clark, 13 June 1928, Port 1–3.
5. Lloyd to Norton, 9 May 1928, Port 1–4.
6. Barbur to City Council, 5 February 1926, Public Works, City Engineer Subject Files, box 32, CPA.
7. Barbur to City Council, 5 February 1926, Public Works, City Engineer Subject Files, box 32, CPA; "Police and Fire Pay, Bridge Bills Carry," *Oregonian*, 22 May 1926, I:1; "City Spends $4,686,829 on Street Work and 1927 Projects Call for $5,750,000," *Oregonian*, 1 January 1927, I:26.
8. Petition to Mayor and City Council, undated [March 1928], Port 1–3; Apperson to Laurgaard, 21 November 1927; Barbur to City Council, 17 December 1927; Barbur to City Council, 23 February 1928; McClure to Laurgaard, 19 March 1928; Apperson to Laurgaard, 1 May 1928 (all found in Public Works, City Engineer Subject Files, box 32, CPA); City Planning Commission to Mayor and Council, 26 June 1928, Minutes, City of Portland Council, 5 July 1928, 87:696–711, CPA.
9. City Planning Commission to Mayor and Council, 26 June 1928, Minutes, City of Portland Council, 5 July 1928, 87:696–711, CPA; East Side Commercial Club et al. to the Mayor and City Council, for presentation at 11 July 1928 City Council meeting, Port 1–2; Minutes, City of Portland Council, 11 July 1928, 88:734–5, CPA.
10. Laurgaard to Barbur, 24 July 1928, Minutes, City of Portland Council, 1 August 1928, 88:31–41, CPA; Laurgaard to City Council, 17 September 1928, filed 1 October 1928, Council Document, CPA; Barbur quoted in "Street Widening Backed by Clubs," *Oregonian*, 2 August 1928, I:1.
11. Lloyd to Norton, 1 August 1928, Port 1–4; Norton to Lloyd, 7 August 1928, Port 1–4; Lloyd to Laurgaard, 25 July 1928, Public Works, City Engineer Subject Files, box 32, CPA.

12. Report of Committee of Sixteen, Street Widening Program, 29 September 1928, Council Document per Calendar 6959–1, CPA; Minutes, City of Portland Council, 3 October 1928, 90:342–9, CPA; Ainsworth to Lloyd, 9 October 1928, JCA, box 1; John C. Ainsworth, Minutes, Special Meeting of the Planning Commission, 10 October 1928, CPA.
13. Lloyd to Ainsworth, 5 October 1928, JCA, box 1 (quoted); Lloyd to Mayor and City Council, cable, 2 October 1928, Council Document 7231, CPA; Barbur to City Council, 12 October 1928, Council Document per Calendar 7066–1, CPA.
14. Lloyd quoted in "Ralph B. Lloyd Visitor," *Oregonian*, 7 November 1928; Lloyd to Hauser, 6 August 1928, Port 1–3; "Lloyd Outlines Project Dreams," *Oregonian*, 4 April 1929, I:1, 12; Fred Lockley, "Eric V. Hauser Sr.," *History of the Columbia River Valley from the Dalles to the Sea*, vol. 2 (Chicago: S. J. Clarke Publishing, 1928), 455–6. In little more than a decade, Ralph Lloyd would qualify his position on widening streets in the context of his efforts to develop properties along Wilshire Boulevard in Los Angeles, as we shall see. In 1940, he wrote to commercial real estate brokers Ritter, Lowe & Co.: "The more we study the effect of rapid traffic on Wilshire Boulevard, Los Angeles, the more it becomes apparent that too much traffic is almost as bad as too little. Recently the pressure has become so great on business men on the City Council to slow up traffic on Wilshire Boulevard that the City Council has done away with the four-way lanes moving into the city in the morning and out of the city at night." Retailers promoted Olympic Boulevard, which ran parallel to Wilshire Boulevard to the south, as the "fast-traffic" route, funneling traffic in and out of downtown. Lloyd observed: "It has been proven that fast traffic hurts business while the slow traffic gives the purchaser the chance to come to a rest without the hazard of fast traffic. Slow traffic builds up the suburbs. Fast traffic builds up the business at either end of the fast traffic lane." On this evidence, he and the Lloyd Corporation were reconsidering "some of [their] opinions at which [they] had arrived by earlier study of the traffic situation." Applying the Wilshire Boulevard experience to the East Side of Portland would mean that "high speedways," including Broadway, Union Avenue, and Sandy Boulevard, would be configured "to bring traffic to a rest under attractive conditions for parking" (Lloyd to Ritter, Lowe & Co., 29 March 1940, LCL 13–3).
15. "Baker Appears Certain Winner," *Oregonian*, 7 November 1928, I:8; Abbott, *Portland*, 97–98.
16. Lloyd to Cox, 3 December 1928, Port 1–3; Lloyd to Norton, 4 January 1929, LCL 4–5; Lloyd to Norton, 15 January 1929, LCL 4–5; Lloyd to Norton, 26 January 1929, LCL 4–5; Lloyd to Killingsworth, 13 February 1929, LCL 4–4.

17. Norton to Lloyd, 7 January 1929, LCL 4–5.
18. Lloyd to Rosenwald, 19 December 1928, Port 1–4.
19. Norton to Lloyd, 13 April 1929, LCL 4–5.
20. Norton to Lloyd, 15 February 1929, LCL 4–5; Lloyd to Norton, 28 February 1929, LCL 4–5 (quoted).
21. Lloyd to Norton, 28 February 1929, LCL 4–5; Lloyd to McGowan, 19 March 1929, LCL 4–5 (quoted).
22. Minutes, City of Portland Council, 3 April 1929, 90:375, CPA.
23. "Lloyd Outlines Project Dreams."
24. Lloyd quoted in "Lloyd Outlines Project Dreams."
25. Lloyd quoted in "Lloyd Outlines Project Dreams."
26. Lloyd quoted in "Lloyd Outlines Project Dreams."
27. "Lloyd Outlines Project Dreams."
28. MacNaughton to Lloyd, 3 April 1929, LCL 4–4; MacNaughton to Baker, 3 April 1929, LCL 4–4 (quoted); East Side Commercial Club to City Council, 4 April 1929, LCL 4–3; A. E. Doyle & Associates to Lloyd, 4 April 1929, LCL 4–3 (quoted); Hollinshead to Mayor and City Commissioners, 9 April 1929, LCL 4–4; Progressive Business Men's Assn. to Lloyd, 10 April 1929, LCL 4–5; Shull to Lloyd, 13 April 1929, LCL 4–6 (quoted). A. E. Doyle had died prematurely of Bright's Disease in January 1928. His partner, William Hamblin Crowell, carried on the firm in Doyle's name. On the career of A. E. Doyle, see Philip Niles, *Beauty of the City: A. E. Doyle, Portland's Architect* (Corvallis: Oregon State University Press, 2008).
29. Minutes, City of Portland Council, 3 April 1929, 90:375; Minutes, City of Portland Council, 10 April 1929, 90:380–1; Minutes, City of Portland Council, 17 April 1929, 90:419–20 (all found in CPA).
30. Class A stores were large, full-fledged department stores that typically contained more than 100,000 square feet.
31. Lloyd to Kittle, 7 May 1926, LCL 1–4.
32. Boris Emmet and John E. Jeuck, *Catalogues and Counters: A History of Sears, Roebuck and Company* (Chicago: University of Chicago Press, 1950), 338–45; Longstreth, *City Center to Regional Mall*, 119–21; Richard Longstreth, "Sears, Roebuck and the Remaking of the Department Store, 1924–42," *Journal of the Society of Architectural Historians* 65 (June 2006): 238–46, 269–72. In 1927, Sears introduced smaller, so-called Class B and Class C retail stores. By the end of 1928, the company had opened 150 Class B and five Class C establishments.

33. Longstreth, *City Center to Regional Mall*, 119–20; Longstreth, "Sears, Roebuck and the Remaking of the Department Store, 1924–42," 244–6.
34. Norton to Lloyd, 28 April 1928, Port 1–4; Lloyd to Norton, 30 April 1928, Port 1–4 (quoted).
35. Lloyd to Rosenwald, second letter, 8 August 1928, Port 1–4.
36. Lloyd to Rosenwald, first letter, 8 August 1928, Port 1–4; Lloyd to Norton, 9 August 1928, Port 1–4; Lloyd to Hoffman, 20 November 1928, Port 1–3; Lloyd to Hoffman, 19 December 1928, Port 1–3; Lloyd to Veach, 2 January 1945, LCR 3–6.
37. Longstreth, "Sears, Roebuck and the Remaking of the Department Store, 1924–42," 249.
38. "Big Sears, Roebuck Store Opens Today," *Oregonian*, 16 May 1929, II:1.
39. Longstreth, *City Center to Regional Mall*, 120; Longstreth, "Sears, Roebuck and the Remaking of the Department Store, 1924–42," 249–51, 271. On Ralph Lloyd's interest in the store's exterior, see Lloyd to Norton, 30 April 1928, Port 1–4; Lloyd to Hoffman, 20 November 1928, Port 1–3.
40. "Big Sears, Roebuck Store Opens Today" (quoted); Lloyd to Rosenwald, 7 May 1929, LCL 4–6 (quoted).
41. Lloyd to Green, 24 May 1929, LCL 4–3 (quoted); MacNaughton to Lloyd, 17 July 1929, LCL 4–3 (quoted); Norton to Lloyd, 24 October 1929, LCL 4–5 (quoted).
42. Lloyd to Sensenich, 16 January 1929, LCL 4–7.
43. Norton to Lloyd, 13 April 1929, LCL 4–5; Norton to Lloyd, 25 May 1929, LCL 4–5.
44. Thatcher quoted in "Lloyd Plan Attacked," *Oregonian*, 25 May 1929.
45. Norton to Lloyd, 25 May 1929, LCL 4–5; Lloyd to Corbett, 29 May 1929, LCL 4–3 (quoted); Lloyd to Shull, 29 May 1929, LCL 4–6; Shull to Lloyd, 3 June 1929, LCL 4–6.
46. Blackford, *A History of Small Business in America*, 124–6.
47. Hemingway quoted in Gregory H. Hemingway, *Papa: A Memoir* (Boston, MA: Houghton Mifflin, 1976), 4. On creating luck as it relates to the success of business ventures, see, for example, Jim Collins and Morten T. Hansen, *Great by Choice: Uncertainty, Chaos, and Luck: Why Some Thrive Despite Them All* (New York: Harper Business, 2011); Peter Morgan Kash, *Make Your Own Luck: Success Tactics You'll Never Learn in B-School* (Upper Saddle River, NJ: Prentice Hall, 2002); John D. Krumboltz and Al S. Levin, *Luck Is No Accident: Making the Most of Happenstance in Your Life and Career*, 2nd ed. (Atascadero,

CA: Impact Publishers, 2010); Michael J. Mauboussin, *The Success Equation: Untangling Skill and Luck in Business, Sports, and Investing* (Boston, MA: Harvard Business Review Press, 2012); Aubrey K. Tjan and Richard T. Harrington, *Heart, Smarts, Guts, and Luck: What It Takes to be an Entrepreneur and Build a Great Business* (Boston, MA: Harvard Business Review Press, 2012).

48. Ertz to Lloyd, 23 January 1930, LCL 5–1.
49. Norton to Lloyd, 19 July 1929, LCL 4–5; Lloyd to Kane, 11 December 1929, LCL 4–4.
50. Ertz to Lloyd, 13 November 1929, LCL 4–3 (quoted); Lloyd to Ertz, 23 November 1929, LCL 4–3 (quoted).
51. C. W. Norton reported the views of the Portland real estate community (Norton to Lloyd, 27 November 1929, LCL 4–5). The *Los Angeles Times*, for one, predicted a rosy 1930 for builders ("Great Building Program Visioned Here," *LAT*, 1 December 1929, V:1; "Great Building Year Predicted," *LAT*, 22 December 1929, V:1; "National Survey Indicates Extensive Building Operations During Coming Year," *LAT*, 29 December 1929, V:2; "Hollywood Closes Year Marked by Great Building Progress," *LAT*, 5 January 1930, V:3).
52. Lloyd to Ertz, 23 November 1929, LCL 4–3.
53. Lloyd to Kane, 11 December 1929, LCL 4–4.
54. Barbur to Council, 27 July 1929, Council Document [illegible], CPA (quoted); Lloyd to Norton, 14 November 1929, LCL 4–5; Norton to Lloyd, 18 November 1929, LCL 4–5; Norton to Lloyd, 15 November 1929, LCL 4–5; Norton to Lloyd, 17 February 1930, LCL 5–3; Ertz to Lloyd, 18 February 1930, LCL 5–1; Lloyd to Ertz, 24 February 1930, LCL 5–1.
55. Lloyd to Mayor and City Council, 19 February 1930, LCL 5–5.
56. Lloyd to Norton, 20 February 1930, LCL 5–5 (quoted); "Burnside Street Plans Defeated," *Oregonian*, 9 February 1930, I:1; Lansing, *Portland*, 318. In 1932, Olaf Laurgaard would become the subject of a grand jury investigation into charges that he and other officials had taken bribes in connection with a planned new municipal market on the waterfront and had overpaid for the site (Lansing, *Portland*, 322–3).
57. Lloyd to Mayor and City Council, 19 February 1930, LCL 5–5.
58. Lloyd to Norton, 20 February 1930, LCL 5–5: See, also, Lloyd to Wilcox, 6 March 1930, LCL 5–3.
59. Norton to Lloyd, 27 February 1930, LCL 5–3.
60. Barbur to Council, 22 July 1930, Council Document 4702, referred to Public Works, 19 March 1930, adopted 30 July 1930, CPA.

61. Robert M. Fogelson, *Downtown: Its Rise and Fall, 1889–1950* (New Haven, CT: Yale University Press, 2001), 219–21; James S. Warren, "The Present Status of the Hotel Business," *AF* 51 (December 1929): 711–14.
62. Lloyd to Hauser, 6 August 1928, Port 1–3; Lloyd to Stern, 6 February 1929, LCL 4–6; "Eric V. Hauser Sr.," obituary, *New York Times*, 18 January 1929; "Lloyd Outlines Project Dreams."
63. Stern to Lloyd, 26 January 1929. LCL 4–6; Lloyd to Stern, 6 February 1929, LCL 4–6; On the financing of the Los Angeles Biltmore, see "Mammoth Hotel for Downtown," *LAT*, 18 April 1921, II:1 (quoted); "Launch Seven Million Dollar Hotel Project, *LAT*, 6 November 1921, V:1; Chapin Hall, "Biltmore Hotel Project," *LAT*, 23 November 1921, I:10.
64. Lloyd to Norton, 4 February 1929, LCL 4–5; Lloyd to MacNaughton, 9 April 1929, LCL 4–4 (quoted).
65. Lloyd to Huckins, letters of 13 July 1929 and 12 August 1929, LCL 4–4; Lloyd to Maples, 13 December 1929, LCL 4–4; *Moody's Manual of Investments and Security Rating Service: Foreign and American Government Securities: Industrials* (New York: Moody's, 1928), 1878–9; *Moody's Manual of Investments and Security Rating Service: Foreign and American Government Securities: Industrials* (New York: Moody's, 1929), 2700–1; "S. B. Hotel Not in Huge Merger of N. Y. Biltmore Chain," *Santa Barbara Daily News*, 9 February 1929; "Confirms Hotel Merger," *New York Times*, 21 February 1929; "Hotels," *Time*, 4 March 1929.
66. Lloyd to East Side Commercial Club, 21 March 1930, LCL 5–1 (quoted); "20-Story Hotel Will Rise Soon," *Sunday Oregonian*, 13 April 1930, I:1; Lloyd to Ertz, 1 May 1930, LCL 5–1.
67. "Lloyd Proves Faith in Portland," *Oregonian*, 14 April 1930, I:8 (quoted); Hollinshead to Lloyd, 15 April 1930, LCL 5–2 (quoted).
68. Lloyd to Ainsworth, 10 May 1930, LCL 5–1 (quoted); Ainsworth to Lloyd, 15 May 1930, LCL 5–5; Lloyd to McNaughton, 15 May 1930, LCL 5–1; Lloyd to Ainsworth, 7 June 1930, JCA, box 1; Ainsworth to Lloyd, 12 June 1930, LCL 5–5 (quoted).
69. Lloyd quoted in "Lloyd Hotel Plan Nearing Fruition," *Sunday Oregonian*, 22 June 1930, II:1.
70. Lloyd Corporation to Morgan, Walls & Clements, 9 August 1930, LCL 5–2; Lloyd Corporation to *Oregonian*, 15 August 1930, LCL 5–3; "Lloyd Hotel Start Big News of Week," *Sunday Oregonian*, 17 August 1930, II:1 (quoted).
71. Lloyd to Merwin, 28 August 1930, LCL 5–2; "Building Experts View Lloyd Plan," *Oregonian*, 9 September 1930, I:1, 14; Lloyd Corporation to *Oregonian*,

13 September 1930, LCL 5–3; Lloyd to Talbot, 27 September 1930, LCL 5–4. The project's "unusual size and unusual features" were the source of deviations from the building code sufficient to fill a six-page memo from the City's building inspector (Barbur to Norton, 18 November 1930, LCL 5–3).

72. Lloyd quoted in "Building Experts View Lloyd Plan," *Oregonian*, 9 September 1930, I:1, 14; Lloyd Corporation to *Oregonian*, 13 September 1930, LCL 5–3.

73. Lloyd to Dent, 15 July 1930, LCL 5–1; Lloyd to Bard, 28 August 1930, LCL 5–1; Lloyd to Ainsworth, 25 September 1930, JCA, box 1; Lloyd to Hauser, 26 September 1930, LCL 5–2; Lloyd to Talbot, 27 September 1930, LCL 5–4; Lloyd to Ainsworth, 6 October 1930, JCA, box 1; Lloyd to Ainsworth, telegram, 11 October 1930, JCA, box 1; MacColl, *The Growth of a City*, 334.

74. Lloyd to Ainsworth, 10 May 1930, LCL 5–1; Lloyd to Ainsworth, 25 September 1930, JCA, box 1; Lloyd to Hauser, 26 September 1930, LCL 5–2; Lloyd to Ainsworth, 20 October 1930, JCA, box 1.

75. Lloyd to Ainsworth, 25 September 1930; Ainsworth to Lloyd, 3 October 1930; Hemphill to Ainsworth, 16 October 1930; Ainsworth to Lloyd, 17 October 1930 (all found in JCA, box 1).

76. Lloyd to Norton, 13 December 1929, LCL 4–5; Norton to Lloyd, 16 December 1929, LCL 4–5; Strong to Ainsworth, 10 March 1930, JCA, box 2; Hauser to Lloyd, 23 September 1930, LCL 5–2.

77. Roger M. Olien and Diana Davids Olien, *Oil and Ideology: The Creation of the American Petroleum Industry* (Chapel Hill: University of North Carolina Press, 2000), 188–93; Daniel Yergin, *The Prize: The Epic Quest for Oil, Money, and Power* (New York: Simon & Schuster, 1991), 244–8.

78. Hawkins to Lloyd, 14 November 1930, LCL 5–1; Lloyd to Ainsworth, 20 December 1930; Ainsworth to Lloyd, 29 December 1930 (all found in JCA, box 1).

79. Dudley to Ainsworth, 6 December 1930; Ainsworth to Strong, 10 December 1930; Ainsworth to Dudley, 17 December 1930; Strong to Dudley, 17 December 1930; Ainsworth to Dudley, 19 December 1930; Dudley to Ainsworth, 23 December 1930; Dudley to Ainsworth, 13 February 1931; Ainsworth to Dudley, 17 February 1931; Strong to Ainsworth, 10 March 1930 (all found in JCA, box 2).

80. Lloyd to Norton, 13 December 1929, LCL 4–5 (quoted); Norton to Lloyd, 16 December 1929, LCL 4–5 (quoted).

81. Hollinshead to Lloyd, 2 January 1931, LCL 6–2; Minutes, City of Portland Council, 7 January 1931, 92:326, CPA.

82. Hoffman to Lloyd, telegram, 7 January 1931, LCL 6–2; "Contract Given for Lloyd Hotel," *Oregonian*, 11 January 1931, I:1, 7. As Ralph Lloyd told the Oregon Public Utility Information Bureau, the changes to the hotel included separating both the power plant and the laundry from the hotel to eliminate the need for a sub-basement (Lloyd to Crawford, 12 November 1930, LCL 5–1). On the Meier and Frank department store building, see King, *An Architectural Guidebook to Portland*, 21–23.
83. Ainsworth to Lloyd, telegram, 12 January 1931, JCA, box 1; Norton to Lloyd, 21 January 1931, LCL 6–4 (quoted); Clark to Lloyd, 21 January 1931, LCL 6–1 (quoted); Merwin to Lloyd, 13 January 1931, LCL 6–3 (quoted).
84. Lloyd to Ainsworth, 17 January 1931, JCA, box 1; Lloyd to Ainsworth, 19 January 1931, JCA, box 1; Barry to Lloyd, 26 January 1931, LCL 6–6; Ainsworth to Lloyd, 30 January 1931, JCA, box 1; MacColl, *The Growth of a City*, 335–7.
85. Lloyd to Ainsworth, 17 January 1931, JCA, box 1; Lloyd to Hall, 9 March 1931, JCA, box 1.
86. Olien and Olien, *Oil and Ideology*, 188–93.
87. Lloyd to Ainsworth, 10 March 1931, JCA; Lloyd to McGonigle, 4 April 1931, LCL 6–3.
88. Lloyd to Strong, 24 June 1931, LCL 6–1 (quoted); Lloyd to Ainsworth, 13 August 1931; Lloyd to Ainsworth, 26 August 1931; Lloyd to Ainsworth and Dick, 10 September 1931 (all found in JCA, box 1); MacColl, *The Growth of a City*, 337–8. Plans for a new downtown hotel died as well.
89. Norton to Lloyd, 29 April 1931, LCL 6–4 (quoted); Lloyd to Ainsworth 11 July 1931, LCL 6–6; Ainsworth to Lloyd, 18 July 1931, JCA, box 1; Lloyd to Ainsworth, 23 July 1931, LCL 6–6; Lloyd to Ainsworth, 24 July 1931, JCA, box 1; Lloyd to Dudley, 28 July 1931, LCL 6–5.
90. Lloyd to Ainsworth, 13 August 1931, JCA, box 1; Lloyd to Ainsworth, 26 August 1931, JCA, box 1 (quoted).
91. Lloyd to Ainsworth, 26 August 1931, JCA, box 1.
92. Ainsworth to Lloyd, 27 August 1931, JCA, box 1.
93. Lloyd to Dinwiddie, 30 January 1932, LCL 7–2.
94. Lloyd to Allen, 27 November 1931, LCL 6–1.
95. Barbur to Lloyd, 21 August 1931, LCL 6–1.
96. Lloyd to Barbur, 26 August 1931, LCL 6–1; Lansing, *Portland*, 318, 321–4. On Ralph Lloyd's conditions for starting construction of the hotel, see, also, Lloyd to Bates, 9 October 1931, LCL 6–1.

97. William Leach, *Land of Desire: Merchants, Power, and the Rise of a New American Culture* (New York: Vintage, 1993), 173–6.
98. Lloyd Corp., Portland office, to the Portland and Multnomah County RFC Committee, draft of memorandum, undated (May 1933), LCL 8–4.
99. Bartholomew quoted in MacColl, *The Growth of a City*, 284–5. On Portland's interwar street improvement program, see also, Earl O. Mills, "Portland's Planning Perseverance Pays," *Western City* 8 (May 1932): 13–16.

CHAPTER 5. THE LLOYD CORPORATION BECOMES AN INDEPENDENT OPERATOR

1. Beaton, *Enterprise in Oil*, 782 (table 1); Petroleum World, *Annual Review: California Oil Industry* (Los Angeles, 1934) (compiled by the author). Shell was relatively less dependent than Associated on Ventura Avenue production, which accounted for 29 percent of its California output.
2. Lloyd to Associated, 24 December 1927, LCL 2–1.
3. 140 Fed. 814 (8th Cir. 1905), quoted in Bruce M. Kramer, "The Interaction between the Common Law Implied Covenants to Prevent Drainage and Market and the Federal Oil and Gas Lease," *Journal of Energy, Natural Resources, and Environmental Law* 15 (1995): 7.
4. Lloyd to Robinson, "Statement as to Financial Standing of Lloyd Corporation," 13 September 1934, LCL 9–1.
5. Lloyd to Rosemary Lloyd, 1 July 1932, not sent, LCL 7–3; *Ira Gosnell and Lena Bowyer v. Ralph B. Lloyd and Lloyd Corporation*, 244–8.
6. *Ira Gosnell and Lena Bowyer v. Ralph B. Lloyd and Lloyd Corporation*, 246–7.
7. *Ira Gosnell and Lena Bowyer v. Ralph B. Lloyd and Lloyd Corporation*, 248–9. T. B. Gosnell's first wife, Caroline, died in December 1913, just two months after Ralph Lloyd and Joseph Dabney leased the property. Gosnell soon remarried. In 1917, however, he and Ethel divorced and she moved to Los Angeles.
8. *Ira Gosnell and Lena Bowyer v. Ralph B. Lloyd and Lloyd Corporation*, 249–50; "Battle Rages Over $10,000,000 Oil Field," *LAE*, 26 June 1927, I:4.
9. *Ira Gosnell and Lena Bowyer v. Ralph B. Lloyd and Lloyd Corporation*, 250; Crotty, "Biography of Ralph B. Lloyd," 102–4.
10. *Ira Gosnell and Lena Bowyer v. Ralph B. Lloyd and Lloyd Corporation*, 250–1; "Battle Rages Over $10,000,000 Oil Field"; "Gosnell Loses $5,000,000 Oil Battle," *LAE*, 23 May 1928; "Court Battle Over Cal. Oil 'Strike' Ends," *LAH*, 22 May 1928. From 1925 to 1932, Ralph Lloyd earned some $1 million in royalties from the Gosnell lease (Lloyd to Rosemary Lloyd, 1 July 1932, not sent, LCL 7–3).

11. Lloyd to Rosemary Lloyd, 1 July 1932, not sent, LCL 7–3.
12. For an explanation of the API scale, see Chapter 1 n40 on page 290.
13. Beaton, *Enterprise in Oil*, 171–85; Sabin, *Crude Politics*, 114–20; Williamson et al., *The American Petroleum Industry: The Age of Energy*, 305–7, 463–5.
14. Lloyd to Hartman Ranch Company, 18 March 1927, LCL 2–3.
15. "Ventura Draws a Pair of Aces," *The Record* 6 (March 1925): 7; untitled note, *The Record* 7 (May 1926): 12; Lloyd to Grubb, 17 August 1928, LCL 3–3; *The Record* 11 (July 1930): 18; Beaton, *Enterprise in Oil*, 111.
16. Bain, *The Economics of the Pacific Coast Petroleum Industry, Part I*, 36–8; White, *Formative Years in the Far West*, 7–24.
17. Kincaid to Lloyd, 13 September 1935, LCL 9–5.
18. Lloyd to Grubb, 16 October 1925, RBL 4–4; Lloyd to Grubb, 2 November 1925, RBL 4–4; Lloyd to Gosnell, Bowyer, and Gosnell, 2 November 1925, RBL, 4–2; Lloyd to Hartman Ranch Company, 18 March 1927, LCL 2–3.
19. Lloyd to Shell, 8 October 1925, RBL 4–4; Lloyd to Roberta Lloyd Dobbins, 27 October 1925, RBL 4–1 (quoted); Lloyd to Gosnell, Bowyer, and Gosnell, 2 November 1925, RBL, 4–2 (quoted); Shell to Lloyd, 3 November 1925, RBL 4–4; Lloyd to Gosnell, Bowyer, and Gosnell, 17 November 1925, RBL, 4–2; Lloyd to Shell, 16 December 1925, LCL 1–4.
20. Lloyd to Shell, 16 December 1925; Shell to Lloyd, 22 December 1925; Lloyd to Shell, 29 January 1926; Shell to Lloyd, 12 February 1926; Shell to Lloyd, 20 February 1926; Lloyd to Shell, 9 February 1926; Shell to Lloyd, 15 March 1926 (all found in LCL 1–4); Humphery to Lloyd, 17 January 1928, LCL 3–1; Lloyd to Associated, 21 January 1928, LCL 3–1; "Tide Water Associated Oil Company: A Major Unit in the Petroleum Industry," *World Petroleum* (August 1937): 20–1.
21. Humphery to Lloyd, 17 January 1928; Lloyd to Associated, 21 January 1928; Humphery to VL&W, South Basin Oil, Lloyd Corp., and State Consolidated Oil Co., 2 February 1928; Lloyd to Humphery, 6 February 1928; Humphery to Lloyd, 14 February 1928 (all found in LCL 3–1). Tide Water Associated Oil Company owned 94.6 percent of Associated's stock ("The Associated Oil Company," *Mining and Oil Bulletin* [November 1922], 646–50; Moody's Investor Service, *Moody's Manual of Industrial Securities, 1928* [New York: Moody's, 1929], 1660–2, 2920).
22. Humphery to Lloyd, 14 March 1928, LCL 3–1.
23. Humphery to Lloyd, 25 February 1928, LCL 3–1; Lloyd to Grubb, 17 August 1928, LCL 3–3 (quoted); Lloyd to Grubb, 20 September 1928, LCL 3–3 (quoted); W. Kenneth Hayes, "Standard Advances Prices Paid Here for Crude

Oil," *LAE*, 15 August 1928, II:1; idem, "Other Companies Meet Standard's Crude Advance," *LAE*, 16 August 1928, II:6.

24. Lloyd to Humphery, 26 September 1928, LCL 3–1. Statewide output for 1928 was 231.8 million barrels, slightly higher than the 231.2 million barrels produced in 1927 (Williamson et al., *The American Petroleum Industry: The Age of Energy*, 302).
25. T. E. Swigart, "Ultimate Oil Production in Ventura District Increased by Controlling Gas Output," *Petroleum World* (October 1928): 78–80, 142. Swigart served as Shell's superintendent at Ventura Avenue.
26. Swigart, "Ultimate Oil Production in Ventura District Increased by Controlling Gas Output."
27. Swigart, "Ultimate Oil Production in Ventura District Increased by Controlling Gas Output."
28. Swigart, "Ultimate Oil Production in Ventura District Increased by Controlling Gas Output"; "More Oil Fields Will Soon Be Curbed in War on Waste," *LAE*, 10 March 1928, I:2.
29. Sabin, *Crude Politics*, 120–1.
30. See the discussion in Gary B. Conine, "The Prudent Operator Standard: Applications beyond the Oil and Gas Lease," *Natural Resources Journal* 41 (Winter 2001): 33–9; Kramer, "The Interaction Between the Common Law Implied Covenants to Prevent Drainage and Market and the Federal Oil and Gas Lease," 6–16.
31. F. W. Hertel, oral history, 15.
32. Lloyd to Hartman, letters of 24 November 1924 (quoted), 8 December 1924, and 30 December 1924 (all found in RBL 3–1); Lloyd to Hartman, 6 January 1925, RBL 4–2; Lloyd to Katherine Hartman and Hartman Ranch Company, 7 January 1925, RBL 4–2.
33. Lloyd to Katherine Hartman and Hartman Ranch Company, 19 August 1925; Lloyd to Hartman Ranch Company, 28 October 1925 (quoted) (all found in RBL 4–2); Lloyd to King and McLaughlin, 28 October 1925, RBL 4–1; *Hartman Ranch Co. v. Associated Oil Co.*, 10 Cal. 2d 238–9.
34. Lloyd to King and McLaughlin, 28 October 1925, RBL 4–1.
35. Lloyd to Taff, 15 December 1925, RBL 4–1.
36. Lloyd to Humphery, 2 October 1928; Lloyd to Humphery, 4 October 1928; Lloyd to Humphery, 3 November 1928 (all found in LCL 3–1); Lloyd to Associated, 4 January 1929; Lloyd to Associated, 18 January 1929; King to Lloyd, 20 January 1929; Humphery to Lloyd, 22 January 1929; Lloyd to Associated, 23 January 1929; Humphery to Lloyd, 23 January 1929; Lloyd to

Associated, 24 January 1929; Humphery to Lloyd, 24 January 1929; Lloyd to Associated, 25 January 1929; Humphery to Lloyd, 25 January 1929; Lloyd to Associated, 28 January 1929 (quoted); King to Lloyd, 29 January 1929; Lloyd to King, 29 January 1929 (all found in LCL 4–2).

37. E. R. Head, "State to Wield Big Stick If Law Disobeyed," *California Oil World*, 5 September 1929, 1, 17.
38. Lloyd to Associated, 20 February 1929, LCL 4–2; Lloyd to Humphery and Jenkins, 15 June 1929, LCL 4–2.
39. Humphery to Lloyd, 20 June 1929, LCL 4–2.
40. *Hartman Ranch Co. v. Associated Oil Co.*, 10 Cal. 2d 232; Associated Oil Company, *Annual Report, 1933* (San Francisco: the company, 1934); Associated Oil Company, *Annual Report, 1935* (San Francisco: the company, 1936).
41. "Bush Asks Operators' Criticism of Gas Law Enforcement Plan," *California Oil World*, 29 August 1929; "Full Text of State Plan of Enforcement," *California Oil World*, 29 August 1929; "State Fires First Gun at S. F. Springs," *California Oil World*, 12 September 1929, 1, 14; "No Curtailment in Field Is Indicated," *VCS*, 12 September 1929, 1; "Ventura Meet Saturday to Discuss Gas," *California Oil World*, 19 September 1929; "Ventura Gas Waste Discussed," *California Oil World*, 26 September 1929; "Interesting Data at Ventura Hearing," *California Oil World*, 3 October 1929, 1, 5.
42. Lloyd to Norton, 7 October 1929, LCL 4–5; "Remedy for Gas Wastage Is Offered," *VCS*, 24 September 1929, 1, 3; "Avenue Field Different, Is Expert View," *VCS*, 25 September 1929, 1; "Landowners Join Fight on Gas Law," *California Oil World*, 10 October 1929, 1, 16.
43. McDuffie quoted in "Avenue Field Different, Is Expert View," 1; Jensen quoted in "Ventura to Fight Gas Cut," *LAT*, 7 October 1929, I:15; "Ventura Meet Saturday to Discuss Gas"; "Ventura Gas Waste Discussed"; "Interesting Data at Ventura Hearing." Notwithstanding the presence of many former Shell employees in the new Los Angeles–based company, incorporated on 21 November 1928, neither Shell nor its affiliates had any connection to Pacific Western Oil Corporation, which John Paul Getty would control by late 1931 (Reinhardt to Lloyd, 19 March 1929, LCL 4–6).
44. "Gas Order for Avenue Field Issued," *VCS*, 3 October 1929, 1; "Ventura Gas Under Curb," *LAT*, 4 October 1929, I:1; "Ventura to Fight Gas Cut"; "Companies Protest Order Issued for Ventura Ave.," *California Oil World*, 10 October 1929, 1, 17; "Gas Notices Are Posted in Oil Field," *VCS*, 11 October 1929, 1.

45. "Independents Are in Session Wednesday," *California Oil World*, 3 October 1929, 1–2; "No Chance of Agreement of Independents," *California Oil World*, 3 October 1929, 1, 8; Howard Kegley, "Oil Field News," *LAT*, 4 October 1929, I:19.
46. Beek and Guio quoted in "Companies Protest Order Issued for Ventura Ave.," 17; Beek quoted in "State Firm on Ventura Gas Matter," *California Oil World*, 21 November 1929, 5; "Gas Notices Are Posted in Oil Field"; "Appeal from Gas Order Is Now on File," *VCS*, 17 October 1929, 1.
47. Kincaid quoted in "Landowners Join Fight on Gas Law," 1; "State Firm on Ventura Gas Matter"; Lloyd to Jenkins, 19 February 1930, LCL 5–1; Sabin, *Crude Politics*, 122.
48. Bennett quoted in "State Firm on Ventura Gas Matter," 1, 5.
49. Lloyd to Rochester, 16 February 1932, LCL 7–2 (quoted); Lloyd to Norton, 28 October 1929, LCL 4–5 (quoted).
50. Quoted in Sabin, *Crude Politics*, 123.
51. "Drilling," *California Oil World*, 13 March 1930, 6.
52. Lloyd to McGonigle, 30 April 1929, LCL 4–4; Lloyd to Humphery, 21 November 1929, LCL 4–2; "Cal. Curtailment Makes Progress," *California Oil World*, 6 March 1930, 1, 8; "State Allowable Held at 610,000," *California Oil World*, 8 May 1930, 1; Arthur M. Johnson, "California and the National Oil Industry," *Pacific Historical Review* 39 (1970): 165–6; Sabin, *Crude Politics*, 128–32.
53. Lloyd to Ralph M. Smith, 6 May 1930, LCL 5–4; "Ventura Avoids Halting of Drill," *California Oil World*, 8 May 1930, 1; "Allowable Now Set at 550,000 Bbls.," *California Oil World*, 18 September 1930, 1, 4.
54. Lloyd to Humphery, 2 October 1928; Lloyd to Humphery, 4 October 1928; Lloyd to Humphery, 3 November 1928 (all found in LCL 3–1).
55. Bayer to Lloyd, 11 December 1929; Lloyd to Bayer, 13 December 1929; Bayer to Lloyd, 16 December 1929 (all found in LCL 4–2); Jenkins to Lloyd, 29 January 1930; Lloyd to Jenkins, 19 February 1930 (quoted) (all found in LCL 5–1).
56. Lloyd to McGonigle, 4 April 1931, LCL 6–3 (quoted); "Lloyd Made President of New Agency," *California Oil World*, 9 April 1931, 1, 9; "Sales Agency Membership Is Growing," *California Oil World*, 23 April 1931, 1, 8; E. R. Head, "460,000-Bbl. Daily Quota May Be Set," *California Oil World*, 23 April 1931, 1, 13.
57. Lloyd to Ainsworth 24 July 1931, JCA, box 1 (quoted); Sabin, *Crude Politics*, 140. On Oklahoma's and Texas's path toward mandatory proration, see Williamson et al., *The American Petroleum Industry: The Age of Energy*, 542–4.

58. On the dissatisfaction of the Hartman family, see, for example, Lloyd to Associated, 5 October 1931, LCL 6–1; Jenkins to Lloyd, 9 October 1931, LCL 6–1; Lloyd to Humphery, 13 July 1933, LCL 8–1; Humphery to Lloyd, 17 July 1933, LCL 8–1.
59. VL&W, Annual Stockholders Meeting, 18 March 1929, VLW 1–2; Crotty, "Biography of Ralph B. Lloyd," 104–5.
60. As soon as Associated took responsibility for drilling operations on the Lloyd lease, as provided for under the agreement concluded in February 1920, Joseph Dabney turned his attention to the Los Angeles Basin, in time to participate in the unprecedented oil boom. In February 1924, Dabney incorporated South Basin Oil Company with capitalization of $1 million. In September 1926, he transferred his interest in the Ventura Avenue field to the company, of which he was the sole shareholder. Ultimately, various companies in which Dabney was interested would pump oil from six dozen or more wells from Long Beach and the nearby Huntington Beach field, where derricks co-mingled with beachgoers, provoking concerted action on the part of civic and business leaders to ban oil development on California's beaches. See "Deep Wells Indicate Southern Limit to Signal Hill Production," *Oil Weekly*, 11 March 1921, 78; M. H. Soyster and M. van Couvering, "Notes on the Long Beach Oil Field, Los Angeles County," *Annual Report of the State Oil and Gas Supervisor* 7 (April 1922): 5–8, 7; "Directory of California Oil Operators," *Annual Report of the State Oil and Gas Supervisor* 10 (August 1924): 119, 171; Lloyd, Dabney, Miley, Lloyd Corp., and South Basin Oil Co. to Associated Oil Company, 15 September 1926, LCL 1–1; Agreement between Lloyd Corp., South Basin Oil Co., and E. J. Miley and Associated Oil Company, 31 December 1926, LCL 1–1; "Directory of California Oil Operators," *Annual Report of the State Oil and Gas Supervisor* 13 (August 1927): 123, 151; "Directory of California Oil Operators," *Annual Report of the State Oil and Gas Supervisor* 16 (July-August-September 1930): 209, 239; "Directory of California Oil Operators," *Annual Report of the State Oil and Gas Supervisor* 18 (July-August-September 1933): 162; Kevles, "Joseph B. Dabney"; *Dabney v. Philleo*, 38 Cal. 2d 60 (1951).
61. Lloyd to Lagomarsino, 30 April 1934, LCL 9–1; Lloyd to Robinson, 30 April 1934, LCL 9–1; Hugh T. Dobbins, President's Report to the Annual Meeting of Stockholders, 15 March 1937, VLW 9–4; Crotty, "Biography of Ralph B. Lloyd," 105–7. Ralph Lloyd remained on the VL&W board of directors until September 1936.
62. Lloyd to Rosemary Lloyd, 1 July 1932, not sent, LCL 7–3.
63. Crotty, "Biography of Ralph B. Lloyd," 107.

64. Henderson quoted in Crotty, "Biography of Ralph B. Lloyd," 108; Lloyd to Humphery, 10 March 1936, LCL 10–1; Hugh T. Dobbins, President's Report to the Annual Meeting of Stockholders, 16 March 1936, VLW 9–4; Lloyd to VL&W, 24 July 1936, Crotty Papers, box 3.
65. Lloyd to Humphery, 10 March 1936, LCL 10–1; Lloyd to Crotty, 10 July 1936, LCL 10–1 (quoted).
66. Olien and Hinton, *Wildcatters*, 56–82.
67. Lloyd to Bayer, 13 December 1929, LCL 4–2; J. R. Pemberton, "Ventura Avenue Field Allocation Schedule, December 1933," LCL 8–1; Ralph B. Lloyd, "Comparative Analysis of Curtailment of Operators in Coastal District," 10 July 1934, LCL 9–1.
68. Humphery to Byles, 31 October 1932, box 7; Humphery to Byles, 2 December 1932, box 7; "Associated Oil Company: Explanation of Changes in 1933 Capital Budget," 8 February 1933, box 5 (all found in Shoup Papers).
69. Shoup to Humphery, 24 December 1934, Shoup papers, box 2. Paul Shoup served as a director and president of Associated from July 1918 to August 1926.
70. Shoup to Humphery, 24 December 1934, Shoup papers, box 2 (quoted); Humphery to Shoup, 4 January 1935, Shoup papers, box 1; "The Associated Oil Company," *Mining and Oil Bulletin*, 650.
71. Lloyd to Humphery, 17 August 1936, LCL 8–1.
72. Associated had spudded two of the three wells that it was obligated to drill under the lease.
73. In fact, the well cost more than $450,000 to complete at a depth of more than 9,500 feet (Lloyd to Crotty, 10 July 1936, LCL 10–1).
74. Jenkins to Lloyd, 26 April 1929, LCL 4–2 (quoted); "Notice of Default," 10 July 1929, LCL 4–2; "Notice of Default," 20 May 1930, LCL 5–1; Lloyd to Associated, 13 November 1933, LCL 8–1; Lloyd to Humphery, 27 December 1933, LCL 8–1 (quoted); Lloyd to Bard, 22 May 1934, LCL 9–1; William F. Humphery, letter to shareholders, 8 March 1933, Associated Oil Company, *Annual Report, 1932* (San Francisco: the company, 1933) (quoted); Benton Turner, oral history, 2–3.
75. Lloyd to Humphery and Jenkins, 8 October 1934, LCL 9–1; Humphery to Lloyd, 3 November 1934, LCL 9–1.
76. Lloyd to Stair, 15 October 1934, LCL 9–3; Lloyd to Stair, 29 December 1934, LCL 9–3.
77. Humphery to Lloyd, 3 November 1934, LCL 9–1.
78. Lloyd to Hudson, 19 October 1936, LCL 10–4.
79. Lloyd to Mears, 14 January 1938, LCL 11–6.

80. Lloyd to Jenkins, 5 January 1935, LCL 9–4; Jenkins to Lloyd, 14 January 1935, LCL 9–4; Lloyd to Ventura Office, 9 May 1938, LCL 12–1; Lloyd to Tide Water Associated, 8 May 1939; Humphery to Lloyd, 8 June 1939; Lloyd to Humphery, 8 November 1939; Lloyd to Humphery, 24 June 1939 (all found in LCL 12–4).
81. Lloyd to Crotty, 10 July 1936, LCL 10–1.
82. Lloyd to Hudson, 19 October 1936, LCL 10–4; Lloyd to Stair, 16 December 1936, LCL 10–4; Rehm to Lloyd, 1 July 1937, LCL 11–1.
83. Shoup to De Forest, 2 April 1936, Shoup papers, box 3; Shoup to Mills, 14 April 1936, Shoup papers, box 3 (quoted); Humphery to Lloyd Corp., South Basin Oil Co., and VL&W, 6 March 1936, LCL 10–1 (quoted).
84. "Tide Water Associated Oil Company: A Brief Discussion of This Company and the Oil Industry," 26 March 1926, LCL 1-4; Tidewater Associated Oil Company, *Annual Report, 1936* (San Francisco: the company, 1937); "Tide Water Associated Oil Company: A Major Unit in the Petroleum Industry," *World Petroleum* (August 1937). Before 1934, federal income tax of corporation and its subsidiaries was based on consolidated net income. Under changes in the law adopted in 1934, each company was required to file a separate return and pay taxes on its separate net taxable income. A holding company could no longer offset losses of one subsidiary against the profits of another subsidiary that it held, and so its potential tax liability might increase substantially. In addition, effective 1 January 1936, 15 percent of dividends received by a corporation from another corporation had to be included as taxable income.
85. Humphery to Lloyd Corp., South Basin Oil Co., and VL&W, 6 March 1936, LCL 10–1; Humphery to Lloyd, 4 August 1936, LCL 10–1.
86. Humphery to Lloyd Corp., South Basin Oil Co., and VL&W, 6 March 1936, LCL 10–1 (quoted); Lloyd to Crotty, 15 April 1937, Crotty Papers, box 3.
87. Lloyd to Humphery, 10 March 1936, LCL 10–1.
88. Lloyd to Humphery, 20 July 1936, LCL 10–1; Lloyd to VL&W, 24 July 1936, Crotty Papers, box 3.
89. Hugh T. Dobbins, Report to Stockholders at a Meeting Held January 13, 1937, VLW 9–4.
90. Humphery to Lloyd, 4 August 1936, LCL 10–1; Humphery to Shoup, 13 March 1937, Shoup papers, box 1; Lloyd to VL&W, 12 April 1937; Lloyd to Crotty, 15 April 1937; Humphery to VL&W, Lloyd Corp., and South Basin Oil Co., 14 September 1937 (all found in Crotty Papers, box 3).
91. Lloyd to VL&W, 12 April 1937; Lloyd to Humphery, 12 April 1937 (quoted); Lloyd to Crotty, 15 April 1937; Crotty to Lloyd, 21 April 1937; Crotty to Lloyd,

20 May 1937; Templeton to Lloyd Dobbins, Lloyd Smith, Dobbins, Crotty, and Montgomery, 29 May 1937; Humphery to VL&W, Lloyd Corp., and South Basin Oil Co., 14 September 1937 (Lloyd quoted) (all found in Crotty Papers, box 3); Pease to Edward M. Lloyd, 29 July 1937, LCL 11–1; Kincaid to Philleo, 26 August 1937, LCL 11–1; Humphery to Shoup, 24 September 1937, Shoup papers, box 1; Shoup to Humphery, 7 April 1938, Shoup papers, box 3.

92. Memorandum of Telephone Conversation, 12 December 1936, LCL 10–5; Lloyd to Hollingsworth, 20 November 1937, LCL 10–6; Kincaid to Lloyd, 2 June 1938, LCL 11–4; Modification of Lease, 1 January 1938, executed 7 June 1938, Crotty Papers, box 3; Agreement and Assignment of Modification of Lease to Associated Oil Company, 1 January 1938, Crotty Papers, box 3; Hugh T. Dobbins, President's Report to Annual Meeting of Stockholders, 22 March 1939, VLW 9–4; *Hartman Ranch Co. v. Associated Oil Co.*, 10 Cal. 2d 232. Associated prevailed on the portion of the judgment relating to the forfeiture of the lease.

93. W. Elliot Brownlee, *Federal Taxation in America: A History*, 3d ed. (New York: Cambridge University Press, 2016), 124–33; Mark Leff, *The Limits of Symbolic Reform: The New Deal and Taxation, 1933–1939* (New York: Cambridge University Press, 1984), 129–47, 202; Christina D. Romer and David H. Romer, "A Narrative Analysis of Interwar Tax Changes," University of California, Berkeley (2012).

94. R. G. Blakely and G. C. Blakely, *The Federal Income Tax* (Clark, NJ: Lawbook Exchange), 428–35; J. S. Seidmen, *Seidmen's Legislative History of Federal Income Tax Laws, 1938–1961* (Clark, NJ: Lawbook Exchange), 155–76; "Troubled Real Estate Leasing Companies Trapped Within the Personal Holding Company Income Tax Provisions," *Brigham Young University Law Review* (November 1978): 986–8. Ralph Lloyd was a majority owner, holding 30,115 of 50,000 shares. His immediate family held all but one of the remaining shares (Lloyd Corporation, List of Stockholders, 11 June 1935, LCL 9–5).

95. Lloyd Corp., tax worksheet (handwritten), 20 June 1938, LCL 11–4. The calculation did not factor the savings enjoyed in avoiding a California Corporation Franchise Tax of four percent.

96. A former university instructor in Japan and district manager for Shell, Mark Justin Dees married the Lloyds' eldest daughter, Eleanor, on 23 January 1932. In April 1933, Ralph Lloyd put him in charge of his company's new Ventura field office (Crotty, "Biography of Ralph B. Lloyd," 123).

97. Lloyd to Jones, 8 January 1935, LCL 9–4; Lloyd to Dees, 4 May 1936, LCL 10–5; Lloyd to Ventura Office, 9 May 1938, LCL 12–1; Lloyd to Ventura Office,

7 June 1938, LCL 12–1; Lloyd to Phillips, 25 October 1938, LCL 11–5 (quoted); Benton Turner, oral history, 2–3. Lloyd Corporation No. 1 was the fourth well on the former Dabney-Lloyd lease. The lease was renamed Shell-Lloyd Lease A at the request of Dabney's heirs. When Shell relinquished its interest in the property, the lease was renamed Lloyd Corporation. The existing Shell-Lloyd Lease A wells retained their names, however. Subsequent wells were named Lloyd Corporation, beginning from number one. South Basin Oil Company maintained its interest in the lease. See, for example, the South Basin Oil Company file for 1938 found in LCL 11–6.

98. Lloyd to Humphery, 24 June 1939, LCL 12–4.
99. McGonigle to Lloyd, 30 December 1940, LCL 13–2.
100. Lloyd Corp. to Security-First National Bank of Los Angeles, 25 May 1936, LCL 10–4; Lloyd Corporation, Ltd., Statement of Assets and Liabilities, 31 December 1950, LCR 12–2; Pasini to Haskins & Sells, 18 April 1951, LCL 24–5; Benton Turner, oral history, 3–4, 7–8; *Petroleum World, Annual Review of California Crude Oil Production* (Los Angeles: Palmer Publications, 1941): 64; CCCOP, *Thirty-Sixth Annual Review of California Crude Oil Production* (Los Angeles: CCCOP, 1951); Lester P. Stockman, "Lloyd's Discovery in Ventura May Prove Most Important in Years," *Petroleum World* 47 (July 1947): 11–12; California Division of Oil and Gas, *Annual Report of the State Oil and Gas Supervisor* 39 (January–June 1953): 29; California Division of Oil and Gas, *Annual Report of the State Oil and Gas Supervisor* 39 (July–December 1953): 39; D. H. Stormont, "Ventura Avenue Sets New High," *Oil & Gas Journal*, 25 January 1954. The scale of exploration and production at Oxnard paled in comparison with activity at Ventura Avenue. At its peak, in 1959, the Oxnard field yielded only 3.2 million barrels of crude oil (California Division of Oil and Gas, *Annual Report of the State Oil and Gas Supervisor* 45 [July–December 1959]: 55).

CHAPTER 6. DEPRESSION-ERA COMMERCIAL REAL ESTATE DEVELOPMENT AND MANAGEMENT

1. Lloyd to LeBaron, 3 March 1931; Corporation Grant Deed, 14 January 1932; Yule to Lloyd, 22 July 1935 (all found in LCR 4–2); Morrison to Lloyd, 7 December 1931 (quoted); Morrison to Lloyd and Kincaid, 18 December 1931; Kincaid to Morrison, 31 December 1931; Gore to Lloyd Corp., 2 January 1932; Escrow Instructions, 4 January 1932; Lloyd, Memorandum, Lloyd Corp. Board of Directors Meeting, 5 January 1932 (all found in LCR 4–3); Lloyd to Robinson, 12 August 1938, LCL 11–6 (quoted); "Boulevard Transfers

Announced," *LAT*, 28 June 1931, V:1; "Prominent Corner in Transfer," *LAT*, 24 January 1932, V:1.

2. W. Elliot Brownlee, *Dynamics of Ascent*, 2d ed. (New York: Knopf, 1979), 395–9.
3. U.S. Bureau of the Census, *Statistical Abstract of the United States, 1957* (Washington, DC: GPO, 1957), 752 (table 971), 759 (table 980).
4. Eberle Economic Service, weekly letter, 17 February 1930, 37–38; "National Survey Indicates Extensive Building Operations During Coming Year," *LAT*, 29 December 1929, V:2.
5. "National Survey Indicates Extensive Building Operations During Coming Year."
6. Eberle Economic Service, weekly letter, 17 February 1930, 37.
7. Eberle Economic Service, weekly letter, 19 January 1931, 13–14.
8. Eberle Economic Service, weekly letter, 15 February 1932, 25–26.
9. Lloyd to Drinker, 21 December 1935, LCL 9–6.
10. James L. Davis, "Profits Beckon Investors in Realty," *LAT*, 24 January 1932, V:1.
11. Sartori quoted in "Bankers Laud Los Angeles Real Estate as Investment and Cite Unusual Opportunities in Present Market," *LAT*, 24 January 1932, V:1.
12. Monnette quoted in "Bankers Laud Los Angeles Real Estate as Investment and Cite Unusual Opportunities in Present Market."
13. "Graphic Record of Los Angeles' New Construction Increase," *LAT*, 9 January 1938, V:1.
14. Lloyd to Anderson, 9 July 1932, LCL 7–1.
15. *Wilshire Topics*, November 1930, quoted in Ralph Hancock, *Fabulous Boulevard* (New York: Funk & Wagnalls, 1949), xi.
16. Lloyd to Dinwiddie, 30 January 1932, LCL 7–2.
17. Morrison to Lloyd, 7 December 1931, LCR 4–3; Anderson to Lloyd Corp., 20 June 1932, LCL 7–1.
18. Longstreth, *City Center to Regional Mall*, 180–1.
19. Morrison to Lloyd, 7 December 1931, LCR 4–3.
20. *Wilshire Topics*, quoted in Hancock, *Fabulous Boulevard*, xii.
21. Lloyd to Title Guarantee & Trust Co., 15 November 1923, LCR 2–2; S. M. Cooper, Builder's Statement, 16 April 1924, LCR 2–2; Longstreth, *City Center to Regional Mall*, 104–5. Ralph Lloyd's uncle, Charles W. Bramel, was John De Lario's grandfather.
22. Longstreth, *City Center to Regional Mall*, 105–27, 106 (quoted); "City's Zoning Law Sustained," *LAT*, 17 May 1927, II:1–2. In 1904, Los Angeles imposed a

thirteen-story limit on buildings to limit urban density. It remained in force until 1958.

23. Longstreth, *City Center to Regional Mall*, 127–34; James W. Elliott, "Wilshire Paces Growth of City," *LAT*, 20 January 1929, V:1 (quoted).
24. Lloyd to Norton, 5 February 1931, LCL 6–4.
25. Longstreth, *City Center to Regional Mall*, 131. Myer Siegel, a women's apparel shop that was established downtown in 1886, opened its third store outside the central business district in the Dominguez-Wilshire Building in October 1931. The opening of the Wilshire Boulevard store was preceded by expansion into Hollywood and Pasadena ("New Wilshire Store to Open," *LAT*, 25 October 1931, III:17).
26. Longstreth, *City Center to Regional Mall*, 133 (quoted).
27. "Wilshire Boulevard Business Growth in Five Year Period Declared Phenomenal," *LAT*, 24 November 1929, V:2; "Many Firms Locating on Boulevard," *LAT*, 23 February 1930, V:3.
28. Steve Vaught, "Lost Hollywood: A John De Lario in the Hills of Beverly," *Paradise Leased*, http://paradiseleased.wordpress.com/2012/05/13/lost-hollywood-a-john-de-lario-in-the-hills-of-beverly. Photographs of the estate may be found in *Architectural Digest* 8:3 (1931): 23–8. Publisher John C. Brasfield wrote that the issue "reproduces in detail some of the outstanding types of fine California homes which stand as the realization of the best creative thought of their designers." The mansion was constructed for about $56,000. Ralph Lloyd valued the property at $500,000—an indication of the substantial wealth that petroleum extraction at Ventura Avenue was generating for the holders of mineral rights (Lloyd Corp., Bids for Residence on Alpine Drive, undated, LCR 6–1; Lloyd to Security-First National Bank of Los Angeles, 2 June 1936, LCL 10–4). Ralph and Lulu Lloyd would live in this home for the rest of their lives.

    John De Lario was lead architect for Hollywoodland, a subdivision of the Sherman & Clark Ranch in the Hollywood Hills that broke ground in 1923. Developer Sidney H. Woodruff envisioned the project as a Mediterranean Riviera that would attract wealthy Easterners to Los Angeles in winter and selected De Lario to execute it. The real estate syndicate behind the project included Woodruff, Harry Chandler, publisher of the *Los Angeles Times*, and Moses H. Sherman, co-founder of the town of Sherman, now known as West Hollywood.
29. Crandall to Dykes, 20 November 1939, Crandall Papers, box 5.
30. On the relative economic decline of Los Angeles nationally, see, also, William H. Mullins, *The Depression and the Urban West Coast, 1929–1933: Los Angeles,*

San Francisco, Seattle, and Portland (Bloomington: Indiana University Press, 1991), 9–13.

31. Lloyd to Dick, 7 December 1933, LCL 8–6; Lloyd to Drinker, 19 December 1935, LCR 7–2; Lloyd to Drinker, 21 December 1935, LCL 9–6; Lloyd to Burns, 7 August 1940, LCL 13–1.
32. See, also, Mullins, *The Depression and the Urban West Coast, 1929–1933*, 15–16, 48–53, 87–9.
33. Wilshire Boulevard Association of Beverly Hills to Lloyd, 12 May 1933, LCL 8–6. Among its directors, the association counted actresses Lilyan Tashman and Mary Pickford, director Clarence Brown, Burton Green, president of the Rodeo Land and Water Co.—the owners and subdividers of the ranch that became the city of Beverly Hills, Walter G. McCarthy, owner of the Beverly-Wilshire Hotel, and G. D. Robertson, a former president of the Los Angeles Realty Board and a director of the California Real Estate Association.
34. Lloyd to Dinwiddie, 30 January 1932, LCL 7–2; Read to Kincaid, 4 March 1932, LCL 7–5; Kincaid to Coldwell, Cornwell & Banker, 15 March 1932, LCL 7–1; Wilshire Boulevard Association of Beverly Hills to Lloyd, 29 August 1932, LCL 7–6; Lloyd to Hewitt, 6 September 1932, LCL 7–6; Lloyd to Wood, 14 September 1932, LCL 7–5; Wood to Lloyd, 15 September 1932, LCL 7–5; Wilshire Boulevard Association of Beverly Hills to Lloyd, 28 September 1932, LCL 7–6; Yule to Hewitt, 29 September 1932, LCL 7–6; Wilshire Boulevard Association of Beverly Hills to Lloyd, 12 May 1933, LCL 8–6.
35. Wilshire Boulevard Association of Beverly Hills to Lloyd, 12 May 1933, LCL 8–6 (quoted); David C. Sloane, "Medicine in the (Mini) Mall: An American Health Care Landscape," in *Everyday America: Cultural Landscape Studies After J. B. Jackson*, ed. Chris Wilson and Paul Groth (Berkeley: University of California Press, 2003), 294.
36. Lloyd to Shinn, 21 January 1932, LCL 7–1; Norton to Lloyd, 17 May 1933, LCR 4–2; Buckley to Lloyd, LCL 8–1; Lloyd to Shinn, 25 April 1940, LCL 13–1. As all of Wilshire Boulevard within Beverly Hills was zoned for commercial use, the Lloyd Corporation did not have to take similar action on behalf of its Business Triangle property.
37. Reeder to Lloyd, 13 October 1933; Statement of Account, 23 October 1933; Deed, 9 December 1933; Statement of Account, 31 March 1934 (all found in LCR 4–4); A. S. Benson to Lloyd Corp., grant deed, 19 May 1934, LCR 5–1; Escrow Instructions, 10 December 1934; Statement of Account, 5 February 1935; Statement of Account, 21 February 1935 (all found in LCR 6–6);

Deposit Receipt, 15 August 1935; Option to Purchase Real Property, 23 October 1935; Statement of Account, 29 October 1935 (all found in LCR 7–1); Lloyd to Security-First National Bank of Los Angeles, 2 June 1936, LCL 10–4.

38. Lloyd to Dick, 18 December 1933, LCL 8–6.
39. Lloyd to Hoffman, 7 February 1928, Port 1–3; Contract with L. H. Hoffman, 2 March 1928, LCR 3–4, Lease, 1 July 1928, LCR 3–4. The building cost a maximum of $32,500 to construct.
40. Fisk Tire Company, Agreement to Lease, 4 June 1928, LCR 3–5; Builder's Contract, 22 August 1928; Notice of Completion, 17 December 1928; Kincaid to Hartford Accident and Indemnity Co., 7 January 1929 (all found in LCR 3–4). The building cost $87,900 to construct.
41. Lease, 18 March 1929, LCR 2–2. The building cost $13,455 to construct. It was the second Willard's restaurant in Los Angeles. In 1928, the first restaurant opened at 9625 West Pico Boulevard.
42. Trombly to Lloyd, 2 May 1929, LCR 4–1; Builder's Schedule, 18 April 1929, LCR 4–2.
43. Charles W. Ertz, "Estimate of Automotive Building for Ralph B. Lloyd," 8 April 1930, LCR 3–1; Lloyd to Lyon, 12 April 1930, LCR 3–1. Ertz estimated the cost of the building to be $22,766.
44. Of the tenants for whom Lloyd constructed buildings between 1926 and 1931, only Sears and Union State Bank, in which Lloyd was interested, maintained their lease payments without modification.
45. Lloyd to Hennessy, 3 August 1932, LCR 2–2.
46. Mercer to Drinker, 7 March 1933, LCR 3–4; Agreement, 14 April 1933, LCR 3–4. By the spring of 1932, International Harvester's local business was 25 percent of normal. A year later, it had fallen again by half.
47. Hennessy to Lloyd, 27 July 1932; Lloyd to Hennessy, 3 August 1932; Hennessy to Lloyd, 4 January 1933; Lloyd to Hennessy, 16 January 1933; Hennessy to Lloyd, 26 May 1933; Lloyd to Hennessy, 13 June 1933; Hennessy to Lloyd, 5 January 1934; Lloyd to Hennessy, 18 July 1934; Lloyd to Hennessy, 22 April 1935; Hennessy to Lloyd, 1 May 1935; Kincaid to Hennessy, 6 May 1935 (all found in LCR 2–2). In 1940, Cecil B. DeMille bought the Willard's restaurant at auction and built a Brown Derby restaurant on the site.
48. Drinker to Kincaid, 8 August 1932; Winfree to Lloyd Corp., 15 August 1932; Drinker to Kincaid, 30 August 1932 (all found in LCR 3–4).
49. Lyon to Lloyd, 14 January 1931; Kincaid to Lyon, 11 May 1932; Drinker to Kincaid, 3 December 1932; Lyon to Lloyd Corp., 10 May 1933 (all found in

LCR 3–1). This chapter and those that follow use the modern addresses for buildings in Portland, per the renaming and renumbering system adopted by the City on 2 September 1931. See Chapter 3 n34 on page 310.

50. Kincaid to Lloyd, 21 October 1932; Eggleston to Kincaid, 7 November 1932; Agreement, 7 November 1932 (all found in LCR 3–5); *Moody's Industrial Manual, 1933* (New York: Moody's, 1934), 2827–8. The company emerged from reorganization on 10 May 1933.
51. Lease, 13 October 1924, LCR 1–3. On Studebaker's struggles during the Great Depression, see Critchlow, *Studebaker*, 101–15.
52. Lloyd to Lloyd's Automotive Service, 3 February 1933, LCL 8–3.
53. Levy to Lloyd Corp., 21 August 1934, LCR 1–3; Yule to Sheperd Tractor and Equipment, 23 January 1940, LCL 13–3; Yule to GE Appliances, 20 February 1947, LCL 20–2.
54. Drinker to Lloyd, 7 August 1933; Lease, 11 October 1933; Latture and Pierce to Lloyd Corp., 31 August 1934; Agreement, 1 October 1934; Lloyd to Drinker, 27 April 1936; Drinker to Lloyd, 6 May 1936 (quoted) (all found in LCR 3–4). C. W. Norton moved to Los Angeles and took a position with W. I. Hollingsworth & Co., a real estate and insurance firm with its own building on the corner of Hill and Sixth Streets (Ralph B. Lloyd, "Statement on Portland Activities," 18 June 1932, LCL 7–5).
55. Paul Shoup, speech delivered in Portland, 18 August 1936, Shoup Papers, box 9.
56. Barry Eichengreen, *Golden Fetters: The Gold Standard and the Great Depression, 1919–1939* (New York: Oxford University Press, 1992); Thomas E. Hall and J. David Ferguson, *The Great Depression: An International Disaster of Perverse Economic Policies* (Ann Arbor: University of Michigan Press, 1998), 115–60; Christina D. Romer, "What Ended the Great Depression?" *Journal of Economic History* 52 (December 1992): 757–61; Herbert Stein, *The Fiscal Revolution in America* (Chicago: University of Chicago Press, 1969), 37–56, 91–128; Peter Temin, *Did Monetary Forces Cause the Great Depression?* (New York: Norton, 1976); Peter Temin, *Lessons from the Great Depression* (Cambridge, MA: MIT Press, 1991).
57. Longstreth, *City Center to Regional Mall*, 107–9.
58. Maris to Lloyd, 19 November 1935; Drinker to Lloyd, 9 December 1935; Drinker to Lloyd, 14 December 1935 (all found in LCR 7–2).
59. Lloyd to Drinker, 29 November 1935, LCR 7–2; Lloyd to Drinker, 29 November 1935; Lloyd to Drinker, 19 December 1935; Lloyd to Drinker, 11

December 1935 (all found in LCR 7–2); Lloyd to Crotty, 10 July 1936, LCL 10–1.

60. Drinker to Lloyd, 19 November 1935, LCR 7–2.
61. Lloyd to Drinker, 23 November 1935, LCR 7–2.
62. Drinker to Lloyd, 14 December 1935, LCR 7–2.
63. Lloyd to Drinker, 19 December 1935, LCR 7–2.
64. Lloyd Corp., Portland, to Lloyd Corp., Los Angeles, cable, 24 December 1935, LCR 7–2; Lloyd Corp., Portland, to Lloyd, cable, 27 December 1935, LCR 7–2; Lease, 10 March 1936, LCR 7–2; Lloyd to Rucker, 24 November 1948, LCL 20–3.
65. "One Firm's Activity Illustrates Brisk Upward Trend in City's Building," *LAT*, 11 October 1925, V:1; "$75,000,000 in New Buildings Planned for 1926," *LAE*, 2 January 1926, IV:1; "Roosevelt Building Enhances City Skyline," *LAE*, 15 June 1927, II:3; "Skyscraper to Rise Downtown," *LAT*, 16 December 1928, V:1; "Two Major Structures Announced," *LAT*, 26 May 1929, V:1; Patricia Bayer, *Art Deco Architecture: Design, Decoration, and Detail from the Twenties and Thirties* (New York: H. N. Abrams, 1992), 87–115; Carla Breeze, *American Art Deco: Modernistic Architecture and Regionalism* (New York: Norton, 2003), 13–34, 223–53; Suzanne Tarbell Cooper, Amy Ronnebeck Hall, and Frank E. Cooper Jr., *Los Angeles Art Deco* (Chicago: Arcadia Publishing, 2005): 9–24.
66. Ertz to Lloyd, 21 May 1935, LCL 9–5; Lloyd to Drinker, 29 November 1935, LCR 7–2; Lloyd to Drinker, 19 December 1935, LCR 7–2; Lloyd to Drinker, 26 December 1935, LCR 7–2; Lloyd Corp., Portland, to Lloyd, cable, 27 December 1935, LCR 7–2; Drinker to Lloyd, 11 March 1936, LCR 7–2; Lloyd to Ertz, Burns & Co., 19 March 1936, LCL 10–2; Lloyd to Rucker, 24 November 1948, LCL 20–3. On the 1930s as a decade of technological innovation, see Alexander J. Field, *A Great Leap Forward: 1930s Depression and U.S. Economic Growth* (New Haven, CT: Yale University Press, 2011).
67. Frank to Lloyd, 25 July 1938, LCL 11–4; Lloyd to Frank, 5 August 1938, LCL 11–4.
68. Frank to Lloyd, 8 August 1938, LCL 11–4.
69. Lease, 15 July 1946, LCL 20–2; Construction Contract, 13 August 1946, LCL 19–4; Lloyd to McIver, 27 July 1947, LCL 20–1; Yule to Frank, 13 October 1947, LCL 20–2; Lloyd to Lerner, 21 January 1949, LCL 22–2.
70. Lloyd to Gray, 23 December 1936, LCL 10–4 (quoted); Longstreth, *City Center to Regional Mall*, 201–3.

71. Merwin to Kincaid, 1 October 1934, LCR 5–2; Lloyd to *Architectural Forum*, 5 February 1936, LCL 10–1; "Syndicate Buys Site Downtown," *LAT*, 26 May 1929, V:1; "History of Office Buildings in Los Angeles Recorded," *LAT*, 13 November 1949, V:6. On the Parkinsons' design of Bullock's Wilshire, see Longstreth, *City Center to Regional Mall*, 112–17. On City Hall as inspiration for the design of contemporary tall buildings, see Longstreth, *City Center to Regional Mall*, 129–32.
72. Tenants in Sixth and Olive Building, table, 17 September 1934, LCR 5–2; Merwin to Kincaid, 1 October 1934, LCR 5–2; Gillette to Wolff, 27 July 1934, LCR 5–1; Lloyd, handwritten spreadsheet, 27 September 1935, LCR 5–2.
73. Merwin to Kincaid, 1 October 1934; Property Service Corp., Application for $400,000 loan, 16 October 1934; Promissory Note, 1 November 1934; Kincaid to Boyle 13 November 1934; Boyle to Kincaid, 14 November 1934 (all found in LCR 5–2).
74. Lloyd to Carver, 10 December 1934, LCL 9–1; Beelman to Lloyd, 5 September 1935, LCR 5–2; Beelman to Lloyd, 16 September 1935, LCR 5–2.
75. Lloyd to Boyle, 1 October 1935, LCR 5–2; Lloyd to Yates, 6 December 1935, LCL 9–6; Agreement, 19 December 1935, LCR 5–2; Lloyd to *Architectural Forum*, 5 February 1936, LCL 10–1; Lloyd to Drinker, 31 July 1936, LCR 3–4.
76. Lloyd to Crotty, 10 July 1936, LCL 10–1; Bernstein's Fish Grotto to W. Ross Campbell Co., 4 November 1936, LCL 10–2; Lloyd to Title Insurance & Trust Co., 18 March 1937, LCR 5–2.
77. Lloyd to Gray, 23 December 1936, LCL 10–4; Lloyd to Title Insurance & Trust Co., 18 March 1937, LCR 5–2; Minutes of Directors Meeting, Lloyd Corp., 29 March 1937, LCL 10–6.
78. Lloyd to Quinlan, 21 October 1935, LCL 9–6.
79. See, for instance, Crandall to Lick, 5 February 1936, Crandall Papers, box 21; Crandall to Stone, 14 April 1939, Crandall Papers, box 12.
80. Charles C. Cohan, "Year's Structural Gain Tremendous," *LAT*, 10 January 1937, V:3.
81. "Graphic Record of Los Angeles' New Construction Increase."
82. Crandall to Barth, 17 May 1937, box 7 (quoted); Crandall to Dyke, 20 May 1936, box 21 (quoted); Crandall to Wiepert, 24 August 1937, box 7 (all found in Crandall Papers).
83. Crandall to Lacey, 5 January 1940, box 1; Crandall to Lacey, 30 January 1940, box 1; Crandall to Roeder, 24 April 1940, box 2; Crandall to Lyons, 30 September 1940, box 1 (all found in Crandall Papers); Romer, "What Ended the Great Depression?" 757.

84. Lloyd to Strong, 24 June 1931, LCL 6–1; Lloyd to Blodgett, 3 February 1933, LCL 8–1. Robert Higgs argues that private investment did not recover to 1929 levels by 1940 because of "a pervasive uncertainty among investors about the security of their property rights in their capital and its prospective returns" ("Regime Uncertainty," 563).
85. Lloyd to Ertz, 3 February 1933, LCL 8–2.
86. Lloyd Corp. to Los Angeles County Board of Supervisors, 19 August 1937, LCL 10–6.
87. Lloyd to Ford, 7 August 1939, LCL 12–2.
88. Lloyd to Drinker, 2 December 1939, LCL 12–3.
89. Crandall to Roeder, 24 April 1940, Crandall Papers, box 2.
90. Shannon Crandall, annual address, 44th Annual Convention of the National Wholesale Hardware Association, Atlantic City, New Jersey, 18 October 1938, Crandall Papers, box 7.
91. Paul Shoup, speech delivered in Portland, 18 August 1936, Shoup Papers, box 9. On Shoup's promotion of his ideas on taxation, see W. Elliot Brownlee, "Carl S. Shoup: Formative Influences," in *The Political Economy of Transnational Tax Reform: The Shoup Mission to Japan in Historical Context*, ed. W. Elliot Brownlee, Eisaku Ide, and Yasunori Fukagai (New York: Cambridge University Press, 2013), 21–26.
92. Lloyd to May, President, 3 February 1936, LCL 10–4; Lloyd to Meehan, 8 May 1936, LCL 10–4; Lloyd to Drinker, 8 May 1936, LCL 10–3; Brownlee, *Federal Taxation in America*, 129–31.
93. Lloyd to Garland, 29 January 1937, LCL 10–6; Lloyd to Carson, 21 April 1936, LCL 10–4.
94. Lloyd to Hubbard, Westervelt & Mottelay, 8 March 1938, LCL 11–4.
95. Lloyd to Hutchings-Hudson, 24 March 1938, LCL 11–4.
96. Lloyd to Bates, 25 July 1939, LCL 12–1.
97. Lloyd to Wright, 1 October 1940, LCL 13–4.
98. Lloyd to Beelman, 28 November 1933, LCL 8–1; Business Property Lease, 28 November 1933, LCR 4–4; Builder's Contract (Pozzo Construction), 25 April 1934, LCR 4–4; Notice of Completion (Victor Hugo Restaurant), 4 December 1934, LCR 4–4; Lloyd to Beverly Auto Park, 5 December 1934, LCL 9–1; Statement of Account, 21 March 1935, LCR 4–4; Lloyd to Philp, 12 July 1935, LCL 9–4 (quoted); Charles C. Cohan, "New Construction Spurs Growth of Los Angeles," *LAT*, 27 May 1934, I:14; Stephanie Stein Crease, *Gil Evans: Out of the Cool: His Life and Music* (Chicago: Chicago Review Press, 2003), 55–7;

Drew Page, *Drew's Blues: A Sideman's Life with the Big Bands* (Baton Rouge: Louisiana State University Press, 1999), 121.

99. Morrison to Lloyd Corp., 7 February 1935, LCL 9–5; Lloyd to Philp, 12 July 1935, LCL 9–4; Longstreth, *City Center to Regional Mall*, 86–8, 89 (quoted).
100. Morrison to Lloyd Corp., 14 February 1935, LCL 9–4; Lloyd to Philp, 12 July 1935, LCL 9–4; Lloyd to Philp, 12 September 1935, LCL 9–4.
101. Longstreth, *City Center to Regional Mall*, 89.
102. Lloyd to St. John, 3 May 1937, LCL 10–6; Lloyd to St. John, 13 May 1937, LCL 10–6; Lloyd to The May Co., 16 June 1938, LCL 11–5.
103. Longstreth, *City Center to Regional Mall*, 139–41.
104. Lloyd to J. W. Robinson Co., 12 August 1938, LCL 11–6.
105. Cohan, "New Construction Spurs Growth of Los Angeles," I:13; Longstreth, *City Center to Regional Mall*, 201–3.
106. Lloyd to J. W. Robinson Co., 11 June 1937, LCL 11–6.
107. Lloyd to J. W. Robinson Co., 12 August 1938, LCL 11–6. Billed by management as "a store of tomorrows, not yesterdays," Coulter's department store was demolished in 1980 (Cooper, Hall, and Cooper Jr., *Los Angeles Art Deco*, 125).
108. Lloyd to Snyder, 6 September 1938, LCL 11–6.
109. Norton to Lloyd, 25 September 1938, LCL 11–6.
110. Lloyd Corp. to Brown Derby Cafe, 19 October 1939, LCL 12–1.
111. Agreement to Exchange, 9 September 1944, LCR 4–4.
112. J. J. Sugarman-Rudolph Co., "Final Statement of Victor Hugo Auction, June 9, 1941," LCR 4–5; Lloyd to Ball, 6 June 1941, LCL 14–1; Lloyd to George, 20 June 1941, LCL 14–1; Yule to Lloyd and Von Hagen, 20 June 1941, LCL 14–2; Lloyd to Title Insurance & Trust Co., 5 October 1944, LCR 4–4; "Confirm Report of Wilshire Deal," *LAH*, 11 May 1935; "Sloane Company Plans Additional Building Unit," *LAT*, 19 May 1935, I:12; "New Building Project Marks Business Growth," *LAT*, 24 May 1936, V:4; "New York Shop to Open in West," *LAT*, 16 November 1937, I:9.
113. Anonymous to Lloyd, postcard, 12 May 1935, LCL 9–7.
114. Longstreth, *City Center to Regional Mall*, 136.

CHAPTER 7. PUBLIC CAPITAL AND THE DEVELOPMENT OF PORTLAND'S EAST SIDE

1. Higgs, "Regime Uncertainty," 577.
2. Lloyd to Dick, 28 August 1933, LCL 8–2; Lloyd to Ertz, 5 October 1933, LCL 8–2; Sabin, *Crude Politics*, 144. In August 1933, Ralph Lloyd had traveled to

Washington, D.C., as chairman of California's Central Proration Committee, a group of oil industry operators, to convince President Roosevelt and NRA "czar" Hugh Johnson to invest responsibility for carrying out the industry's production code for California in the committee.

3. Lloyd to Holbrook, 23 December 1930, LCL 5–3; Lloyd to Weil, 9 October 1931, LCL 6–6; Kincaid to Bigelow, 8 April 1933, LCL 8–4; Lloyd Corp. to Portland and Multnomah County RFC Committee, draft of memorandum, undated (May 1933), LCL 8–4; Lloyd to Hoffman, 12 May 1933, LCL 8–2; Lloyd to Los Angeles Board of Education, 4 August 1936, LCL 10–2; Abbott, *Portland,* 110. The proposed hotel would contain between 300 and 400 rooms and cost half as much to build as the 600-room behemoth designed by Morgan, Walls & Clements. The Lloyd Corporation would donate the excavated site and contribute as much as $200,000 of the project's estimated $1–2 million cost.
4. "Lloyd Proposes Big Development," *Oregonian*, 7 May 1933, I:12.
5. James S. Olson, *Saving Capitalism: The Reconstruction Finance Corporation and the New Deal, 1933–1940* (Princeton, NJ: Princeton University Press, 1988), 13–21.
6. Olson, *Saving Capitalism*, 42–62; Abbott, *Portland,* 108; Lansing, *Portland,* 325; James S. Olson, *Herbert Hoover and the Reconstruction Finance Corporation, 1931–1933* (Ames: Iowa State University Press, 1977), 76–80.
7. Dick to Lloyd, 9 May 1933, LCL 8–2.
8. Ainsworth to Lloyd, 6 June 1933, LCL 8–6.
9. Hemphill to Lloyd, 13 May 1933, LCL 8–2.
10. Murphy to Lloyd, 22 May 1933, LCL 8–4.
11. Portland Chamber of Commerce to Lloyd, 15 May 1933, LCL 8–4; Smith to Lloyd, 15 May 1933, LCL 8–4.
12. "Lloyd Proposes Big Development," I:13 (quoted).
13. "Lloyd Proposes Big Development," I:13 (quoted); Lansing, *Portland*, 322–3.
14. Marks to Swinerton, 12 May 1933, LCL 8–3; Lloyd Corp., Portland office, to Lloyd, cable, 16 June 1933, LCL 8–3; "Lloyd Proposes Big Development," I:13; MacColl, *The Growth of a City*, 340.
15. Boss to Lloyd, 14 July 1933, LCL 8–1.
16. Marks to Swinerton, 12 May 1933, LCL 8–3; "Lloyd Proposes Big Development," I:13; MacColl, *The Growth of a City*, 340.
17. Beelman to Lloyd, 19 June 1933, with attachment, Wilkinson to Beelman, undated, LCL 8–1; Drinker to Yule, 25 July 1933, LCL 8–4; Lloyd Corp. to Richanbach & Co., 12 September 1933; Richanbach & Co. to Board of County

Commissioners, 13 September 1933; Drinker to Lloyd, 20 October 1933; Drinker to Lloyd, 23 October 1933; Drinker to Lloyd, 25 October 1933 (all found in Port 2–2); Olson, *Herbert Hoover and the Reconstruction Finance Corporation*, 88–89.

18. Kincaid to Lloyd, 21 October 1932; Eggleston to Kincaid, 7 November 1932; Agreement, 7 November 1932; Lease, 1 May 1934; Deed of Trust, 23 October 1946; Whitney to Von Hagen, 8 June 1948; Whitney to Lloyd Corp., 3 February 1949 (all found in LCR 3–5); Yule to DMV, 23 January 1940, LCL 13–1; Von Hagen to McGaughey, 19 November 1946, LCL 19–1; Yule to DMV, 20 February 1947, LCL 20–1.
19. Drinker to Lloyd, 29 May 1937; Drinker to Lloyd, 3 June 1937; Lloyd to Drinker, 17 June 1937; Lloyd to the Honorable Governor of Oregon and the State Board of Control, 25 June 1937 (quoted) (all found in BI 3–3).
20. Drinker to Lloyd, 3 July 1937; Drinker to Lloyd, 18 October 1937; Lloyd to Drinker, 4 August 1938; Drinker to Lloyd, 9 August 1938 (all found in BI 3–3). On labor actions in the auto and steel industries during the first half of 1937, see, for instance, Lizabeth Cohen, *Making a New Deal: Industrial Workers in Chicago, 1919–1939*, 2d ed. (New York: Cambridge University Press, 2008), 292–359; Michael Dennis, *The Memorial Day Massacre and the Movement for Industrial Democracy* (New York: Palgrave Macmillan, 2010); David M. Kennedy, *Freedom from Fear: The American People in Depression and War, 1929–1945* (New York: Oxford University Press, 1999), 308–19; Nelson Lichtenstein, *Walter Reuther: The Most Dangerous Man in Detroit* (Urbana: University of Illinois Press, 1997), 63–87, 109–11; Jack Metzger, *Striking Steel, Solidarity Remembered* (Philadelphia: Temple University Press, 2000), 23–30.
21. Drinker to Lloyd, 28 January 1938; Lloyd to Drinker, 1 February 1938; Yule to Lloyd, 9 February 1938; Lloyd to Drinker, 4 August 1938, BI 3–3; Drinker to Lloyd, 9 August 1938 (all found in BI 3–3).
22. Drinker to Lloyd, 3 November 1945, LCR 10–4; Lloyd to Davis, 13 December 1945, LCL 18–1; Drinker to Lloyd, cable, 21 January 1946, LCR 10–4; Lloyd to McIver, 23 January 1946, LCL 19–2; Drinker to Lloyd, 24 January 1946, LCR 10–4; Lease, draft, 1 May 1946, LCR 10–4; Von Hagen to Lloyd, 11 March 1947, LCL 20–2 (quoted); Yule to State Board of Control, 20 January 1949, LCL 22–3.
23. Lloyd Corp., Minutes of Directors Meeting, 29 March 1937, LCL 10–6; Lloyd to Archibald, 8 June 1937, LCL 10–5; Lloyd to Drinker, 20 August 1940, LCL 13–2; FHA, *Seventh Annual Report* (Washington, DC: GPO, 1941), 89.

24. Architectural Resources Group, *Garden Apartments of Los Angeles: Historical Context Statement*, October 2012, 3–15; Miles L. Colean, "Multiple Housing under FHA," *AR* 84 (September 1938): 96.
25. Richard W. Bartke, "The Federal Housing Administration," *Wayne Law Review* 13 (1966): 651–2, 660; Catherine Bauer, "Planned Large-Scale Housing: Advance Sheet of Progress," *AF* 74 (May 1941): 93; FHA, *The FHA Story in Summary, 1934–1959* (Washington, DC: GPO, 1959), 5–14; Leo Grebler, David M. Blank, and Louis Winnick, "Long-Term Changes in Cost and Terms of Mortgage Financing," in *Capital Formation in Residential Real Estate: Trends and Prospects*, ed. Grebler, Blank, and Winnick (Princeton, NJ: Princeton University Press, 1956), 228; Kenneth T. Jackson, *Crabgrass Frontier: The Suburbanization of the United States* (New York: Oxford University Press, 1985), 203–5.
26. Colean, "Multiple Housing under FHA," 96.
27. FHA, *Rental Housing as Investment* (Washington, DC: FHA, 1938), 30, quoted in Jackson, *Crabgrass Frontier*, 208.
28. "Garden Apartments," *AF* 72 (May 1940): 309; Bauer, "Planned Large-Scale Housing," 93.
29. Colean, "Multiple Housing under FHA," 97–109, 104 (quoted); FHA, *Property Standards: Requirements for Mortgage Insurance under Title II of the National Housing Act*, circular No. 2 (Washington, DC: GPO, 1936).
30. Andrew H. Whittemore, "How the Federal Government Zoned America: The Federal Housing Administration and Zoning," *Journal of Urban History* 39 (July 2013): 621–4.
31. Drinker to Lloyd, 12 May 1938, LCL 14–3.
32. FHA, *Seventh Annual Report*, 96.
33. Yule to Lloyd, 18 July 1938; Yule to Ertz, 19 July 1938; Drinker to Lloyd, 3 October 1938 (all found in LCL 14–3).
34. Yule to Lloyd, 18 July 1938; Yule to Ertz, 19 July 1938; Drinker to Lloyd, 3 October 1938; Yule to Drinker, 27 October 1938; Drinker to Yule, 28 October 1938 (all found in LCL 14–3); Lloyd to Drinker, 20 August 1940, LCL 13–2. The typical three-room unit consisted of a living room, bedroom, combined kitchen and dining alcove, and bathroom. The three-and-a-half room unit either separated the dining alcove from the kitchen or added a small bedroom. The four-room unit was similar to the three-room unit in plan but added a second bedroom. The four-and-a-half-room unit was similar to the three-and-a-half-room unit in plan but added a second bedroom. A five-room unit offered two bedrooms and a dining room fully separated from the kitchen. The FHA did not include bathrooms in its room count. Walk-up projects

insured by the FHA averaged 3.7 rooms (FHA, *Seventh Annual Report*, 95–96).

35. Drinker to Lloyd, 18 January 1939; Lloyd to Dick, 2 August 1939; Platt to Lloyd, 9 August 1939; Drinker to Lloyd, 9 August 1939 (all found in LCL 14–3).
36. FHA, *Property Standards*, 10.
37. "Garden Apartments," 310.
38. Lloyd to Hubbard, Westervelt & Mottelay, 17 October 1938, LCL 11–4; Drinker to Lloyd, 18 January 1939, LCL 14–3; "Garden Apartments," 318–19; FHA, *Fifth Annual Report* (Washington, DC: GPO, 1939), 127.
39. Yule to Lloyd, 27 February 1939; Drinker to Lloyd, 21 March 1939; Lloyd to Dick, 2 August 1939 (quoted); Platt to Lloyd, 9 August 1939; Drinker to Lloyd, 9 August 1939 (all found in LCL 14–3); Colean, "Multiple Housing under FHA," 100–4.
40. Yule to Lloyd, 18 July 1938, LCL 14–3; Hubbard, Westervelt & Mottelay to Lloyd, 24 October 1938, LCL 11–4; Lloyd to Hubbard, Westervelt & Mottelay, 1 November 1938, LCL 11–4; Drinker to Lloyd, 7 December 1938; Lloyd to Drinker, 12 December 1938; Drinker to Lloyd, 18 January 1939; Drinker to Lloyd, 1 August 1939; Lloyd to Drinker, 3 August 1939; Lloyd to Platt, 3 August 1939 (all found in LCL 14–3).
41. Platt to Lloyd, 18 July 1939, LCL 14–3; Lloyd to Platt, 3 August 1939, LCL 14–3.
42. Drinker to Lloyd, 1 August 1939; Dick to Lloyd, cable, 1 August 1939; McIntyre to Lloyd, cable, 1 August 1939; Lloyd to Dick, 2 August 1939; Lloyd to McIntyre, 2 August 1939; McIntyre to Lloyd, 4 August 1939 (quoted); Platt to Lloyd, 9 August 1939 (all found in LCL 14–3).
43. Platt to Lloyd, 9 August 1939, (quoted); Drinker to Lloyd, 9 August 1939; Lloyd to McGovern, cable, 10 August 1939; McGovern to Lloyd, cable, 11 August 1939; Lloyd to McGovern, cable, 14 August 1939; McGovern to Lloyd, 14 August 1939 (all found in LCL 14–3).
44. Platt to Lloyd, 18 July 1939; Colton to Snyder, 28 August 1939; Snyder to Von Hagen, 12 September 1939; Von Hagen to Snyder, 14 September 1939; Von Hagen to Snyder, 9 October 1939 (quoted); Von Hagen to Drinker, 11 October 1939; Von Hagen to Platt, 11 October 1939; Von Hagen to Drinker, 23 October 1939; Von Hagen to Snyder, 23 October 1939 (all found in LCL 14–3).
45. McGovern to Johnson, 29 April 1940, PVA 8–3; McGovern to Lloyd Corp., 14 May 1940, PVA 8–3 (quoted).

46. Ertz & Burns to Johnson, 17 May 1940, enclosure, Burns to Lloyd Corp., 17 May 1940; Lloyd to McGovern, cable, 22 May 1940; Lloyd to McGovern, 23 May 1940; McGovern to Lloyd Corp., 29 May 1940; McGovern to Powell, 29 May 1940; McGovern to Lloyd Corp., 13 June 1940; McGovern to Lloyd Corp., 17 June 1940; McGovern to Johnson, 25 June 1940; McGovern to Johnson, 27 June 1940; Ertz & Burns to Johnson, 2 July 1940 (all found in PVA 8–3).
47. Lloyd to Hoffman, cable, 10 May 1940, LCL 14–3; Lloyd to Burns, 7 August 1940, LCL 13–1; Lloyd Corp., press release, attached to Lloyd to Drinker, 20 August 1940, LCL 13–2 (quoted); Lloyd to Drinker, cable, 28 August 1940, LCL 14–3; Yule to Lloyd and Von Hagen, 16 July 1945, LCL 18–2; "Lloyd Apartment Project Started," *Oregon Journal*, 1 September 1940, II:1; FHA, *Seventh Annual Report*, 88–9.
48. In Ralph Lloyd's lifetime: while the Lloyd Center may be attributed to his vision, it would not break ground until five years after his death.
49. Karl Boyd Brooks, *Public Power, Private Dams: The Hell's Canyon High Dam Controversy* (Seattle: University of Washington Press, 2006), 9–11, 39–41; Philip J. Funigiello, *Toward a National Power Policy: The New Deal and the Electric Utility Industry, 1933–1941* (Pittsburgh, PA: University of Pittsburgh Press, 1973), 216; Richard Lowitt, *The New Deal in the West* (Bloomington: University of Indiana Press, 1984), 139–63; Richard White, *"It's Your Misfortune and None of My Own": A New History of the American West* (Norman: University of Oklahoma Press, 1991), 483–7; Richard White, *The Organic Machine: The Remaking of the Columbia River* (New York: Hill & Wang, 1995), 64–72.
50. Brooks, *Public Power, Private Dams*, 33 (quoted), 34–5; Gerald D. Nash, *World War II and the West: Reshaping the Economy* (Lincoln: University of Nebraska Press, 1990), 27–66; White, *The Organic Machine*, 72–74.
51. Drinker to Lloyd, 15 June 1938; Lloyd Corp., Portland office, to Lloyd Corp., Los Angeles office, cable, 22 June 1938 (quoted); Lloyd Corp., Portland office, to Lloyd Corp., Los Angeles office, cable, 22 June 1938; Lloyd Corp., Portland office, to Lloyd Corp., Los Angeles office, cable, 25 June 1938; Lloyd Corp., Portland office, to Lloyd Corp., Los Angeles office, cable, 25 June 1938; Lloyd Corp., Portland office, to Lloyd Corp., Los Angeles office, cable, 12 July 1938; Lease, 1 August 1938; Drinker to Lloyd, 2 August 1938; Statement of Fair Market Value, n.d. [1938]; Stewart to Drinker, 16 September 1938 (all found in LCR 8–2); Lloyd to Ickes, 12 October 1944, Port 3–4 (quoted).
52. Drinker to Lloyd, 21 November 1939, LCR 8–3.

53. Drinker to Lloyd, 21 November 1939, LCR 8–3; Drinker to Lloyd, 25 November 1939, LCR 8–3.
54. Drinker to Lloyd, cable, 24 November 1939, LCL 12–3; Drinker to Lloyd, 25 November 1939, LCR 8–3 (quoted); Lloyd to Drinker, 5 February 1940, LCL 13–2.
55. Burns to Lloyd Corp., Portland office, 25 November 1939, LCR 8–3; Lease, 5 April 1940, LCR 8–3; Drinker to Lloyd, 10 April 1940, LCR 3–4; Lloyd to Stewart, 14 May 1940, LCR 8–3; Stewart to Lloyd, 5 July 1940. LCR 8–3; Lloyd to Drinker, 9 July 1940, LCR 3–4; Beelman to Von Hagen, 30 July 1946, LCL 19–1.
56. Drinker to Von Hagen, 21 January 1943, LCL 16–2.
57. The leases on the administration building at 811 NE Oregon Street and Bonneville Engineering Building No. 2 expired in 1943. Ralph Lloyd insisted on capitalizing the cost of servicing the buildings in a new lease. (Maintenance costs had increased because of the need to pay overtime to existing maintenance workers; mobilization had created a labor shortage that made hiring new workers all but impossible for the Portland office.) The Office of Price Administration resolved the issue when it decided that services provided under BPA leases were exempt under the Emergency Price Control Act of 1942. On 1 July 1943, BPA signed a two-year lease for both buildings with an expiration date that coincided with the expiration of the lease on Bonneville Engineering Building No. 1. With lease in hand, Franz Drinker outsourced building maintenance to a local contractor (Drinker to Lloyd, 20 May 1943; Drinker to Lloyd, 12 June 1943; Drinker to Lloyd, 21 June 1943; Lloyd to Drinker, 23 June 1943; Veness to Drinker, 20 July 1943; Drinker to Lloyd, 20 July 1943 [all found in LCR 3–4]; Yule to Stewart, 15 January 1946, LCL 19–4).
58. Bradeen to Lloyd Corp., 16 April 1954, LCR 12–1; Brooks, *Public Power, Private Dams*, 39–59.
59. Paul F. Ewing, "Sparkling City View Features Controversial Lloyd, BPA Buildings," *Oregonian*, 14 March 1954.
60. Lloyd to Ickes, 12 October 1944, Port 3–4.
61. Lloyd to Raver, 28 September 1944, Port 3–4 (quoted); Lloyd Corp. Board of Directors, resolution, n.d., attachment to Lloyd to Ickes, 12 October 1944, Port 3–4.
62. Lloyd to Ickes, 12 October 1944, Port 3–4 (quoted); Lloyd to Ickes, 13 November 1944, Port 3–4; Drinker to Lloyd, 1 March 1945, LCL 18–3; Ewing, "Sparkling City View Features Controversial Lloyd, BPA Buildings."

63. Lloyd to Drinker, 16 November 1944, Port 3–4; Drinker to Lloyd, 15 December 1944, Port 3–4; Drinker to Lloyd, 1 March 1945, LCL 18–3; Drinker to Lloyd, 26 April 1945, LCL 18–3.
64. Lloyd to Drinker, 9 May 1945 (quoted), LCL 18–3; Lloyd to Drinker, 4 October 1945, LCL 18–3; Drinker to Lloyd, 30 July 1947, LCR 10–4. To ensure that they retained the office space in which their employees were working presently as they discussed plans for additional office space, BPA officials signed a five-year lease, covering the administration building and both engineering buildings, for a total of $4,876 in monthly rental payments (Lease, 1 July 1945, LCR 8–2).
65. Lloyd to Davis, 13 December 1945, LCL 18–1; Lloyd to McIver, 23 January 1946, LCL 19–2; Von Hagen to Lloyd, 25 March 1946; Statement of Fair Value, "Bonneville West Building," undated; Statement of Fair Value, "Bonneville East Building," undated; Builder's Contract (2), 1 July 1946, LCL 19–1; Drinker to Lloyd, 30 July 1947 (all found in LCR 10–4). The difference in size between buildings with similar footprints was accounted for in part by the inclusion of a basement in the Bonneville West Building.
66. Von Hagen to Beelman, 7 May 1946, LCL 19–1; Beelman to Von Hagen, 30 July 1946, LCL 19–1; Lloyd to McIver, 27 July 1947, LCL 20–1. The Bonneville East and Bonneville West Buildings cost $252,000 and $495,000 to build, respectively.
67. King, *An Architectural Guide to Portland*, 256.
68. Drinker to Ertz, 10 April 1947, LCL 20–1 (quoted); Lloyd to McIver, 27 July 1947, LCL 20–1; Drinker to Lloyd, 30 July 1947, LCR 10–4; Statement of Fair Value, "Bonneville West Building," undated, LCR 10–4; Statement of Fair Value, "Bonneville East Building," undated, LCR 10–4.
69. Under separate five-year leases, BPA paid $12,202 per month in total to rent the Bonneville East and Bonneville West Buildings (Lease, "Bonneville West Building," 15 July 1947; Lease, "Bonneville East Building," 16 October 1947; Statement of Fair Value, "Bonneville West Building," undated; Statement of Fair Value, "Bonneville East Building," undated [all found in LCR 10–4]).
70. Beginning 1 November 1947, Pacific Telephone & Telegraph Company leased former Bonneville Engineering Building Nos. 1 and 2 for $2,100 per month for ten years (Lease, 3 October 1947, LCL 20–3).
71. Paul J. Raver to the Secretary of the Interior, letter of transmittal, 1 December 1947, in U.S. Department of the Interior, *1947 Report on the Columbia Power System* (Washington, DC: GPO, 1948).

72. Raver to Drinker, 17 October 1947, LCR 10–6.
73. Drinker to Lloyd, 12 February 1947, LCR 10–6.
74. Lloyd to Drinker, 21 February 1947, LCR 10–6.
75. Abbott, *Portland*, 147–58, 206–22. On postwar concern regarding the obsolescence of downtown generally, see Fogelson, *Downtown*, 381–94; Alison Isenberg, *Downtown America: A History of the Place and the People Who Made It* (Chicago: University of Chicago Press, 2004), 166–83.
76. "Portland Recast in Larger Mold," *Oregon Journal*, 7 November 1946, enclosure, Drinker to Lloyd, 12 November 1946, LCL 19–3.
77. Von Hagen to Lloyd, 11 March 1947, LCL 20–2; Drinker to Lloyd, 9 May 1947, LCR 10–6; Drinker to Von Hagen, 26 May 1947, LCL 20–3; Lloyd to Drinker, 28 May 1947, LCR 10–6; Drinker to Von Hagen, 3 July 1947, LCR 10–6; Lloyd to Drinker, 25 September 1947, LCL 20–3 (quoted); Drinker to Lloyd, 24 October 1947, LCR 10–6.
78. Raver to Drinker, 17 October 1947, LCR 10–6.
79. Drinker to Raver, 10 November 1947, LCR 10–6.
80. Earlier in 1949, General Petroleum occupied eight floors of its height-limit building at 612 South Flower Street in Los Angeles. Designed by Walter Wurdeman and Welton Becket, the Late Moderne structure featured a ceramic veneer exterior and aluminum fin sunshades from the third to the thirteenth floor. Upon its completion, it was the largest office building in Southern California. Together with the Mirror and Prudential Buildings, the General Petroleum Building broke a fifteen-year lull in the construction of tall office buildings in Los Angeles ("Oil Firm Dedicates 13-Story Building," *LAT*, 2 April 1949, I:5; "History of Office Buildings in Los Angeles Recorded").
81. Drinker to Lloyd 19 August 1948; Lease, "Bonneville North Building," 16 October 1949; Statement of Fair Market Value, "Bonneville North Building," undated (all found in LCR 10–6). General Petroleum moved into its building under a 15-year lease agreement. It agreed to lease two-thirds of the building for $2,500 per month, with an option to lease the remaining space at $1.50 per square foot. The shorter lease period (over GP's willingness initially to consider a 25-year lease) may have owed in part to the fact that Ralph Lloyd did not obtain the terms that he had sought: two-thirds of the building for $3,750 per month, with an option to lease the remaining space at $2.50 per square foot (Lloyd to Clark, 3 January 1949, with enclosure, Lease, "General Petroleum Building," draft, 31 December 1948; Lease, "General Petroleum Building," 17 June 1949 [all found in LCR 10–6]).

82. Bradeen to Von Hagen, 3 May 1951, LCR 12–1 (quoted); Bessey to Drinker, 3 May 1951, LCR 12–1; Doyle to Drinker, 4 May 1951, LCR 12–1; Searles to Fowler, 31 January 1952, LCR 12–1; Bradeen to Lloyd Corp., 16 April 1954, LCR 12–1; Lawrence Barber, "Foreign Exports, River Traffic Decline, Other Shipping Holds About Average," *Oregonian*, 31 December 1950, I:11; "Zone Shifted for Building," *Oregonian*, 9 May 1951; Ewing, "Sparkling City View Features Controversial Lloyd, BPA Buildings."
83. Lloyd Corp., "Proposed Changes for Bonneville Power Administration," memorandum, 24 February 1950, LCR 8–2; Lloyd Corp. to GSA, 14 November 1950, LCL 23–4; Bradeen to Lloyd Corp., 16 April 1954, LCR 12–1 (quoted); Lease, 20 March 1956, LCR 10–4; "Zone Shifted for Building."
84. Lloyd Corp., Proposed Changes for Bonneville Power Administration, memorandum, 24 February 1950, LCR 8–2; Yule to Lloyd and Von Hagen, 10 October 1950, LCL 23–2; Lloyd Corp to Public Buildings Service, GSA, 14 November 1950, LCL 23–4; Wright to Von Hagen, 16 April 1951, LCR 12–2; Bessey to Drinker, 3 May 1951, LCR 12–1; NPA to Lloyd Corp., Portland office, 25 June 1951, LCR 12–1; Agreement, GSA and Lloyd Corp., 12 April 1951, LCR 12–1; Elmer E. Gunnette quoted in "High, 50' Wide and Handsome," *Pacific Architect and Builder* 61 (June 1955): 10.
85. Lloyd Corp., Petition [for vacation of NE 10th Avenue], 8 December 1950; Drinker to Von Hagen, 21 March 1951; Lloyd Corp., Petition [for change of zone from II to III], 21 March 1951; Drinker to Von Hagen, 24 April 1951; City of Portland, Ordinance 94209, 17 May 1951; Deed from Lloyd Corp. to the City of Portland, 26 June 1951; City of Portland, Ordinance 94076, 13 September 1951 (all found in LCR 12–1).
86. DePuy to Von Hagen, 29 August 1952, with attachment, Agreement, 28 August 1952, LCR 12–2.
87. "High, 50' Wide and Handsome"; "Welded Building, Two Bays Wide, Designed to Resist Seismic Loads," *Engineering News-Record*, 13 October 1953, 30–31; "Welded Steel Construction Puts Interior Bldg. in Spotlight," *Pacific Architect and Builder* 61 (June 1955): 12.
88. Bradeen to Lloyd Corp., 16 April 1954, LCR 12–1.
89. Brooks, *Public Power, Private Dams*, 176–216.
90. Lloyd Corp. to GSA, 20 April 1954, LCR 12–1; Drinker to Von Hagen, 9 July 1954, LCR 12–1; "Government Agencies Now Located in Old Bonneville Buildings," undated, LCR 10–4. Half of the basement housed BPA's reproduction department; the other half was devoted to parking. It is unclear if Lloyd

Corporation properties were able to accommodate all of the Interior Department's Portland employees upon completion of the building. At the time, five agencies occupied some 72,000 square feet of space at Swan Island. Only about 35,000 square feet was available in the new building. According to the Portland Chamber of Commerce, GSA officials planned to convert space in the basement to office use to make up the difference (Clark to Condon, 26 March 1954, LCR 12–2).

91. Yule to Von Hagen, 11 January 1960, LCR 12–1.
92. Wright to Von Hagen, 16 April 1951, LCR 12–2.
93. Bradeen to Lloyd Corp., 16 April 1954, LCR 12–1.
94. Lloyd to Shafer, 3 December 1947, LCR 3–4; Lloyd to Rowold, 1 April 1948, LCR 11–1; Gailliland to Yule, 28 February 1949, LCR 3–4; "Mack Truck Internal Building," worksheet, 9 September 1963, LCR 11–1.
95. From 1936 to 1950, the value of buildings and improvements on the Lloyd Corporation's balance sheet increased from $1,278,971 to $5,957,760 (Lloyd Corp. to Security-First National Bank of Los Angeles, 25 May 1936, LCL 10–4; Lloyd Corporation, Ltd., Statement of Assets and Liabilities, 31 December 1950, LCR 12–2).

CHAPTER 8. THE SUBURBANIZATION OF URBAN SPACE: THE LLOYD CENTER

1. Richard Longstreth, *The American Department Store Transformed, 1920–1960* (New Haven, CT: Yale University Press, 2010), 242 (quoted), 241 (quoted).
2. Victor Gruen quoted in "Cities in Trouble—What Can Be Done," *U.S. News & World Report*, 20 June 1960, 86; M. Jeffrey Hardwick, *Mall Maker: Victor Gruen, Architect of an American Dream* (Philadelphia: University of Pennsylvania Press, 2004), 162–209; Alex Wall, *Victor Gruen: From Urban Shop to New City* (Barcelona: Actar, 2005), 116–97. On the postwar crisis facing America's downtown districts and the efforts of municipal leaders to address it, see, for instance, Lisabeth Cohen, "Buying into Downtown Revival: The Centrality of Retail to Postwar Urban Renewal in America Cities," *Annals of the American Academy of Political and Social Science* 611:1 (2007): 82–95; Isenberg, *Downtown America*, 166–202.
3. Victor Gruen, "Dynamic Planning for Retail Areas," *Harvard Business Review* 32 (November-December 1954): 53–62.
4. Hardwick, *Mall Maker*, 162–209; "Master Plan for Revitalizing Ft. Worth's Central Core," *Business Week*, 17 March 1956, 70–74; Wall, *Victor Gruen*, 136–57, 198–222.

5. Lloyd Corp., Lloyd Center Information Kit, Lloyd Center General Release, undated [March 1960], LC 10–3.
6. Richard G. Horn, "Trends from Regional Center to 'Complete City,'" *Urban Land* 21 (September 1962): 9–10.
7. "BPA's New Home Part of Lloyd Center," *Pacific Architect and Builder* 61 (June 1955): 12.
8. Larry Smith & Co., "Lloyd's in Portland," 10 January 1951, LC 12–1.
9. Smith to Lloyd, 2 August 1950, LC 12–3.
10. Larry Smith & Co., "Lloyd's in Portland," 10 January 1951, LC 12–1.
11. On the overwhelmingly suburban location of first-generation regional shopping centers, see Lizabeth Cohen, "From Town Center to Shopping Center: The Reconfiguration of Community Marketplaces in Postwar America," *American Historical Review* 101 (October 1996): 1050–81; Howard Gillette Jr., "The Evolution of the Planned Shopping Center in Suburb and City," *Journal of the American Planning Association* 51 (Autumn 1985): 449–60; Hardwick, *Mall Maker*, 91–161; Kenneth T. Jackson, "All the World's a Mall: Reflections on the Social and Economic Consequences of the American Shopping Center," *American Historical Review* 101 (October 1996): 1113–4.
12. Lloyd to Hubbard, Westervelt & Mottelay, 16 June 1937, LCL 10–6 (quoted); Lloyd to Hubbard, Westervelt & Mottelay, 8 March 1938, LCL 11–4 (quoted).
13. Lloyd to Drinker, 10 January 1945, LCL 18–3; Lloyd to Winnett and Arnett, 18 January 1945, LCL 18–1 (quoted). Both Bullock's and I. Magnin, a high-end retailer, were rumored to be considering downtown Portland locations for stores (Drinker to Lloyd, 6 January 1945, LCL 18–3).
14. See, for instance, Lloyd to Frank, 20 April 1948, LCL 21–3; Lloyd to Hoppe, 9 December 1949, LCL 22–2.
15. Lloyd to McIver, 23 January 1946, LCL 19–2.
16. Lloyd to Frank, 5 December 1947, LCL 20–2; Lloyd to Aaron Frank, 20 April 1948, LCL 21–3; Lloyd to Frank, 16 September 1949, LCL 22–3; Lloyd to Adams, 12 October 1949, LCL 22–3; Von Hagen to Barnett, 7 December 1956, LC 10–5; Smith to Barnett, 7 December 1956, LC 10–5; King, *Portland*, 22. Under B. Earle Puckett, who reorganized the former Hahn Department Stores in 1934, Allied Stores added more than a dozen stores to a growing empire by 1947, when it reaped some $20 million in net profit (Longstreth, *The American Department Store Transformed*, 50).
17. Lloyd to Statler, Hennessy, and Douglas, 31 March 1948, LC 8–5 (quoted); Irwin to Lloyd, 26 October 1949, LC 8–4; Irwin to Lloyd, 8 December 1949, LC 8–4; Irwin to Lloyd, 15 December 1949, LC 8–4; Lloyd to Executive

Department, Statler Hotels, 7 August 1950, LC 8–5; Douglas to Lloyd, 16 August 1950, LC 8–5; Lloyd to Douglas, 19 September 1950, LC 8–5; Larry Smith & Co., "Second Interim Report on the Lloyd Properties," October 1950, LC 12–1.

18. Indeed, controls imposed on raw materials in conjunction with defense mobilization delayed the start of construction of the U.S. Department of the Interior Building (Drinker to Von Hagen, 28 June 1951, with enclosure, NPA to Lloyd Corp., 25 June 1951; Freiburg to Graham, 5 September 1951; NPA to Lloyd Corp., 10 September 1951; Olverson to Lloyd Corp., Portland office, 6 November 1951; Drinker to NPA, 20 December 1951; Lloyd Corp. to NPA, 24 January 1952; Searles to Fowler, 31 January 1952; Fowler to Searles, 6 February 1952; NPA, Authorized Construction and Allotment of Controlled Materials, 29 February 1952 [all found in LCR 12–1]). On the Controlled Materials Plan implemented by the Truman Administration on 1 July 1951 to allocate aluminum, copper, and steel according to requirements and production schedules developed by defense mobilization planners, see Elliot V. Converse III, *Rearming for the Cold War, 1945–1960* (Washington, DC: Historical Office, Office of the Secretary of Defense, 2012), 96–7; Robert D. Cuff, "Organizational Capabilities and U.S. War Production: The Controlled Materials Plan of World War II," *Business and Economic History* 2 (1990): 105–8; Paul G. Pierpaoli, *Truman and Korea: The Political Culture of the Early Cold War* (Columbia: University of Missouri Press, 1999), 120; William Stueck, *The Korean War: An International History* (Princeton, NJ: Princeton University Press, 1987), 199–200.
19. Lloyd to Bergey, 8 December 1950, LCL 23–3 (quoted); Lloyd to Carr, 8 December 1950, LCL 23–4; Lloyd to Holman, 8 December 1950, LCL 23–2; Lloyd to Pedersen, 8 December 1950, LCL 23–3.
20. Lloyd to Executive Department, Statler Hotels, 7 August 1950, LC 8–5.
21. "A Report on the Work and Organization of John Graham & Co., Architects and Engineers, Seattle, Washington, 1950," LC 7–2; Meredith L. Clausen, "Northgate Regional Shopping Center—Paradigm from the Provinces," *Journal of the Society of Architectural Historians* 43 (May 1984): 144–61. On the development of the local shopping center in Los Angeles, see Longstreth, *The Drive-In, The Supermarket, and the Transformation of Commercial Space in Los Angeles*, 32–75. On the interwar emergence of the branch department store in Los Angeles, see Longstreth, *City Center to Regional Mall*, 81–141. On Northgate Shopping Center, see, also, Arthur W. Priaulx, "Northgate—Suburban Shopping Center, Seattle, Washington," *Architect & Engineer* 182

(September 1950): 14–21; "Shopping Centers that Offer New Ideas," *Urban Land* 10 (December 1951): 1, 3; "The Architect's Place in the Suburban Retail District," *AF* 93 (August 1950): 116.

22. "A Report on the Work and Organization of John Graham & Co., Architects and Engineers, Seattle, Washington, 1950," LC 7–2; Clausen, "Northgate Regional Shopping Center," 150–5; "Shopping Centers that Offer New Ideas," 3 (quoted).
23. Clausen, "Northgate Regional Shopping Center," 157; "Shopping Centers that Offer New Ideas," 3.
24. "Shopping Centers that Offer New Ideas," 3 (quoted); "A Report on the Work and Organization of John Graham & Co., Architects and Engineers, Seattle, Washington, 1950," LC 7–2 (quoted); Clausen, "Northgate Regional Shopping Center," 153–7. On Broadway-Crenshaw Center, see Longstreth, *City Center to Regional Mall*, 227–38.
25. Clausen, "Northgate Regional Shopping Center," 150, 159; Longstreth, *City Center to Regional Mall*, 232–3.
26. Graham to Lloyd, 8 May 1950, LC 7–2; Smith to Lloyd, 27 June 1950, LC 12–3; Smith to Lloyd, 2 August 1950, LC 12–3; Larry Smith & Co., "Second Interim Report on the Lloyd Properties," October 1950, LC 12–1; David W. Yule, timetable, undated, in folder, "Lloyd Center Financing: Prudential Ins. Co.—Preliminary Papers," LC 10–4; Von Hagen to Barnett, 7 December 1956, LC 10–5; Smith to Barnett, 7 December 1956, LC 10–5. Smith established his real estate consultancy in 1939 ("He's Built More Than 75 Malls," *Syracuse Post-Standard*, 25 March 1976, special section; Hardwick, *Mall Maker*, 114–6).
27. Larry Smith & Co., "Second Interim Report on the Lloyd Properties," October 1950, LC 12–1.
28. Smith to Lloyd, 25 October 1950, LC 12–3. Ralph Lloyd had written to John Graham on 18 October (letter not found).
29. Larry Smith & Co., "Lloyd Center, Portland," undated, LC 12–2; Larry Smith & Co., "Lloyd's in Portland," 10 January 1951, LC 12–1; Larry Smith & Co., "Lloyd's in Portland: Supplement to General Report of January 10, 1951," 22 January 1951, LC 12–2 (quoted). On Portland's efforts at downtown renewal during this period, see Abbott, *Portland*, 149–64.
30. Larry Smith & Co., "Memorandum Concerning the Opportunity for a Hotel at Lloyd Center, Portland," undated [1951], LC 12–2.
31. Abbott, *Portland*, 151–3.
32. Smith to Lloyd, 25 October 1950, LC 12–3 (quoted); Larry Smith & Co., "Second Interim Report on the Lloyd Properties," October 1950, LC 12–1;

Larry Smith & Co., "Lloyd's in Portland," 10 January 1951, LC 12–1, 24 (quoted).

33. Graham to Lloyd, 8 May 1950, LC 7–2 (quoted); Richard R. Von Hagen, "Résumé of Trip to Portland," 31 July 1951, LCL 24–2; Larry Smith & Co. to Von Hagen, 30 August 1951, LCL 24–4; David W. Yule, timetable, undated, in folder, "Lloyd Center Financing: Prudential Ins. Co.—Preliminary Papers," LC 10–4.
34. Smith to Von Hagen, with attachment, "Outline of Proposal to Allied," 26 May 1952, LCL 25–4 (quoted); Crabb to Meek, 9 June 1952, LC 12–3.
35. Larry Smith & Co. to Von Hagen, with attachment, "Memorandum for Allied Stores Corporation, New York," 28 August 1952, LCL 25–4.
36. Rebber to Lloyd, 15 January 1953, LCL 26–2.
37. Rebber to Lloyd, 9 February 1953, LCL 26–2; Rebber to Lloyd, 12 February 1953, LCL 26–2; Crabb to Von Hagen, 20 February 1953, with attachment, "Portland—Lloyd: Work Schedule—Tentative," LC 12–3; Rebber to Lloyd, 5 March 1953, LCL 26–2; Robert J. Crabb, "Memorandum Re Portland-Lloyd," 4 March 1953, LC 9–4; Rebber to Lloyd, 20 March 1953, LCL 26–2; David W. Yule, timetable, undated, in folder, "Lloyd Center Financing: Prudential Ins. Co.—Preliminary Papers," LC 10–4.
38. Smith to Von Hagen, 11 January 1956, LC 12–3; Smith to Beggs, 18 February 1956, LC 12–3.
39. David W. Yule, timetable, undated, in folder, "Lloyd Center Financing: Prudential Ins. Co.—Preliminary Papers," LC 10–4.
40. In 1948, the Lloyd Corporation constructed a three-story office building on the northeast corner of Olympic Boulevard and Beverly Drive in Beverly Hills (9441 Olympic Boulevard), which served as the company's headquarters upon its completion (Lloyd Corp. to Rohkam, 2 November 1948, LCR 7–4; Cribbs to Von Hagen, 22 July 1986, LCR 7–3).
41. Von Hagen to Moore, 1 March 1954, LCL 27–3; Robert J. Crabb, "Memorandum Re Portland-Lloyd," 13 April 1954, LC 9–4; Von Hagen to Barnett, 7 December 1956, LC 10–5; Smith to Barnett, 7 December 1956, LC 10–5 (quoted); David W. Yule, timetable, undated, in folder, "Lloyd Center Financing: Prudential Ins. Co.—Preliminary Papers," LC 10–4.
42. Robert J. Crabb, "Memorandum Re Lloyd Center," 14 December 1954, LC 9–4; Minutes of General Meeting, 24 February 1955, LC 9–4; Richard G. Horn, Minutes of Staff Meeting, 6 and 7 April 1955, LC 9–4; Horn to Von Hagen, 20 June 1955, LCL 28–2; Smith to Beggs, 18 February 1956, LC 12–3; Von Hagen to Barnett, 7 December 1956, LC 10–5; Smith to Barnett, 7

December 1956, LC 10–5 (quoted); David W. Yule, timetable, undated, in folder, "Lloyd Center Financing: Prudential Ins. Co.—Preliminary Papers," LC 10–4.

43. Horn to Von Hagen, 20 June 1955, LCL 28–2; Smith to Beggs, 18 February 1956, LC 12–3; "Table," 4 April 1956, LC 10–4; Von Hagen to Barnett, 7 December 1956, LC 10–5; Smith to Barnett, 7 December 1956, LC 10–5; David W. Yule, timetable, undated, in folder, "Lloyd Center Financing: Prudential Ins. Co.—Preliminary Papers," LC 10–4; "M&F Gets," *Sunday Oregonian*, 29 July 1956, I:1. In September 1953, a quarter century after Bullock's Wilshire opened in Los Angeles, Meier & Frank announced the "first decentralization by a major downtown department store" in Portland with its plan to open a branch store some eight miles to the northeast of Holladay Park. It would anchor a shopping center at NE Sandy Boulevard and NE 122nd Avenue that would offer consumers clothing, shoe, and variety stores, banks, supermarkets, and restaurants (Horn to Von Hagen, 11 August 1954, LC 9–4; "Meier & Frank Acquires 50-Acres Parkrose Site for Shopping Center," *Journal of Commerce*, 2 September 1953; "Meier & Frank to Build at NE Sandy and 122d," *Oregonian*, 2 September 1953 [quoted]).
44. Horn to Fast, 30 July 1956, LC 12–3 (quoted).
45. Von Hagen to Barnett, 7 December 1956, LC 10–5.
46. Leland L. Rebber, "Preliminary Plans—Lloyd Center," 5 June 1953, LCL 26–2.
47. Robert J. Crabb, "Memorandum Re Portland-Lloyd," 15 April 1953, LC 9–4; Yule to Lloyd, 13 May 1953, LCL 26–2; Robert J. Crabb, "Memorandum Re Portland-Lloyd," 12 May 1953, LC 9–4; Rebber to Lloyd, 7 August 1953, LCL 26–2; G. R. Cysewski, "Portland's Lloyd Center," *Traffic Engineering* 29 (December 1958): 11–12.
48. Rebber to Lloyd, 5 June 1953, LCL 26–2.
49. Rebber to Lloyd, 19 February 1953; Rebber to Lloyd, 20 March 1953 (quoted); Rebber to Lloyd, 17 April 1953; Rebber to Lloyd, 24 April 1953 (quoted); Rebber to Lloyd, 5 June 1953 (all found in LCL 26–2).
50. Richard G. Horn and W. Joseph McFarland, "Forum and Seminar, Portland Chapter, American Institute of Banking," undated, LC 10–3.
51. Robert J. Crabb, "Memorandum Re Portland-Lloyd," 4 March 1953, LC 9–4.
52. Through August 1955, Lloyd Corporation would spend more than $2.5 million to acquire several dozen residential properties that lay north of NE Multnomah Street (Yule to Janeck, 24 January 1957, LC 10–5).
53. Drinker to Von Hagen, 10 September 1954, LC 12–5; Bartholomew to Graham, 26 October 1954, LC 12–5; W. Joseph McFarland, Notes of Meeting,

1 and 2 December 1954, LC 9–4; Robert J. Crabb, "Memorandum Re Lloyd Center," 14 December 1954, LC 9–4; Fast to Graham, 3 December 1954, LC 9–4; Drinker to Graham, 26 January 1955, LC 7–2; Horn to Von Hagen, 18 October 1954, LCL 27–2; Horn to Von Hagen, 15 February 1955, LC 12–5; Drinker to Von Hagen, 24 April 1956, LC 12–5; Drinker to Von Hagen, 6 March 1957, LC 12–5; Cysewski, "Portland's Lloyd Center," 13, 38.

54. State Highway Commission, Footage Estimates for The Lloyd Corporation, 6 December 1949; Lloyd to State Highway Commission, 30 January 1950; Drinker to Lloyd, 21 February 1950; Drinker to Lloyd, 24 March 1950; Lloyd to Drinker, 14 April 1950; "Recommendations to the Board of Directors, Lloyd Corporation," for meeting of 17 October 1950; Lloyd to Cooper, 25 October 1950; Cooper to Lloyd, 13 November 1950; Devers to Von Hagen, 18 July 1951; Drinker to Von Hagen, 18 April 1952; Drinker to Von Hagen, 15 May 1952; "Real Estate Option," 29 July 1952; Warranty Deed, 16 September 1952 (all found in LCR 2–5).
55. Cysewski, "Portland's Lloyd Center," 13.
56. Von Hagen to Barnett, 7 December 1956, LC 10–5.
57. Miles L. Colean quoted in Thomas W. Hanchett, "U.S. Tax Policy and the Shopping-Center Boom of the 1950s and 1960s," *American Historical Review* 101 (October 1996): 1095.
58. Hanchett, "U.S. Tax Policy and the Shopping-Center Boom of the 1950s and 1960s," 1092–7, 1095 (quoted).
59. Horn to Von Hagen, 27 July 1954, LC 11–3; McFarland to Von Hagen, 31 August 1954, LC 11–3; McFarland to Von Hagen, 4 October 1954, LCL 27–2; Horn to Von Hagen, 10 November 1955, LC 11–3; Horn to Fast, 23 February 1956, LC 11–3; Horn to Von Hagen, 6 March 1956, LC 11–3; Horn to Von Hagen, 26 September 1956, LC 11–3; Homer Hoyt, "Impact of Suburban Shopping Center in September, 1956," *Urban Land* 15 (September 1956): 4–5.
60. McFarland to Von Hagen, 31 August 1954, LC 11–3; Horn to Von Hagen, 10 November 1955, LC 11–3.
61. Hoyt, "Impact of Suburban Shopping Center in September, 1956," 3.
62. Horn to Von Hagen, 27 July 1954, LC 11–3.
63. Horn to Von Hagen, 10 November 1955, LC 11–3.
64. Horn to Von Hagen, 10 November 1955, LC 11–3.
65. Horn to Von Hagen, 10 November 1955, LC 11–3.
66. Horn to Von Hagen, 27 July 1954, LC 11–3.
67. McFarland to Von Hagen, 31 August 1954, LC 11–3.
68. Drinker to Von Hagen, 19 October 1954, LC 11–3.

69. Cohen, "From Town Center to Shopping Center," 1073–5.
70. Larry Smith & Co., "Lloyd Corporation Progress Report 3: Financial Prospectus Draft," 28 October 1953, LC 11–6.
71. Larry Smith & Co., "Lloyd Corporation Progress Report 6," 23 and 24 September 1954, LC 11–6; John Graham & Co. to Von Hagen, 8 October 1954, LCL 27–2; Wilkinson to Graham, 15 October 1954, LC 9–4; W. Joseph McFarland, Notes of Meeting, 1 and 2 December 1954, LC 9–4; Robert J. Crabb, "Memorandum Re Lloyd Center," 14 December 1954, LC 9–4.
72. Larry Smith & Co., "Lloyd Corporation Progress Report 6," 23 and 24 September 1954, LC 11–6; John Graham & Co. to Von Hagen, 8 October 1954, LCL 27–2; Wilkinson to Graham, 15 October 1954, LC 9–4; W. Joseph McFarland, Notes of Meeting, 1 and 2 December 1954, LC 9–4; Robert J. Crabb, "Memorandum Re Lloyd Center," 14 December 1954, LC 9–4; Larry Smith & Co., "Lloyd Corporation Progress Report 10," 24 August 1955, LC 11–6; Richard G. Horn, Minutes of Staff Meeting, 25 and 26 August 1955, LC 9–4; "Lloyd Center Tenants By Store Classification," table, Larry Smith & Co., "Lloyd Center: Terminal Report," 2 December 1957, LC 12–2.
73. Larry Smith & Co., "Lloyd Corporation Progress Report 6," 23 and 24 September 1954, LC 11–6; Drinker to Von Hagen, 19 October 1954, LC 11–3 (quoted); "Suburban Retail Districts," *AF* 93 (August 1950): 107.
74. Minutes of General Meeting, 24 February 1955, LC 9–4; Von Hagen to Drechsel, 15 June 1955, LC 10–5; Von Hagen to Abernathy, 28 June 1955, LC 10–5; Abernathy to Von Hagen, 27 July 1955, LC 10–5; Abernathy to Von Hagen, 19 October 1955, LC 10–5; David W. Yule, "Lloyd Center Financing," 19 December 1956, LC 10–4; Building Loan Agreement, 15 April 1958, LC 10–4; W. J. Randall, "Characteristics of Regional Shopping Centers," *The Real Estate Analyst Appraisal Bulletin* 26 (29 March 1957): 109–14.
75. Bansbach to Von Hagen, 8 April 1957, LC 10–5; Fast to Von Hagen, 30 April 1957, LC 10–5; Cohen, "From Town Center to Shopping Center" 1072–3; Cysewski, "Portland's Lloyd Center," 11–12; John L. Denny, "Record $21,300,000 Permit Issued for Lloyd Center," *Oregonian*, 11 April 1958, I:1; Horn, "Trends from Regional Center to 'Complete City,'" 9; "Portland Gets Nation's Largest Shopping Center," *Engineering News-Record*, 3 September 1959, 41–2.
76. Larry Smith & Co., "Lloyd Center: Terminal Report," 2 December 1957, LC 12–2.
77. Building Loan Agreement, 15 April 1958, LC 10–4; Betty Edgar, "We Helped Build Lloyd Center; Portland, Oregon's 'New Downtown,'" *Prudential*

Mortgage Loan Mirror (February 1961): 7. A copy of the article may be found in LC 10–5.

78. Adamson, *A Better Way to Build*, 3–4.
79. Daniel Hovey Calhoun, *The American Civil Engineer: Origins and Conflict* (Cambridge, MA: MIT Technology Press, 1960); "Dividing Line Between Engineer and Architect," *The Architect and Engineer* 60 (January 1920): 104–5; Otto E. Goldschmidt, "The Owner, the Architect, and the Engineer," *The Architect and Engineer* 47 (November 1916): 88–92; "The Relative Positions of the Engineer and the Architect in Designing Commercial Buildings," *The Architect and Engineer* 39 (November 1914): 108–9.
80. Adamson, *A Better Way to Build*, 6–8, 14.
81. "The Architect's Place in the Suburban Retail District," *AF* 93 (August 1950): 110.
82. "Architectural Profile: John Graham and Company," *This Earth* 13:1 (January 1960): 8–9. *This Earth* was a publication of Permanente Cement Corporation and Kaiser Gypsum. In 1958, John Graham & Co. ranked twenty-fifth in *AF*'s top 100 architectural and engineering firms, with $46.3 million in revenue.
83. "Shopping Centers: More, Bigger, Better," *Printers' Ink*, 25 November 1956, 44–45.
84. John Graham, "Lloyd Center," 10 September 1954, LC 12–3.
85. "How to Plan Successful Shopping Centers," *AF* 100 (March 1954): 139; Hardwick, *Mall Maker*, 116. See, also, Robert J. Crabb "Tenant's Work," attachment to Crabb to McFarland, 25 August 1954, LC 12–3. Victor Gruen and Larry Smith later compiled the "Bible" of regional shopping center planning, *Shopping Centers USA: The Planning of Shopping Centers* (New York: Reinhold, 1960).
86. Smith to Von Hagen, 21 September 1954, LC 12–3.
87. Robert J. Crabb, "Memorandum Re Construction Procedures," 21 September 1954, LC 12–3.
88. Robert J. Crabb "Tenant's Work," attachment to Crabb to McFarland, 25 August 1954, LC 12–3.
89. John Graham & Co., "Exhibit B: Description of Landlord's and Tenant's Work, Lloyd Center, Portland, Oregon," 11 May 1955, LC 7–5.
90. John Graham & Co., "Exhibit B: Description of Landlord's and Tenant's Work, Lloyd Center, Portland, Oregon," 9 April 1957, LC 7–5.
91. Fast to Von Hagen, 17 February 1958, LC 7–2.
92. Minutes of General Meeting, 15, 16 and 17 August 1956, LC 9–4.

93. Minutes of General Meeting, 15, 16 and 17 August 1956, LC 9–4; York to Von Hagen, 4 December 1956, LC 12–3.
94. Holt to Von Hagen, 4 December 1956, LC 12–3.
95. John Graham, "Construction Schedule, Lloyd Center, Portland Oregon," 17 July 1957, LC 8–2; Minutes of Meeting, 14 and 15 August 1957, LC 9–3; "Bidding Procedure: Lloyd Center, Portland, Oregon," 15 October 1957, LC 7–3; Bennett to Von Hagen, 6 November 1957, LC 7–3; Horn to Crear, 7 May 1958, LC 11–3; Brewton to Yule, 23 September 1958, LC 12–7; Von Hagen to Graham, 5 February 1959, LC 7–2.
96. Larry Smith & Co., "Lloyd Center Progress Report," 2 May 1957, LC 11–6.
97. On early construction management, see, for instance, "Amid Controversy, Construction Management Blossoms," *BD&C* 13 (February 1972): 35; "Change: The Building Team Is Getting Together for a Change," *BD&C* 14 (December 1973): 34–36; "The Changing Role of the General Contractor," *BD&C* 12 (April 1971): 43; "Delivery Options: A Wide Range of Choices," *BD&C* 21 (February 1980): 66–69; Jane Edmunds, "The Pendulum Swings Toward Design-Construct: A Committee of 100 Report," *Consulting Engineer* (October 1984): 73.
98. Von Hagen to Graham, 7 January 1957, LC 7–2; Graham to Von Hagen, 17 January 1957, LC 7–2; Minutes of General Meeting, 29 and 30 January 1958, LC 9–3; Von Hagen to Graham, 5 February 1959, LC 7–2; Graham to Von Hagen 18 February 1959, LC 7–2 (quoted).
99. Simon Co., press release prepared for Lloyd Corp., 11 April 1958, LC 8–4; Denny, "Record $21,300,000 Permit Issued for Lloyd Center"; Jack Pement, "$21,000,000 Lloyd Center Here Begins," *Oregon Journal*, 11 April 1958, I:1.
100. David W. Yule, "Suggested Change Order Procedure," 19 May 1958; Brewton to Yule, 22 May 1958; Fast to Von Hagen, 4 June 1958; Aydelott to Von Hagen, 26 May 1958; Yule to Von Hagen, 22 May 1959; John Graham & Co., "Change Orders," office procedure, 26 November 1959 (all found in LC 6–4).
101. See, for example, change orders approved for: overtime payments on Contract 1 from week of 12 May 1960 through the week of 12 September 1960 for $24,205, relocating the elevators in Block H for $14,645 (Contract 2), and completing the finish of Rosenblatt's store for $71,221 (Contract 3) (Smith to Yule, 9 October 1959; Smith to Yule, 11 January 1960; Brewton to Kammer, 12 January 1960). All Lloyd Center change orders may be found in LC 6–4.
102. "Report 24: Final Phase I Disbursement," attachment to Yule to Eldridge, 4 August 1960, LC 10–6.

103. All seven California majors operated bulk petroleum terminals with deepwater dock facilities by 1930. Of the major operators at Ventura Avenue, Associated and GP opened depots at Linnton in 1903 and 1924, respectively; Shell constructed a terminal at Willbridge Cove in 1913 (Beaton, *Enterprise in Oil*, 78–9, 670–1; "General Petroleum Buys Linnton Site," *Oregonian*, 14 June 1924; "Pioneering in Oil: The Story of the Building of Our Organization," *The Record* 3 [December 1922]: 8–13). On the characteristics of the region's isolated market for petroleum products, see Arthur M. Johnson, "California and the National Oil Industry." *Pacific Historical Review* 39 (May 1970): 155–69; Williamson et al., *The American Petroleum Industry: The Age of Energy, 1899–1959*, 26–28, 37–44.
104. Lloyd Corporation, "Lloyd Center: Regional Shopping Center . . . Portland, Oregon," brochure, undated, [1958–1959], Crotty papers, box 33.
105. Larry Smith & Co., "Lloyd Corporation Progress Report 3: Financial Prospectus Draft," 28 October 1953, LC 11–6 (quoted); Drinker to Von Hagen, 19 October 1954, LC 11–3.
106. "Shopping Centers: More, Bigger, Better," 44.
107. Horn, "Trends from Regional Center to 'Complete City,'" 9.
108. Lloyd Corporation, "Lloyd Center: Regional Shopping Center . . . Portland, Oregon," brochure, undated, [1958–1959], Crotty papers, box 33.
109. Longstreth, *The American Department Store Transformed, 1920–1960*, 243.
110. It should be noted that, in the twenty-first century, the Lloyd District is considered to be an extension of Portland's downtown, as Ralph Lloyd envisioned. It includes a number of office buildings constructed by the Lloyd Corporation after 1960, including Lloyd Center Tower, Lloyd Plaza, and the Lloyd 500, Lloyd 847, Lloyd 919, and Lloyd 1500 Buildings.

CONCLUSION

1. Blackford, *A History of Small Business in America*, 122.
2. Dalit Baranoff, "Shaped by Risk: The American Fire Insurance Industry, 1790–1920," Ph.D. diss., Johns Hopkins University, 2003.
3. Olien and Olien, *Oil in Texas*, 150–4.
4. On joint ventures in mining, see Snow and MacKenzie, "The Environment of Exploration," 888–92.
5. Chandler, *The Visible Hand*; Charles E. Edwards, *Dynamics of the United States Automobile Industry* (Columbia: University of South Carolina Press,

1965); Robert Lewis, "Local Production Practices and Chicago's Automobile Industry, 1900–1930," *Business History Review* 77 (2003): 621 (quoted); John B. Rae, *American Automobile Manufacturers: The First Forty Years* (New York: Chilton, 1959).

6. Thorp, "The Merger Movement and the Paint and Varnish Industry," 6.
7. "18,500,000 Gallons of Paint," *Fortune* (August 1935): 70–83; Benjamin Moore & Company, *Foundation for the Future: The Story of Benjamin Moore & Co.* (New York: the company, 1958); Alfred D. Chandler, Jr. and Stephen Salsbury, *Pierre S. Du Pont and the Making of the Modern Corporation* (New York: Harper & Row, 1971), 381–6; Edmond Clowes Jones, "Some Major Problems Confronting a National Paint Manufacturer Entering the California Market and Suggestions for Solving Them," M.B.A. thesis, University of California, Berkeley, 1950; Pittsburgh Plate Glass Corporation, *Seventy-Five Years of Colorful History: PPG's Coatings and Resins Division* (Pittsburgh, PA: the company, 1975); Sherwin-Williams Company, *The Story of Sherwin-Williams* (Cleveland, OH: the company, 1950).
8. California Chamber of Commerce, *An Examination of California's Paint, Varnish, and Lacquer Industry* (Los Angeles: the organization, 1952); Perry Linder, "Retail Marketing Practices of the California Paint Industry," M.B.A. thesis, University of California, Berkeley, 1956.
9. U.S. Department of Commerce, *Census of Manufactures* (Washington, DC: GPO, various years); Charles H. Kline & Co., *Kline Guide to the Paint Industry* (Fairfield, NJ: the company, 1969, 1972, 1975, 1978, 1981); "New Gloss at Sherwin-Williams," *Business Week*, 15 July 1967, 154–6.
10. McClatchy Newspapers, *Consumer Analysis of the Buying Habits and Brand Preferences in Metropolitan Sacramento, Metropolitan Fresno, and Greater Modesto* (Sacramento, CA: the company, 1947–1962); San Jose Mercury-News, *Consumer Analysis of the San Jose ABC City Zone* (San Jose, CA: the company, 1947–1962). The data are not conclusive regarding market share, as consumers did not state the quantity of paint purchased.
11. Theodore H. Erdman, "The Badger Paint & Hardware Stores, Inc., 1918 to 1863," 1988, TS, Badger Paint & Hardware Stores, Inc., MS 2417, Milwaukee County Historical Society; Milwaukee Journal, *Consumer Analysis of the Greater Milwaukee Market* (Milwaukee, WI: the company, 1930–1931, 1939, 1943–1969).
12. Moody's Investor Service, *Moody's Manual of Industrial Securities, 1940* (New York: Moody's, 1941), 2689; Pittsburgh Plate Glass Company, Paint and

Brush Division, Milwaukee, Wisconsin, pamphlet, undated [circa 1955], Milwaukee County Historical Society.

13. Johnson, "California and the National Oil Industry," 155–69; Olien and Olien, *Oil in Texas*, 220–1.
14. Joe Paul, Jr., "Exit the Drilling Rigs," *Ventura County Star-Free Press*, 14 September 1964, I:1, 4.
15. Adamson, *A Better Way to Build*, 253–90.
16. "Introduction," *Energy Capitals: Local Impact, Global Influence*, ed. Joseph A. Pratt, Martin V. Melosi, and Kathleen A. Brosnan (Pittsburgh, PA: University of Pittsburgh Press, 2014), xv.
17. Hornberg quoted in Sandy Smith-Nonini, "The Role of Corporate Oil and Energy Debt in Creating the Neoliberal Era," *Economic Anthropology* 3 (2016): 57.

Index

Page numbers in italics refer to figures and tables.